**LANCHESTER LIBRARY, Coventry University**

Much Park Street, Coventry CV1 2HF Telephone 01203 838292

This book is due to be returned not later than the date and
time stamped above. Fines are charged on overdue books

# SPIN CHOREOGRAPHY

# Spin Choreography

## Basic Steps in High Resolution NMR

**RAY FREEMAN**

OXFORD • NEW YORK • TOKYO
OXFORD UNIVERSITY PRESS
1998

Oxford University Press, Great Clarendon Street, Oxford OX2 6DP

Oxford    New York

Athens    Auckland    Bangkok    Bogota    Buenos Aires    Calcutta
Cape Town    Chennai    Dar es Salaam    Delhi    Florence    Hong Kong    Istanbul
Karachi    Kuala Lumpur    Madrid    Melbourne    Mexico City    Mumbai
Nairobi    Paris    São Paulo    Singapore    Taipei    Tokyo    Toronto    Warsaw

and associated companies in
Berlin    Ibadan

Oxford is a trade mark of Oxford University Press

Published in the United States
by Oxford University Press Inc., New York

First published by Spektrum Academic Publishers, 1997
First published by Oxford University Press, 1998

A catalogue record for this book is available from the British Library

Library of Congress Cataloging in Publication Data
(Data available)

ISBN 0 19 850481 0

Printed in Great Britain by
Biddles Ltd., Guildford & King's Lynn

*To our first grandchild, Alessandro*

# Contents

# *Preface*

The award of the Nobel Prize in Chemistry to Richard Ernst in 1991 has served to highlight the fact that high resolution NMR is not only an essential physical technique for chemists and biochemists, but also offers a fascinating intellectual study in its own right. Nuclear spins can be manipulated in myriad different ways in order to extract information about molecular structure and molecular motion; a whole new science has been created where none existed before. We can lead the spins through an intricate dance, carefully programmed in advance, to enhance, simplify, correlate, decouple, edit or assign the NMR spectra. Yet we are dealing with entities far too small to be seen under any microscope, and this remarkable ballet can only be inferred indirectly, based on our knowledge of the 'music' provided by the radiofrequency pulse sequences and our observation of the ensuing NMR signals.

If we are to practise this new art of 'spin choreography' we must first master the basic steps and then learn how to put them together (*enchaînement*). That is the aim of this book. In the last four decades, high resolution NMR has grown into such a vast undertaking that it is now well-nigh impossible to describe the entire scenario in a single volume, even if we omit solid-state studies and medical applications. A comprehensive classification of all existing pulse sequences and procedures used by today's chemists and biochemists would be like trying to catalogue all of classical dance. Yet there are some basic themes that run through all these experiments, and there are certain ways of visualizing the motion of the spins that are fundamental and essentially universal. So this book concentrates on the basic steps, leaving the really creative work to you, the choreographer. A few exercises are included (Chapter 3) to help ensure that the *répertoire* is smoothly executed.

The first three chapters set out the basic tools required for understanding the rest of the book. The remaining nine chapters are intended as self-contained entities. Although this entails a certain amount of repetition of the subject matter if the stand-alone chapters are to be easily readable, this may be a useful feature because each section examines the material from a different viewpoint. There is a system of cross-referencing of these topics, independent of the subject index. Each chapter examines a specific theme, for example spin echoes, and traces the way it influences our understanding of high resolution NMR methodology. The aim is to approach the subject matter from a fresh perspective. If this leaves some blind spots, there is no lack of textbooks that cover the subject in a more conventional manner.

The question of how to indicate the relative importance of the different topics is a tricky one; some are central to present-day NMR methodology, others certainly border on the esoteric. I leave the reader to select what seems relevant to the task in hand, just as one would use a dictionary or an encyclopaedia. Like the ballet, high resolution NMR follows certain fashions, as particular procedures enjoy the limelight while others are temporarily forgotten. I have sought to strike a balance between present utility and intrinsic interest—the main emphasis is on widely-used techniques, but with some *divertissements*. These more arcane phenomena could one day become crucial to some exciting future extravaganza.

I have waged a deliberate campaign to minimize the use of abbreviations and technical jargon, without carrying this to impossible extremes. The fact that the word jargon is derived from zircon, a common gemstone of little value, warns us to be wary of its overuse. The proliferation of NMR pulse sequences has encouraged the use of acronyms as a shorthand notation, where the full descriptions would be tedious and repetitious. Although acronyms can be quite descriptive, memorable, even witty, they remain jargon, and I have used them sparingly (see the Appendix).

That this book was written at all owes a great deal to the patience and unflagging support of Michael Rodgers, who also guided me through an earlier saga—the compilation of *A Handbook of Nuclear Magnetic Resonance*. I am touched by his loyalty. Gareth Morris and Peter Hore were kind enough to read the first draft of this book and made many shrewd comments, corrections and suggestions, and I am greatly in their debt. Professor Jack Roberts kindly scrutinized a later draft and pointed out many instances where the explanations could be improved and where the text could be made more user-friendly. Computer simulations are often very helpful for understanding NMR phenomena, and I am indebted to Michael Woodley, Jean-Marc Nuzillard and Eriks Kupče, who have contributed many beautiful illustrations. Finally I must thank all the young research students, both at Oxford and at Cambridge, who did the hard work for many of the spectra used as illustrative examples.

*Cambridge*                                                                                    R.F.
*February 1996*

# *Acknowledgements*

I am grateful to the following for permission to reproduce diagrams:

Longman Scientific and Technical for Fig. 1 of Chapter 2. Academic Press for Figs. 10 and 14 of Chapter 4; Figs. 5, 10, 11, 12, 14, 15, 16, 17, 18, 21 and 22 of Chapter 5; Fig 1 of Chapter 6; Figs. 2, 3, 4 and 6 of Chapter 7; Figs. 1, 2, 9, 11 and 14 of Chapter 8; Figs. 3 and 4 of Chapter 10; Figs. 1 and 2 of Chapter 11; Figs. 1, 2, 4, 5, 7, 8 9, 10 and 11 of Chapter 12. John Wiley & Sons for Fig. 20 of Chapter 5. The Royal Society for Fig. 13 of Chapter 8.

# 1

# Energy levels

## 1.1 Introduction

Spectroscopy provides direct and incontrovertible evidence of quantization of energy, and nowhere is this more simply illustrated than in magnetic resonance experiments. Textbooks on quantum mechanics would surely have chosen spin systems as their starting point, rather than atomic spectra, had magnetic resonance been discovered at an earlier date. The resonance lines which we observe in high resolution NMR spectra are transitions between these energy levels. However, in contrast to conventional spectroscopy, the separation between levels is a function of the applied magnetic field. Many modern NMR experiments, for example correlation spectroscopy or multiple-quantum spectroscopy, become clearer if we think in terms of the appropriate energy level diagram. The populations of these energy levels are important in studies of spin-lattice relaxation, the nuclear Overhauser experiment and chemically induced nuclear polarization. On a more mundane level, the general form of the conventional NMR spectrum can be predicted on the basis of the energy level diagram appropriate to that particular spin system.

It is convenient to label the energy levels according to the magnetic quantum number $m_I$ which can take values from $+I$ to $-I$ in integer steps. Each $m_I$ value corresponds to a different orientation of the nuclear magnetic moment with respect to the magnetic field. The energy of each level (in Joules) is given by

$$E(m_I) = -m_I \gamma B_0 h/2\pi \qquad \qquad \textbf{1.1}$$

where $B_0$ is the applied magnetic field (in tesla), $h$ is Planck's constant and $\gamma$ is the gyromagnetic ratio, which has a characteristic value (and sign) for each nuclear species. The allowed magnetic dipole transitions all have an energy

$$\Delta E = \gamma B_0 h/2\pi \qquad \qquad \textbf{1.2}$$

NMR spectroscopists often express such energies in practical frequency units (Hz) by implicitly invoking $\Delta E = h\nu$ and then dropping Planck's constant:

$$\nu_L = \gamma B_0/2\pi \qquad \qquad \textbf{1.3}$$

This is the Larmor precession frequency. When we consider the chemical shift this expression is slightly modified:

$$\nu_L = \gamma B_0(1-\sigma)/2\pi \qquad \qquad \textbf{1.4}$$

(Strictly the chemical shielding parameter $\sigma$ is not necessarily independent of the applied magnetic field, but $\sigma$ is very small, and any variation is much smaller still, so for all intents and purposes we can neglect these higher-order terms and regard the shielding parameter $\sigma$ as a constant.)

## 1.1.1 Two-spin system

The nuclear spins interact with one another by the spin–spin coupling which acts through the electrons in the chemical bonds. In everything that follows we shall consider only nuclei with a spin $\frac{1}{2}$ and we shall represent two coupled spins by $I$ and $S$, where $S$ is the spin to be observed. A nuclear spin $S$ is aware of the presence of a neighbouring spin $I$ since the latter generates a tiny magnetic field at the site of the $S$ spin. This small magnetic field is either positive or negative depending on the sign of $J_{IS}$ and on whether the $I$ spin is aligned parallel ($m_I = +\frac{1}{2}$) or antiparallel ($m_I = -\frac{1}{2}$) to the main field $B_0$. These two spin states are often labelled $\alpha$ and $\beta$ respectively. This splits the $S$-spin resonance frequency into a doublet with splitting $J_{IS}$ (Hz) where $J_{IS}$ is the coupling constant. The interaction is reciprocal; the $I$ spin experiences a similar splitting attributable to the presence of the neighbouring $S$ spin. Unlike the chemical shift, the spin–spin coupling effect is independent of the applied magnetic field because it is a built-in physical interaction within the molecule, whereas the chemical shift arises from a shielding effect which slightly reduces the applied magnetic field at the site of the nucleus. A valuable distinction can be made between spin–spin coupling (a fixed physical interaction) and spin–spin splitting (the quantity observed in the spectrum). Splittings can be changed by decoupling, chemical exchange or second-order effects; couplings remain constant.

Consider the case of two spins $I$ and $S$ with Larmor frequencies $\nu_I$ and $\nu_S$ but (initially) with zero spin–spin coupling:

$$\nu_I = \gamma_I B_0 (1 - \sigma_I)/2\pi$$
$$\nu_S = \gamma_S B_0 (1 - \sigma_S)/2\pi. \qquad \textbf{1.5}$$

We adopt the usual convention, dropping Planck's constant and writing the expressions for energy in frequency units (Hz). We may assume that $\nu_I > \nu_S$ without loss of generality. There are four levels:

| I S | Energy |
|---|---|
| $\beta\beta$ | $E_4 = +\frac{1}{2}\nu_I + \frac{1}{2}\nu_S$ |
| $\beta\alpha$ | $E_3 = +\frac{1}{2}\nu_I - \frac{1}{2}\nu_S$ |
| $\alpha\beta$ | $E_2 = -\frac{1}{2}\nu_I + \frac{1}{2}\nu_S$ |
| $\alpha\alpha$ | $E_1 = -\frac{1}{2}\nu_I - \frac{1}{2}\nu_S$     $\textbf{1.6}$ |

If we now introduce a non-zero spin–spin coupling $J_{IS}$, the energy levels are shifted slightly

| I S | Energy |
|---|---|
| $\beta\beta$ | $E_4 = +\frac{1}{2}\nu_I + \frac{1}{2}\nu_S + \frac{1}{4}J_{IS}$ |
| $\beta\alpha$ | $E_3 = +\frac{1}{2}\nu_I - \frac{1}{2}\nu_S - \frac{1}{4}J_{IS}$ |
| $\alpha\beta$ | $E_2 = -\frac{1}{2}\nu_I + \frac{1}{2}\nu_S - \frac{1}{4}J_{IS}$ |
| $\alpha\alpha$ | $E_1 = -\frac{1}{2}\nu_I - \frac{1}{2}\nu_S + \frac{1}{4}J_{IS}$     $\textbf{1.7}$ |

Note how the sign of the term in $J_{IS}$ depends on the spin states of both nuclei, whereas the sign of the terms in $\nu$ only depends on the spin state of one nucleus. For the moment we assume that $(\nu_I - \nu_S) \gg |J_{IS}|$; this is known as the weak-coupling approximation and will be re-examined later in the chapter. We can now draw an energy level diagram for the coupled two–spin system:

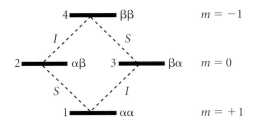

Here $m$ is the total magnetic quantum number. Since $\nu_I$ and $\nu_S$ are very much larger than $|\nu_I - \nu_S|$ and $|J_{IS}|$, the separation of levels (2) and (3) is negligible on this scale.

The allowed transitions have $\Delta m = \pm 1$, and are (1)–(3), (2)–(4), (1)–(2) and (3)–(4). The first two involve a change in the $I$-spin quantum number and are thus $I$ spin transitions:

$$\nu_{13} = \nu_I - \tfrac{1}{2}J_{IS}$$
$$\nu_{24} = \nu_I + \tfrac{1}{2}J_{IS} \qquad \textbf{1.8}$$

the second two are $S$-spin transitions:

$$\nu_{12} = \nu_S - \tfrac{1}{2} J_{IS}$$
$$\nu_{34} = \nu_S + \tfrac{1}{2} J_{IS}$$

**1.9**

Without loss of generality we can assume that $J_{IS} > 0$.

The corresponding four-line spectrum consists of responses at the $I$ and $S$ chemical shifts, split into doublets by the $J$ coupling.

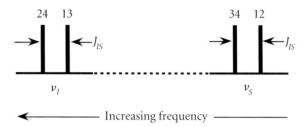

These results can be generalized to the case of $N$ spin-$\tfrac{1}{2}$ nuclei:

$$E = -\sum_{i=1}^{N} m_i \nu_i + \sum_{i>j}^{N} m_i m_j J_{ij}$$

**1.10**

where the second summation is over all pairs of (unlike) spins $i$ and $j$.

We can label the transitions in a rather different manner. The two lines of the $S$ spin response correspond to the two possible spin states of the neighbouring $I$ spin. The $S$ spin experiences a weak magnetic field transmitted from $I$ through the bonding electrons, which either reinforces or diminishes the total field at the $S$-spin site depending on the orientation of the $I$ spin in the field. We can now label the two $S$ lines according to the $I$ spin states and the $I$ lines according to the $S$ spin states:

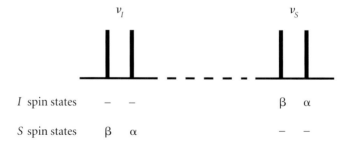

## 1.2 Energy level populations

For a simple two-level system at equilibrium, the energy levels are populated according to the Boltzmann equation:

$$N_{\text{upper}}/N_{\text{lower}} = \exp(-\Delta E/kT)$$

**1.11**

For NMR transitions in realizable magnetic fields and at ambient temperatures, $\Delta E \ll kT$ and the exponential can be expanded to first order in $\Delta E$, giving

$$N_{upper}/N_{lower} = 1 - \Delta E/kT \qquad\qquad \textbf{1.12}$$

which can be rewritten

$$(N_{lower} - N_{upper})/N_{lower} = \Delta E/kT \qquad\qquad \textbf{1.13}$$

That is to say, the population differences are directly proportional to the energy gaps. With this approximation, the equilibrium population differences for all proton (single-quantum) transitions may be put equal, since we can safely disregard the very small changes in $\Delta E$ due to chemical shifts. Other nuclear species (X) would then have population differences reduced in the ratio $\gamma_X/\gamma_H$.

We may write the populations in terms of the small excess ($\Delta \ll 1$):

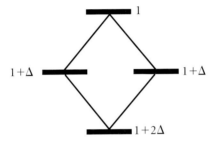

Boltzmann populations

Since the probabilities for stimulated emission and stimulated absorption are equal, the absorption of energy due to upward transitions is largely cancelled by the emission of energy due to downward transitions. Consequently, the observed signal depends only on the population difference $\Delta$.

Spin populations may be perturbed by radiofrequency pulses, the most severe disturbance being a population inversion pulse (often called a 180° pulse or a $\pi$ pulse). A population inversion corresponds to an inverted NMR signal in the spectrum. The excess spins in the upper level return to the ground state by the process of relaxation, characterized by the spin–lattice relaxation time $T_1$.

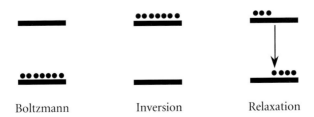

Boltzmann            Inversion            Relaxation

Spin inversion is often used to label the nuclear magnetization arising from a particular chemical site. For example, slow chemical exchange may be monitored[1] by following the fate of an inverted signal created by a $\pi$ pulse applied selectively to a given resonance. If atoms move between molecules, or between different conformations of the same molecule, their nuclei carry with them the population disturbance. The exchange must proceed at a rate that is appreciably faster than, or at least comparable with, the rate of spin–lattice relaxation. Slow chemical exchange can also be studied by two-dimensional spectroscopy (§ 8.2.7).

### 1.2.1 Nuclear Overhauser effect

Probably the most generally useful population experiment is the proton–proton Overhauser effect[2,3] since it provides information about the proximity of two protons within a molecule, and is widely used in NMR studies of macromolecules of biochemical interest. The heteronuclear ($^{13}$C—H) Overhauser effect[4] is a standard technique for enhancing the sensitivity of $^{13}$C spectroscopy. The nuclear Overhauser effect can be described in qualitative terms by considering the interplay between saturation and spin–lattice relaxation. First we assign suitable numbers to the energy levels, proportional to the excess spin populations with respect to the top level. Since the gyromagnetic ratio of $^{13}$C is just a quarter of that of the proton, the population numbers 0, 1, 4 and 5 may be used to represent the equilibrium situation where the proton population differences are four times those of $^{13}$C (left).

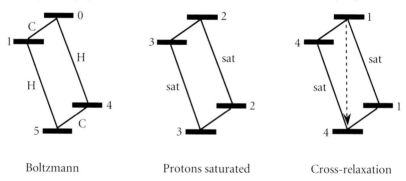

Boltzmann             Protons saturated            Cross-relaxation

The first step is to irradiate the proton resonances to maintain saturation (equal populations). Note that this in itself still leaves the $^{13}$C intensities unchanged (centre). Cross-relaxation then occurs. If the molecular reorientation is fast and if the relaxation mechanism is dipolar, the principal effect is to take spins from the top level (1) to the bottom level (4) but this competes with the weaker influence of natural $^{13}$C relaxation which tries to restore Boltzmann populations across the $^{13}$C transitions. A compromise is reached with the rearrangement of populations shown on the right, leaving the proton transitions still saturated and the $^{13}$C transitions with population differences that have been increased threefold, giving a threefold improvement in the $^{13}$C signal-to-noise ratio. A more rigorous analysis of the population dynamics (§ 9.1.4) predicts the same result. This sensitivity improvement is exploited in a routine manner in most investigations of high resolution $^{13}$C spectroscopy, since it arises as a byproduct of broadband proton decoupling.

In the general case the maximum attainable enhancement (the fractional increase in intensity) is given by

$$\eta_{max} = \tfrac{1}{2}\, \gamma_I/\gamma_S \qquad\qquad \textbf{1.14}$$

where $I$ is the saturated spin and $S$ the observed spin. This maximum effect occurs when there is no leakage as a result of relaxation mechanisms other than the dipole–dipole interaction. For homonuclear systems the maximum enhancement is 50%. Beware of $^{15}$N and $^{29}$Si, which have negative gyromagnetic ratios and thus give a negative value for $\eta$ when protons are saturated; when other relaxation mechanisms compete with the dipole–dipole interaction, the observed signal may have either sign or may even be zero for these nuclei. Negative Overhauser enhancements are also observed for proton–proton systems in very large molecules where the tumbling motion is slow compared with the Larmor frequency (§ 9.3.1).

### 1.2.2 Polarization transfer

The Overhauser effect is not the only population disturbance that is used for enhancing sensitivity. A more advantageous effect can be achieved by inverting populations across a single proton transition while observing the response from, say, a coupled $^{13}$C or $^{15}$N spin. In this situation we need make no assumption about the relaxation mechanism, but the two spin species must be coupled with a resolvable coupling $J_{IS}$, otherwise selective population inversion is not feasible. The inversion can be achieved by a soft radiofrequency pulse (§ 5) or, more generally, by the 'INEPT' polarization transfer sequence described in § 4.2.6:

Protons: $\qquad\qquad (\pi/2)_x - \tau - (\pi)_x - \tau - (\pi/2)_y$
$^{13}$C or $^{15}$N: $\qquad\qquad\qquad\qquad \pi \qquad\qquad\qquad\qquad \pi/2 \text{ acquire} \qquad \textbf{1.15}$

where the delay $\tau = 1/(4J_{IS})$.

For the case that $I = \,^{1}$H and $S = \,^{13}$C, Boltzmann equilibrium is once again represented by the population numbers 0, 1, 4 and 5, proportional to the actual excess populations with respect to the top level.

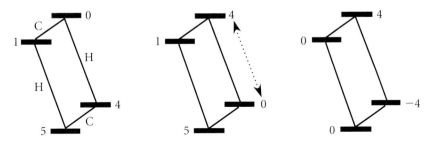

Inversion of one of the two proton transitions immediately creates new population differences across the two $^{13}$C transitions (centre). One $^{13}$C transition increases its intensity by a factor 5 while the other is increased by a factor 3 but is inverted. The usual practice is to suppress the natural $^{13}$C signal and detect only the signal obtained by polarization transfer. This is equivalent to subtraction of the corresponding energy level populations, predicting enhancements of $+4$ and $-4$ (right). The general expression for the enhancement is $\pm\gamma_I/\gamma_S$, which enjoys an advantage over the nuclear Overhauser effect, particularly for the observation of low-$\gamma$ nuclei. No complications arise with a negative gyromagnetic ratio or a competing relaxation mechanism. This is the basis for the INEPT experiment[5] and several heteronuclear two-dimensional techniques.[6] During repetitive-scan experiments, these polarization transfer methods also benefit significantly from the faster spin–lattice relaxation of the protons compared with $^{13}$C or $^{15}$N; we can repeat the excitation sequence as soon as the protons have relaxed.

### 1.2.3  Spin sorting processes

The methods outlined above exploit the slight excess spin population in the ground state, usually no more than about one part in $10^5$ of the total number of spins in the sample. Of course, at very low temperatures the Boltzmann factor favours a higher degree of polarization, but high resolution work is usually performed in solution and it is not feasible to reduce the temperature much below 273 K. However, there are two or three remarkable experiments where much higher excess populations may be created, in principle (though not in practice) as high as 100% polarization. They involve a sorting of the $\alpha$ and $\beta$ spin states.

Certain free radical reactions are influenced at the key moment of recombination by the orientation of a nuclear spin (either $\alpha$ or $\beta$). The nucleus exerts a spin–spin interaction with the electron, and electrons have the power to steer the reaction in one of two possible directions since a stable chemical bond can only be formed if the two valence electrons are antiparallel (a singlet state). Two free radicals generated by photolysis spend some time trapped in a 'cage' of solvent molecules where they undergo many collisions before they can escape from the cage. If they collide while the electron spins are parallel (a triplet state), no bond can be formed and they eventually part company and are scavenged by some other species in the solution. Spin–spin interaction with the nucleus can cause the two electron spins to evolve into an antiparallel configuration. This known as triplet to singlet conversion, as the interaction with the nucleus causes one electron to precess slightly faster than the other.

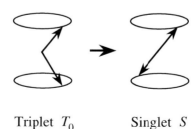

Triplet $T_0$              Singlet $S$

A collision may then result in bond formation. This is known as geminate recombination. The effect depends on the electronic $g$-values and electron–nucleus hyperfine coupling, but for the sake of argument we may assume that the nucleus in the β state favours triplet to singlet conversion. Then:

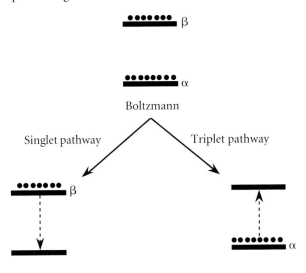

So the feeble nuclear spin, which has such a low energy difference between α and β states, acts as a trigger to control the state of the bonding electrons and thus determines whether the reaction takes the singlet pathway or the triplet pathway. If the sorting process were 100% efficient, then the singlet pathway would generate a product with all the spins in the upper level, giving an enormously enhanced emission spectrum (inverted lines), whereas the product from the triplet pathway would have an enormously enhanced absorption spectrum. Note that it is essential that the α and β nuclear spins end up at sites with different chemical shifts, otherwise the positive and negative enhancements would cancel. These extraordinarily high polarizations are only a transient phenomenon and the normal spin population distribution is eventually restored by spin–lattice relaxation. Note that relaxation for the products of the triplet pathway means taking spins **up** from the ground state to the excited state until Boltzmann equilibrium is restored.

Two years elapsed between the discovery of this **chemically induced nuclear polarization** effect and its explanation by Kaptein and Oosterhoff[7] and by Closs.[8] All previous experience had led spectroscopists to believe that inverted NMR lines only occurred after radiofrequency pumping experiments such as the nuclear Overhauser effect. It was quite hard to accept that the state of the nuclear spin could affect the direction of a chemical reaction. Once the phenomenon had been correctly interpreted, it emerged that several workers had previously observed transient inverted NMR signals but had concluded that the effect was spurious because it seemed irreproducible. This was because these instances involved the accidental generation of free radicals by the ultraviolet radiation in sunlight, before the sample was placed in the (dark) NMR probe. Attempts to reproduce the results failed because it was not realized that the sample must be removed from the probe and re-irradiated in order to create a fresh batch of free radicals just before acquisition of the NMR spectrum.

If we regard the normal nuclear polarization as that which corresponds to a Boltzmann distribution of spin populations, then the Overhauser effect and chemically induced polarization represent extraordinary situations. It is as if a Maxwell demon had somehow sorted out those molecules with spins in the ground state ($\alpha$) from those with their spins in the excited state ($\beta$), so that we could run their spectra separately.

A spin-sorting experiment of a rather different kind has been proposed[9] and demonstrated[10] by Bowers and Weitekamp starting with *para*-hydrogen. The requirement that the overall molecular wavefunction of the $H_2$ molecule be antisymmetric with respect to interchange of the two protons restricts the rotational states of the *ortho* and *para* forms. The former has only odd states and the latter, only even states. Consequently *para*-hydrogen has the lowest rotational energy level of all, and it can be separated from the usual mixture of *ortho*- and *para*-hydrogen molecules by cooling to a low temperature (liquid nitrogen) in the presence of a suitable catalyst. If the catalyst is then removed, the *para*-hydrogen remains stable when allowed to warm up to room temperature. This form of hydrogen has the spins of the two protons antiparallel, as if the Maxwell demon had already been at work. Although *para*-hydrogen generates no abnormally intense NMR signal itself, because it contains equal numbers of spins in $\alpha$ and $\beta$ states, if the two protons are separated by addition across a double bond, we have the possibility for spin sorting and extraordinarily strong signals from the products.

Let the two proton sites of the addition product be $I$ and $S$, coupled by $J_{IS}$, which we shall assume to be positive. For the sake of simplicity, we can ignore any extraneous couplings and consider that the responses from both sites are just doublets of splitting $J_{IS}$. Immediately after addition of *para*-hydrogen across the double bond, both the $I$-spin and $S$-spin responses are antiphase doublets (the up–down pattern) with enormously enhanced intensities. How does this come about ? Basically because only the $\alpha\beta$ and $\beta\alpha$ levels are populated while the $\alpha\alpha$ and $\beta\beta$ levels are empty, a situation which will be treated later under the topic of longitudinal two-spin order (§ 3.2.1).

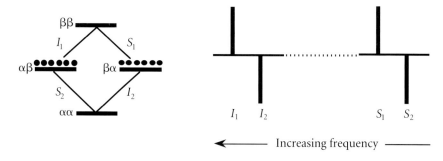

Increasing frequency

The higher-frequency component of the $I$ doublet ($I_1$), which comes from molecules with the $S$ spin in a β state, must have all the $I$ spins in an α state, and gives very intense absorption. The lower-frequency component ($I_2$), which arises from molecules with the $S$ spin in an α state, must have all the $I$ spins in a β state, and gives very intense emission. Identical arguments apply to the $S$-spin response. The demon has obviously been hard at work again. Since there is no "*ortho–para*" symmetry restriction on the $IS$ molecule because it is unsymmetrical, spin-lattice relaxation eventually restores sanity, gradually restoring the normal distribution of spin populations, with only a very small difference between upper and lower levels. All four transitions now have the much weaker intensities expected for Boltzmann equilibrium and they are in-phase doublets again.

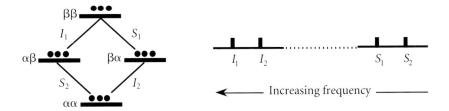

In practice, signal enhancements of two orders of magnitude have been observed[10] in this *para*-hydrogen spin sorting experiment and even higher polarizations are predicted theoretically.

Interestingly, this effect may have been observed on a previous occasion[11] and erroneously ascribed to chemically induced nuclear polarization. The samples had been stored in liquid nitrogen to prevent decomposition, but this may well have allowed some *para*-hydrogen to be formed, which then caused the observed signal enhancements.

Optical pumping provides another path to high nuclear polarizations. Pioneering work by Happer and co-workers[12] has shown how laser pumping of rubidium vapour mixed with xenon gas can generate spin polarizations in $^{129}$Xe that are several orders of magnitude higher than at Boltzmann equilibrium, giving very strong $^{129}$Xe signals. Pines and co-workers[13] have exploited this effect to polarize protons at low temperature on the surface of a polytriarylcarbinol polymer by Hartmann–Hahn cross-polarization (§ 5.5.6) from xenon to protons. This provides a powerful new method for studying surface properties.

## 1.3 **Coherence**

Until about 1970, most high resolution NMR spectra were studied by continuous-wave slow passage methods, and it was sufficient to describe them in terms of transitions between energy levels as in conventional optical spectroscopy. If we knew the energy level separations, the spin populations and the transition probabilities, we could predict the spectrum quite satisfactorily. The selection rule for magnetic dipole transitions is $\Delta m = \pm 1$, and most observed NMR responses were confined to these allowed transitions. However, it was known that the formally forbidden multiple-quantum transitions could be excited by increasing the intensity of the radiofrequency field so that the spin system was forced to absorb several equal quanta simultaneously. (Note that this precludes the excitation of zero-quantum transitions.) Similarly, double resonance experiments could be performed (§ 1.4.4) in which the spin system absorbed quanta from the perturbing radiofrequency field $B_2$ as well as from the observation field $B_1$. Nevertheless, the vast majority of spectra represented transitions that satisfied the selection rule $\Delta m = \pm 1$.

### 1.3.1  Single-quantum coherence

With the advent of pulse excitation methods and Fourier transformation, and particularly after the introduction of two-dimensional spectroscopy,[14,15] it was no longer possible to describe all the observed phenomena in terms of this simple picture of transitions between energy levels. The phase of the precessional motion is important. Spectroscopists began to talk about the concept of coherence between states of the spin system, induced by the application of a radiofrequency pulse (or sequence of pulses). This is sometimes called a coherent superposition of states. The states in question may be separated by $\Delta m = \pm 1$ (single-quantum coherence) or by $\Delta m = 0, \pm 2, \pm 3$, etc. (multiple-quantum coherence). This induced coherence is transient; if left to itself it decays with time as the phase relationship is gradually destroyed by the processes of spin–spin relaxation.

We now use the term state rather than energy level in order to signify that phase is also being taken into account, and we use a slightly different pictorial representation. Formally, the condition of the spin system is represented by a density matrix $\sigma$. Consider the simple case of the coupled two-spin ($IS$) system. It may be represented by a $4 \times 4$ density matrix, in which the diagonal elements, $\sigma_{11}, \sigma_{22}, \sigma_{33}$, and $\sigma_{44}$ correspond to the spin populations.

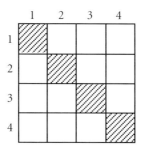

At Boltzmann equilibrium and in the absence of radiofrequency fields, all off-diagonal elements are zero; there are no coherences. A radiofrequency pulse excites single-quantum coherences, represented by the elements $\sigma_{13}, \sigma_{24}, \sigma_{12}$, and $\sigma_{34}$ (and their corresponding symmetrical counterparts $\sigma_{31}, \sigma_{42}, \sigma_{21}$, and $\sigma_{43}$).

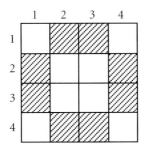

These coherences have the property that they induce a radiofrequency voltage on the receiver coil; together these four oscillating voltages constitute the free induction signal. Single-quantum coherence is therefore an alternative description for precessing transverse magnetization. We may represent the free induction decay as the superposition of two magnetization components from the $I$ spins and two from the $S$ spins. The Fourier transform is identical with the slow-passage continuous-wave spectrum.

### 1.3.2 Multiple-quantum coherence

The need for this new terminology becomes apparent when we consider multiple-quantum effects, the coherent concerted motion of two or more spins. These coherences are represented by the off-diagonal elements $\sigma_{14}$ and $\sigma_{23}$ (together with $\sigma_{41}$ and $\sigma_{32}$).

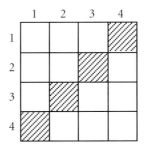

Pulse sequences can be devised (§ 3.4.1) that excite these multiple-quantum coherences, which then persist until damped by spin–spin relaxation processes. For example, the term $\sigma_{14}$ represents double-quantum coherence (where $I$ and $S$ act in the same sense), while $\sigma_{23}$ represents zero-quantum coherence (where $I$ and $S$ act in opposition).

Multiple-quantum coherence differs from the single-quantum coherence described above in that there is no net precessing magnetization, and therefore no signal induced in the receiver coil. Its existence must be detected indirectly by allowing it to precess during the evolution period of a two-dimensional experiment, followed by conversion into observable precessing magnetization. In this manner it is possible to obtain multiple-quantum spectra (§ 8.4.2) derived from the appropriate two-dimensional experiment. Multiple-quantum coherences are unusually sensitive to radiofrequency phase shifts, magnetic field gradients or changes in radiofrequency offset (§ 3.4.4 ).

The general principles of multiple-quantum spectroscopy may be illustrated by considering a weakly-coupled three-spin system ($ISR$) which has energy levels that can be derived from Equation 1.10.

| I S R | Energy | Magnetic quantum number |
|---|---|---|
| βββ | $E_8 = +\frac{1}{2}\nu_I + \frac{1}{2}\nu_S + \frac{1}{2}\nu_R + \frac{1}{4}J_{IS} + \frac{1}{4}J_{IR} + \frac{1}{4}J_{SR}$ | $m = -\frac{3}{2}$ |
| βαβ | $E_7 = +\frac{1}{2}\nu_I - \frac{1}{2}\nu_S + \frac{1}{2}\nu_R - \frac{1}{4}J_{IS} + \frac{1}{4}J_{IR} - \frac{1}{4}J_{SR}$ | $m = -\frac{1}{2}$ |
| αββ | $E_6 = -\frac{1}{2}\nu_I + \frac{1}{2}\nu_S + \frac{1}{2}\nu_R - \frac{1}{4}J_{IS} - \frac{1}{4}J_{IR} + \frac{1}{4}J_{SR}$ | $m = -\frac{1}{2}$ |
| ββα | $E_5 = +\frac{1}{2}\nu_I + \frac{1}{2}\nu_S - \frac{1}{2}\nu_R + \frac{1}{4}J_{IS} - \frac{1}{4}J_{IR} - \frac{1}{4}J_{SR}$ | $m = -\frac{1}{2}$ |
| ααβ | $E_4 = -\frac{1}{2}\nu_I - \frac{1}{2}\nu_S + \frac{1}{2}\nu_R + \frac{1}{4}J_{IS} - \frac{1}{4}J_{IR} - \frac{1}{4}J_{SR}$ | $m = +\frac{1}{2}$ |
| βαα | $E_3 = +\frac{1}{2}\nu_I - \frac{1}{2}\nu_S - \frac{1}{2}\nu_R - \frac{1}{4}J_{IS} - \frac{1}{4}J_{IR} + \frac{1}{4}J_{SR}$ | $m = +\frac{1}{2}$ |
| αβα | $E_2 = -\frac{1}{2}\nu_I + \frac{1}{2}\nu_S - \frac{1}{2}\nu_R - \frac{1}{4}J_{IS} + \frac{1}{4}J_{IR} - \frac{1}{4}J_{SR}$ | $m = +\frac{1}{2}$ |
| ααα | $E_1 = -\frac{1}{2}\nu_I - \frac{1}{2}\nu_S - \frac{1}{2}\nu_R + \frac{1}{4}J_{IS} + \frac{1}{4}J_{IR} + \frac{1}{4}J_{SR}$ | $m = +\frac{3}{2}$ |

**1.16**

Without loss of generality we may assume $\nu_I > \nu_S > \nu_R$ and that all three coupling constants are positive. The energy level diagram can be written:

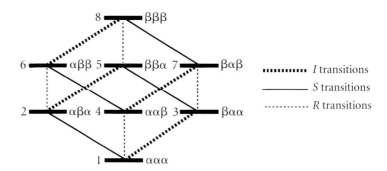

The exact appearance of the single-quantum spectrum depends on the relative magnitudes of the shifts and couplings. Each line of a particular multiplet can be associated with the spin states of the neighbour nuclei:

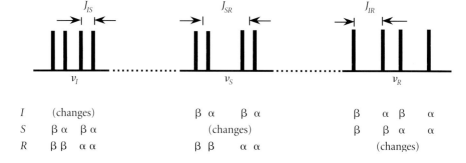

If, for example, the sign of $J_{IS}$ is made negative, the $I$-spin spectrum remains the same but the spin state labelling is different.

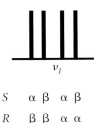

$$
\begin{array}{ccccc}
S & \alpha & \beta & \alpha & \beta \\
R & \beta & \beta & \alpha & \alpha
\end{array}
$$

Consequently, the signs of the coupling constants cannot be deduced from the form of the spin multiplet (unless there are strong coupling effects).

### 1.3.3 Double-quantum coherence

So far we have restricted the discussion to transitions that obey the selection rule $\Delta m = \pm 1$ since these are the only ones which can be easily observed directly as oscillating voltages on the receiver coil. However, it is possible to excite coherence between two energy levels separated by zero, two, three or more quanta. Here we consider double-quantum effects.

The energy level diagram for a three-spin system permits six double-quantum coherences, for example between levels (1) and (7):

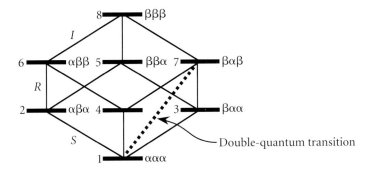

Coherence between states $\alpha\alpha\alpha$ and $\beta\alpha\beta$ involves the $I$ and $R$ spins, but not the $S$ spin, which remains passive. This is an $IR$ double-quantum coherence. The energies of the two states involved are:

$$E_1 = -\tfrac{1}{2}\nu_I - \tfrac{1}{2}\nu_S - \tfrac{1}{2}\nu_R + \tfrac{1}{4}J_{IS} + \tfrac{1}{4}J_{IR} + \tfrac{1}{4}J_{SR} \qquad \textbf{1.17}$$

$$E_7 = +\tfrac{1}{2}\nu_I - \tfrac{1}{2}\nu_S + \tfrac{1}{2}\nu_R - \tfrac{1}{4}J_{IS} + \tfrac{1}{4}J_{IR} - \tfrac{1}{4}J_{SR} \qquad \textbf{1.18}$$

and the double-quantum frequency is:

$$\nu_{17} = +\nu_I + \nu_R - \tfrac{1}{2}J_{IS} - \tfrac{1}{2}J_{SR} \qquad \textbf{1.19}$$

Note how this double quantum frequency lies at the sum of the chemical shift frequencies minus half the sum of the couplings to the passive spin S. In this way we can calculate the form of the double-quantum spectrum:

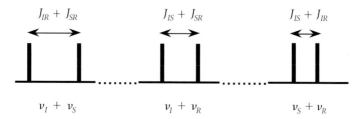

Double-quantum transitions were initially recorded on continuous-wave frequency-sweep spectrometers by increasing the radiofrequency field intensity $(B_1)$ so that the spin system could absorb two equal quanta simultaneously.[16] This brute force method breaks down the selection rule, exciting a normally forbidden transition. The observed frequency is then exactly one half the energy level separation. Nowadays, double-quantum coherences are studied indirectly by two-dimensional spectroscopy, and the actual double-quantum frequencies are obtained.

We can see from the formulae and diagram above that the splittings within the double-quantum spectrum depend on the relative signs of the two couplings to the passive spin. For example the left-hand doublet always has a splitting $|J_{IR} + J_{SR}|$; if these couplings have opposite signs the splitting is smaller than if they have the same sign. This was one of the early methods for relating the signs of proton–proton coupling constants.[17] Double-quantum transitions can also be used as an aid to assignment of spectra.

If we want to redraw an energy level diagram for a more intense applied magnetic field, the energy levels must be shifted in proportion to $mB_0$ where $m$ is the total magnetic quantum number; the $m=0$ levels are only affected by a quite negligible amount, proportional to the difference in chemical shifts, $B_0(\sigma_I - \sigma_S)$. This means that any spatial inhomogeneity of the applied magnetic field $B_0$ broadens the energy levels in proportion to $m$. For example, in a two-spin system the top (4) and bottom (1) levels are broadened but the intermediate levels (2) and (3) are essentially unaffected:

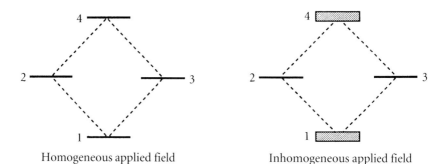

Homogeneous applied field                    Inhomogeneous applied field

(The effect is grossly exaggerated for the purpose of illustration.) Consequently, when double-quantum coherence is detected indirectly in a two-dimensional experiment, it is twice as sensitive to static field inhomogeneity (or applied field gradients) as a single-quantum coherence.

### 1.3.4 Zero-quantum coherence

There are six different zero-quantum ($\Delta m = 0$) coherences. These cannot be observed in continuous-wave experiments and must be detected indirectly by two-dimensional spectroscopy.

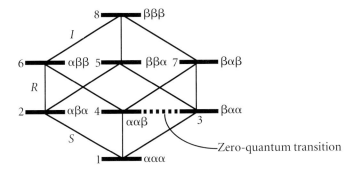

As an example, we can consider the coherence between levels (3) and (4) (*IR* zero-quantum coherence) where the *S* spin remains passive. Its frequency is

$$\nu_{34} = +\nu_I - \nu_R - \tfrac{1}{2}J_{IS} + \tfrac{1}{2}J_{SR} \qquad \textbf{1.20}$$

Homonuclear zero-quantum coherences have some unusual properties; for example, they are essentially unaffected by inhomogeneity of the applied field $B_0$. They may therefore be separated from other orders of coherence by the application of a field gradient pulse. When excited in a pulse sequence, the phase of the zero-quantum coherence is not affected by a radiofrequency phase shift. (A formal treatment of this phenomenon is presented in § 3.4.4) Consequently, phase cycles can be devised to separate zero-quantum coherence from higher-order coherences but not, of course, from longitudinal magnetization.

The spectrum of zero-quantum coherences for a three-spin system is of the form

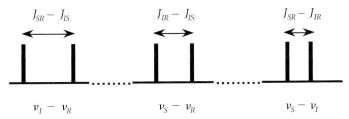

Again we see that the experimental splittings reflect the relative signs of coupling constants to the passive spin.

### 1.3.5 **Three-spin coherence**

Triple-quantum coherence ($\Delta m = 3$) involves the concerted motion of three spins in the same sense. It spans the energy levels (1) and (8) and has the frequency $v_I + v_S + v_R$, but contains no terms in the coupling constants. The triple-quantum transition can be observed in continuous-wave experiments when the radiofrequency field intensity is increased well above its normal setting.[16] It falls at the frequency $\frac{1}{3}(v_I + v_S + v_R)$.

At this stage it is necessary to introduce a further point of nomenclature, since three spins may act in concert in a different manner involving the absorption of two quanta with the simultaneous emission of one quantum. These are sometimes known as combination lines, by analogy with certain forbidden transitions in infrared spectroscopy. They are more properly described as three-spin single-quantum coherences (§ 3.4.2). By extension, systems with larger numbers ($n$) of coupled spins may support $n$-spin $p$-quantum coherences.

The three-spin single-quantum coherences have the frequencies

$$v_{36} = -v_I + v_S + v_R$$
$$v_{27} = +v_I - v_S + v_R$$
$$v_{45} = +v_I + v_S - v_R$$

**1.21**

Note that these frequencies do not involve coupling constants.

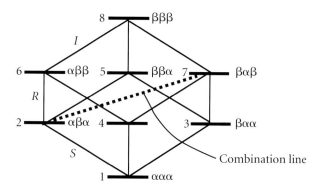

Combination lines appear in strongly-coupled spectra (§ 1.5.7) although they are often quite weak. They may be detected indirectly in a two-dimensional experiment[18] in which the three-spin coherences are allowed to precess during an evolution period $t_1$, and they also appear in three-dimensional coherence transfer experiments. Since the net change in quantum number is $\pm 1$, combination lines behave like ordinary single-quantum coherences as far as magnet field inhomogeneity, applied field gradient pulses, and radiofrequency phase shifts are concerned.

If we sketch out the spectrum of three-spin single-quantum coherences we see that it is stretched with respect to the conventional single-quantum spectrum, the three separations being doubled.

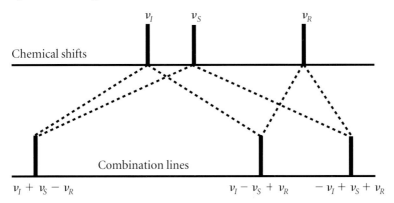

It is as if the conventional spectrum had been stripped of its fine structure, expanded by a factor of two, and reversed left to right. Because the line widths of combination lines are the same as those of ordinary single-quantum transitions, this affords a twofold increase in chemical shift dispersion.[18]

The simplicity of the combination-line spectrum may be put to good use. The introduction of a fourth (passive) spin $P$ that is coupled to $I$, $R$ and $S$, splits all three resonances, giving the multiplets:

$$-\nu_I + \nu_S + \nu_R \pm \tfrac{1}{2}(-J_{IP} + J_{SP} + J_{RP})$$
$$+\nu_I - \nu_S + \nu_R \pm \tfrac{1}{2}(+J_{IP} - J_{SP} + J_{RP})$$
$$+\nu_I + \nu_S - \nu_R \pm \tfrac{1}{2}(+J_{IP} + J_{SP} - J_{RP})$$

**1.22**

while the triple-quantum coherence is split into a multiplet:

$$+\nu_I + \nu_S + \nu_R \pm \tfrac{1}{2}(+J_{IP} + J_{SP} + J_{RP})$$

**1.23**

These four expressions allow us to determine the magnitudes and relative signs of the spin couplings to the passive spin $P$.

## 1.3.6 X approximation

We see from the above that there has been a factorization of the problem of the four-spin system into two related three-spin systems. This is an example of a simplification that has become known as the $X$-approximation and has proved very useful in the analysis of complex high resolution spectra.[19,20] This approximation is justified as long as $X$ is only weakly coupled to all the spins in the subsystem under consideration. In effect we start with a relatively simple subsystem ($IS$) and then introduce an additional spin ($R$) and calculate how it affects the $IS$ energy level diagram. It splits it into two four-level manifolds, one where the $R$ spin is in state $\alpha$, the other where $R$ is in state $\beta$:

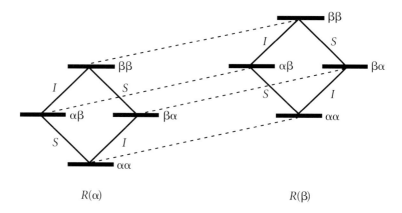

$$R(\alpha) \qquad\qquad\qquad\qquad R(\beta)$$

Note that this is just a different representation of the *ISR* energy level diagram shown earlier. The *X* approximation represents an important simplification in the analysis of strong-coupling effects, treated in § 1.5.4. A very common example is the decomposition of the $^{13}C$ satellite spectra of protons into subspectra from $^{13}C$ in an $\alpha$ or a $\beta$ spin state (§ 1.5.7).

For certain experiments *R* behaves as a spectator, and is not involved in any of the manipulations we may care to impose on one of the *IS* manifolds. For example it is possible to perform selective decoupling or selective coherence transfer in the $R(\alpha)$ manifold without significantly affecting the $R(\beta)$ manifold. The subspectra corresponding to $R(\alpha)$ and $R(\beta)$ are separated in frequency by $|J_{IR} + J_{SR}|$ which may be a larger or smaller quantity depending on the sign of $J_{IR}$ compared with the sign of $J_{SR}$. Coupling to *R* can be thought of as creating effective chemical shifts:

$$\nu_I(\alpha) = \nu_I + \tfrac{1}{2}J_{IR}$$
$$\nu_S(\alpha) = \nu_S + \tfrac{1}{2}J_{SR} \qquad\qquad \textbf{1.24}$$

in one subspectrum, and

$$\nu_I(\beta) = \nu_I - \tfrac{1}{2}J_{IR}$$
$$\nu_S(\beta) = \nu_S - \tfrac{1}{2}J_{SR} \qquad\qquad \textbf{1.25}$$

in the other. The selective decoupling or coherence transfer experiment identifies resonances that belong to a given subspectrum, thus determining the relative signs of the two coupling constants to the spectator spin *R*. Absolute signs of these coupling constants remain ambiguous since, although we can recognize the existence of separate manifolds for $R(\alpha)$ and $R(\beta)$, we cannot identify which is $R(\alpha)$ and which is $R(\beta)$. We shall see in § 8.2.3 that in certain types of two-dimensional correlation experiment, a similar separation of energy level manifolds provides the relative signs of coupling constants by inspection.

## 1.4 **Double resonance**

### 1.4.1 **Connectivity of transitions**

There are some important practical consequences of the topology of the energy level diagram.[21] Consider the two transitions (1)–(2) and (2)–(4) below. These two transitions are called connected transitions as they share a common energy level. There are only two topologically distinct configurations of two connected transitions. In this example the quantum numbers of the outer energy levels differ by two units. The two transitions are said to be progressively connected.

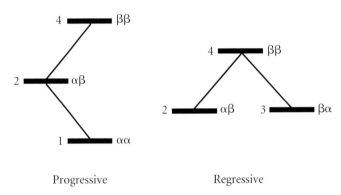

Progressive                        Regressive

In contrast, the connected transitions (2)–(4) and (3)–(4) have the same magnetic quantum number for the outer energy levels. They are said to be regressively connected. This distinction has important consequences for double resonance experiments and two-dimensional correlation spectroscopy. For example, there is a double resonance experiment (spin tickling) in which a single transition is irradiated, generating a small splitting of all connected transitions.[21] For the progressive case this splitting is rather less well resolved than we might expect from the instrumental broadening since there is some double-quantum character in the two lines of the doublet, making them more sensitive to $B_0$ inhomogeneity. By contrast, the regressive case shows a more sharply resolved doublet since it involves some zero-quantum character, which is insensitive to $B_0$ inhomogeneity.

### 1.4.2 **Correlation spectroscopy**

Experiments that create a population disturbance affect a progressively connected transition in one sense and a regressively connected transition in the opposite sense.[22] There is a similar distinction in two-dimensional coherence transfer[15] (COSY) experiments. When coherence is transferred between two connected transitions, the sense of the observed signal is positive for the progressive case and negative for the regressive case. This accounts for the familiar up–down pattern of intensities observed in a COSY cross-peak (§ 8.2.3).

In these coherence transfer experiments, it is useful to define a third kind of relationship between transitions within the energy level diagram—parallel transitions. These are transitions of the same spin; they have no energy level in common.

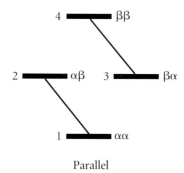

Parallel

In order to transfer coherence between parallel transition we have to flip a third spin ($R$) during the mixing sequence. The probability of flipping $R$ depends on the flip angle of the mixing pulse; if this is small, then this probability is small and the parallel transitions are very weak, if the flip angle is near $\pi$ radians, the parallel transitions are very strong.

For a coupled two-spin system with chemical shifts $\nu_I$ and $\nu_S$, we may define four kinds of coherence transfer, giving rise to four categories of response in the two-dimensional correlation spectrum.

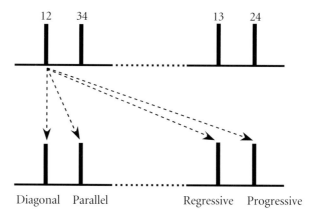

Diagonal   Parallel                  Regressive   Progressive

When no frequency jump occurs at all, the response must necessarily lie exactly on the principal diagonal ($F_1 = F_2$) of Figure 1.1. When the coherence is transferred to a parallel transition ($12 \rightarrow 34$), the corresponding response lies close to the principal diagonal, offset only by the $J$ splitting. When the transfer involves the progressive case ($12 \rightarrow 24$) or the regressive case ($12 \rightarrow 13$), the corresponding responses form a cross-peak in a square pattern with alternation in the intensities (Figure 1.1). Cross-peaks are centred at the coordinates ($\nu_I, \nu_S$) and ($\nu_S, \nu_I$) and are normally recorded in the absorption mode (circular contours in Figure 1.1). The

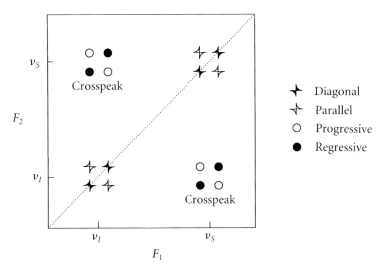

**Figure 1.1** Schematic representation of a COSY spectrum of a two-spin system. The usual phase adjustment sets the coherence transfer peaks (progressive and regressive transitions) in absorption with alternation of the phase. Signals that involve no transfer (diagonal transitions) or only transfer within a multiplet (parallel transitions) are recorded in double dispersion and are all in phase.

square patterns of diagonal and parallel peaks are centred at the coordinates $(v_I, v_I)$ and $(v_S, v_S)$ on the principal diagonal, and are normally recorded as dispersion signals in both dimensions. Various aspects of correlation spectroscopy are treated in more detail in § 2.6.1, § 3.3.5, § 8.2 and § 8.3.

## 1.4.3 **Bloch–Siegert effect**

Energy level separations are determined by the magnetic field experienced by the nuclei—not simply the applied static field $B_0$ but also any radiofrequency field near resonance. This principle was first invoked by Bloch and Siegert[23] before NMR in bulk matter had even been discovered. For reasons of practical convenience, magnetic resonance is almost invariably excited by a linearly oscillating radiofrequency field, although a circularly polarized field would suffice and might seem more logical since it would tell us the sense of the nuclear precession. An oscillating field can be decomposed into two counter-rotating circularly polarized fields:

$$2B_1\cos(\omega t) = B_1[\exp(+i\omega t) + \exp(-i\omega t)]. \qquad \textbf{1.26}$$

We normally neglect the component that rotates in the opposite sense to the nuclear precession since it has a very large offset frequency $2\omega$. Bloch and Siegert calculated the effect of the unused component on the resonance condition and found that it caused a small shift to higher frequency. The nuclei find themselves in an effective field $B_{\text{eff}}$ that is slightly higher than $B_0$.

The effect can be generalized[24] to include a radiofrequency field ($B_2$) deliberately imposed in a double resonance experiment; now the shift can be much larger, because one of the two counter-rotating components of $B_2$ can be quite close to resonance. The Bloch–Siegert effect displaces the energy levels involved in transitions close to resonance with $B_2$. This has repercussions on other transitions sharing the same energy-level manifold and, as we shall see below, can be used to explain many double-resonance effects. We now neglect the component of $B_2$ that rotates in the wrong sense and transform into a frame rotating in synchronism with the other rotating component. In this frame the line of interest has as offset $\Delta B$. The presence of $B_2$ creates a resultant effective field given by

$$B_{\text{eff}}^2 = \Delta B^2 + B_2^2 \qquad \qquad \textbf{1.27}$$

The resonance line experiences a field $B_{\text{eff}}$ instead of $\Delta B$ and thus suffers a shift

$$B_{\text{eff}} - \Delta B = \Delta B \left[1 + B_2^2/\Delta B^2\right]^{\frac{1}{2}} - \Delta B \qquad \qquad \textbf{1.28}$$

If $B_2 << \Delta B$, the square-root expression can be expanded as a power series, retaining only the second-order term:

$$B_{\text{eff}} - \Delta B = \tfrac{1}{2} B_2^2/\Delta B \qquad \qquad \textbf{1.29}$$

or, in practical frequency units (Hz), a Bloch–Siegert shift of

$$\Delta \nu = \tfrac{1}{2} \gamma \, B_2^2/(2\pi\Delta B) \qquad \qquad \textbf{1.30}$$

We may conclude that this also distorts the energy level diagram, moving the respective levels apart by this same amount.

Whenever a continuous radiofrequency field is left switched on during acquisition of the signal, Bloch–Siegert shifts are induced in the spectrum. Water suppression experiments (§ 11.2) often use the proton decoupler for presaturation, but we must be careful to switch it off before data acquisition to avoid this kind of displacement since it could cause problems in difference spectroscopy. Nuclear Overhauser difference-mode experiments (§ 9.2.1) are particularly susceptible to Bloch–Siegert shifts unless the proper precautions are taken. For heteronuclear decoupling experiments we normally disregard the Bloch–Siegert shift on any other nuclear species because the resonance offset is large (MHz). Nevertheless, for very high levels of decoupler power (say $\gamma B_2/2\pi = 10$ kHz) applied at the $^{13}$C frequency, there is a small but significant Bloch–Siegert shift at the deuterium lock frequency, of the order of 1 Hz.

On occasion we may have to deal with a distribution of Bloch–Siegert shifts. If the decoupler field has an appreciable spatial inhomogeneity then there is a Bloch–Siegert broadening of adjacent resonances and a distortion of the line shape.

## 1.4.4  Spin decoupling

The distortion of the energy level diagram by the Bloch–Siegert effect can help to explain the phenomenon of coherent decoupling. Consider a system of two coupled spins $I$ and $S$ where the $I$ spins are irradiated with a field $\gamma B_2/2\pi$ and the $S$ spins are observed by the usual methods. In the absence of $B_2$, the energy levels may be written as

$$
\begin{aligned}
(4) &\quad +\tfrac{1}{2}\nu_I + \tfrac{1}{2}\nu_S + \tfrac{1}{4}J \\
(3) &\quad -\tfrac{1}{2}\nu_I + \tfrac{1}{2}\nu_S - \tfrac{1}{4}J \\
(2) &\quad +\tfrac{1}{2}\nu_I - \tfrac{1}{2}\nu_S - \tfrac{1}{4}J \\
(1) &\quad -\tfrac{1}{2}\nu_I - \tfrac{1}{2}\nu_S + \tfrac{1}{4}J
\end{aligned}
$$

**1.31**

In order to calculate the effect of the irradiation field $B_2$, it is necessary to make a transformation into a rotating frame where $B_2$ is stationary.[25] Let the decoupler be set at some offset $\Delta$ Hz from the exact chemical shift $\nu_I$ of the $I$ spins. The appropriate frame therefore rotates at $\nu_I - \Delta$ Hz.

A shift of NMR transition frequencies in the rotating frame implies that the energy level diagram must also suffer a transformation in that frame of reference. Since all the transition frequencies are reduced by an amount $(\nu_I - \Delta)$ this entails lowering level (4) and raising level (1) by this amount.

$$
\begin{aligned}
(4) &\quad -\tfrac{1}{2}\nu_I + \tfrac{1}{2}\nu_S + \Delta + \tfrac{1}{4}J \\
(3) &\quad -\tfrac{1}{2}\nu_I + \tfrac{1}{2}\nu_S - \tfrac{1}{4}J \\
(2) &\quad +\tfrac{1}{2}\nu_I - \tfrac{1}{2}\nu_S - \tfrac{1}{4}J \\
(1) &\quad +\tfrac{1}{2}\nu_I - \tfrac{1}{2}\nu_S - \Delta + \tfrac{1}{4}J
\end{aligned}
$$

**1.32**

We can represent this transformation schematically, recognizing that the shift of levels (1) and (4) is very large compared with the level separations in the rotating frame:

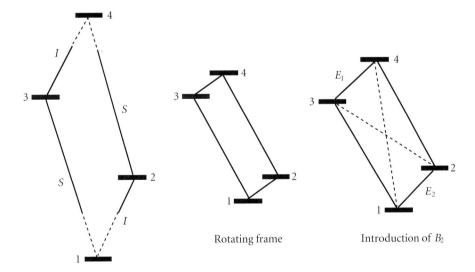

Laboratory frame

Rotating frame

Introduction of $B_2$

So far the transformation into the rotating frame has assumed $B_2 = 0$. We now introduce the decoupling field $B_2$ and evaluate the effective fields acting on the two $I$ spin transitions:

$$E_1 = [(\Delta + \tfrac{1}{2}J)^2 + (\gamma B_2/2\pi)^2]^{\frac{1}{2}}$$
$$E_2 = [(\Delta - \tfrac{1}{2}J)^2 + (\gamma B_2/2\pi)^2]^{\frac{1}{2}} \qquad \textbf{1.33}$$

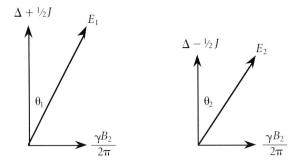

The Bloch–Siegert shift acts on the $I$-spin transitions, increasing the separation of levels (4) and (3) from $(\Delta + \tfrac{1}{2}J)$ to $E_1$ and increasing the separation of levels (2) and (1) from $(\Delta - \tfrac{1}{2}J)$ to $E_2$. These energy levels move apart in a symmetrical fashion:

$$\begin{aligned}
(4) \quad & -\tfrac{1}{2}\nu_I + \tfrac{1}{2}\nu_S + \tfrac{1}{2}\Delta + \tfrac{1}{2}E_1 \\
(3) \quad & -\tfrac{1}{2}\nu_I + \tfrac{1}{2}\nu_S + \tfrac{1}{2}\Delta - \tfrac{1}{2}E_1 \\
(2) \quad & +\tfrac{1}{2}\nu_I - \tfrac{1}{2}\nu_S - \tfrac{1}{2}\Delta + \tfrac{1}{2}E_2 \\
(1) \quad & +\tfrac{1}{2}\nu_I - \tfrac{1}{2}\nu_S - \tfrac{1}{2}\Delta - \tfrac{1}{2}E_2 \qquad \textbf{1.34}
\end{aligned}$$

The perturbation by the $B_2$ field has another important effect—it makes the previously forbidden transitions (1)–(4) and (2)–(3) partly allowed. There are therefore four transitions observable in the $S$-spin spectrum. When we reconvert these frequencies to the laboratory frame by adding $(\nu_I - \Delta)$, we obtain expressions for the $S$-spin transitions under decoupling conditions:

| Levels | Frequency | Intensity | Designation |
|---|---|---|---|
| (1)–(4) | $\nu_S + \tfrac{1}{2}(E_1 + E_2)$ | $\sin^2[\tfrac{1}{2}(\theta_1 - \theta_2)]$ | satellite |
| (2)–(4) | $\nu_S + \tfrac{1}{2}(E_1 - E_2)$ | $\cos^2[\tfrac{1}{2}(\theta_1 - \theta_2)]$ | main |
| (1)–(3) | $\nu_S - \tfrac{1}{2}(E_1 - E_2)$ | $\cos^2[\tfrac{1}{2}(\theta_1 - \theta_2)]$ | main |
| (2)–(3) | $\nu_S - \tfrac{1}{2}(E_1 + E_2)$ | $\sin^2[\tfrac{1}{2}(\theta_1 - \theta_2)]$ | satellite |

The relative intensities are determined by the tilt angles $\theta_1$ and $\theta_2$ of the effective fields in the rotating frame:

$$\tan\theta_1 = (\gamma B_2/2\pi)/(\Delta + \tfrac{1}{2}J)$$
$$\tan\theta_2 = (\gamma B_2/2\pi)/(\Delta - \tfrac{1}{2}J) \qquad \textbf{1.35}$$

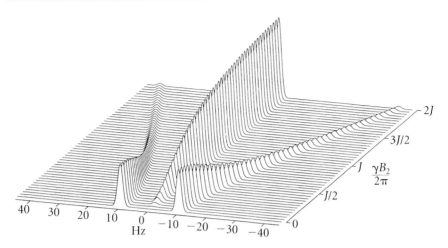

**Figure 1.2** Decoupling at the exact resonance condition ($\Delta = 0$) as a function of the intensity of the decoupler field $\gamma B_2/2\pi$. The normal lines move apart and lose intensity, while the decoupled line appears at the chemical shift frequency and grows in intensity, eventually dominating the decoupled spectrum. The parameters are $J_{IS} = 20$ Hz; Lorentzian linewidth 2 Hz.

If we consider the case of decoupling at exact resonance for the $I$ spins ($\Delta = 0$) then the two main transitions fall exactly at the $S$-spin chemical shift since $E_1 = E_2$. However, there are two other transitions separated by $E_1 + E_2$. For high levels of the decoupler field the latter lines are weak and are normally called the satellites, but note that at very low levels of $B_2$ these become the normal $S$-spin transitions at $\pm\frac{1}{2}J_{IS}$. We may sketch out the $S$-spin spectrum as a function of the intensity of the irradiation field $B_2$ in the form of a stacked-trace diagram (Figure 1.2). As $B_2$ is increased, transitions (1)–(3) and (2)–(4) at the centre of the spectrum take over more and more of the total intensity until, for $\gamma B_2/2\pi = 2J$, the spectrum is virtually decoupled with only very weak satellite lines visible. This is why the decoupling condition is usually quoted as $\gamma B_2/2\pi \gg |J|$. In many decoupling experiments the satellite lines are simply ignored since they are very weak when the decoupling field is intense. Moreover, since the position of the satellite lines is quite sensitive to the intensity of $B_2$, if there is any spatial inhomogeneity of the decoupler field they are preferentially broadened.

Usually the chemist is more interested in the form of the observed spectrum as the decoupling frequency is changed for a constant $B_2$ intensity. Figure 1.3 illustrates the case where $\gamma B_2/2\pi = 20$ Hz, $J = 10$ Hz and the linewidth is 1.5 Hz. At exact resonance for the decoupling frequency ($\Delta = 0$) there is a strong singlet at the $S$-spin chemical shift. As $B_2$ is shifted from exact resonance, a residual splitting appears, increasing as $\Delta$ increases and eventually becoming equal to $J_{IS}$ at large offsets. When $|\Delta| \gg |J|$ the tilt angles of the two effective fields are almost equal and the outer transitions (1)–(4) and (2)–(3) have vanishing intensities while the main lines (1)–(3) and (2)–(4) acquire all the intensity.

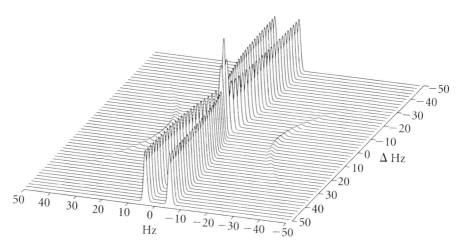

**Figure 1.3** Simulation of the effect of decoupling a two-spin system as a function of the decoupler offset $\Delta$ Hz. The relevant parameters are $J_{IS} = 10$ Hz, $\gamma B_2/2\pi = 20$ Hz and the linewidth is 1.5 Hz. Note the weak satellite lines which are most prominent for small decoupler offsets. The equations governing this behaviour were derived on the basis of the Bloch–Siegert shifts of the energy levels.

More complicated spin systems obey analogous rules.[25] In each case the usual multiplet structure shrinks as the decoupling field $B_2$ is brought closer to resonance, eventually coalescing into a singlet at the chemical shift frequency, while weak satellite lines appear, symmetrically disposed about the chemical shift frequency. Off-resonance coherent proton decoupling is occasionally used to scale down the CH splitting so as to indicate the number of protons attached to a given $^{13}$C site without the complications that arise with overlapping multiplets. The positions of some lines are particularly sensitive to the intensity of the $B_2$ field because it affects the intensity of $B_{\text{eff}}$. For example, the outer lines of a partially decoupled $^{13}$C quartet of a methyl group often show an abnormal broadening because their frequencies are dependent on $B_{\text{eff}}$, and they are therefore three times more sensitive to the spatial inhomogeneity of $B_2$ than the inner lines.[26]

We shall see in § 7.3 that the much more complex theory of broadband hetero-nuclear decoupling is based on the same kinds of considerations about energy-level shifts induced by the decoupler field through the generalized Bloch–Siegert effect.

### 1.4.5 Hartmann–Hahn experiment

Transformation into a rotating frame can be used to understand how to tune-in the frequency of one nuclear species to that of another that has a quite different gyro-magnetic ratio. This is the essence of the Hartmann–Hahn experiment[27] where a spin species $I$ transfers polarization to another spin species $S$, thereby increasing the intensity of the signals in the $S$-spin spectrum. The $I$-spin energy level manifold is transformed into a reference frame rotating in synchronism with the radio-frequency field $B_1$. This involves lowering the upper level and raising the lower level by one-half of the frequency of the $B_1$ field. Similarly the $S$-spin energy level manifold is transformed into a different frame rotating in synchronism with a second radiofrequency field $B_2$. The two radiofrequencies are chosen to make these levels almost degenerate in their respective rotating frames.

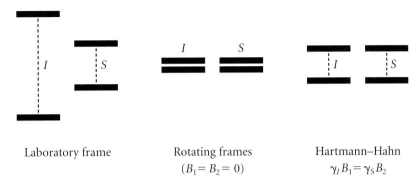

Laboratory frame             Rotating frames             Hartmann–Hahn
                             $(B_1 = B_2 = 0)$             $\gamma_I B_1 = \gamma_S B_2$

(Note that these diagrams are not to scale; energy level separations in the laboratory frame are much higher than in the rotating frame.) The degeneracies are lifted by introducing the fields $B_1$ and $B_2$. The quantities $\gamma_I B_1$ and $\gamma_S B_2$ are chosen to be strong in comparison with the natural spread of $I$ and $S$ precession frequencies caused by dipolar or scalar coupling. In some circumstances this may require specially designed pulse sequences[28–30] (§ 7.4.8) to extend the effective bandwidths covered by $B_1$ and $B_2$. The introduction of these radiofrequencies splits the corresponding levels by $\gamma_I B_1$ and $\gamma_S B_2$. If the Hartmann–Hahn matching condition

$$\gamma_I B_1 = \gamma_S B_2 \qquad \textbf{1.40}$$

is satisfied, then the new energy level separations of the $I$ and $S$ species are equal and the two spins may now communicate. In one form of the experiment the hot $I$ spins are cooled by the relatively cold $S$ spins and this effect is used to detect the presence of the rare $S$ spins indirectly. In a solid there are strong dipolar interactions between the spins, and it is permissible to define a spin temperature for the abundant $I$ spins, which may differ from the lattice temperature.

In liquid-phase NMR the concept of spin temperature is no longer appropriate. The relevant interaction is the spin–spin coupling $J_{IS}$ rather than the dipolar coupling, and it induces an oscillatory exchange of coherence between $I$ and $S$. Applications of this variant of the Hartmann–Hahn experiment are discussed in detail in § 5.5.7 and § 7.4.8.

## 1.5 Strong coupling effects

### 1.5.1 Two-spin system

Everything discussed so far in this chapter has presupposed the condition $|\nu_I - \nu_S| >> |J_{IS}|$, which is known as the weak coupling or first-order approximation.

If this condition does not hold, the simple product wavefunctions are no longer the eigenfunctions of the system. It turns out that wavefunctions with the same total $m$ value are mixed. For example in the two-spin system, the two $m=0$ levels (2 and 3) are slightly shifted in energy and acquire the mixed wavefunctions

$$\Psi_2 = |\alpha\beta> \cos\theta + |\beta\alpha> \sin\theta$$
$$\Psi_3 = |\beta\alpha> \cos\theta - |\alpha\beta> \sin\theta \qquad \textbf{1.41}$$

where $\tan 2\theta = |J_{IS}|/|\nu_I - \nu_S|$. The term $\sin 2\theta$ is called the strong coupling parameter and is very small in the weak coupling limit, where the two wavefunctions revert to the basis functions $|\alpha\beta>$ and $|\beta\alpha>$. The stage where strong coupling effects become appreciable may be taken to be where $\tan 2\theta$ is greater than about 0.1.

The convention[31] for the nomenclature of spin systems is to assign letters close together in the alphabet if the corresponding spins are strongly coupled. Thus an $AB$ spectrum is strongly coupled but an $AX$ system is only weakly coupled. In order to follow this convention, we now change the two-spin nomenclature from $IS$ to $AB$, and write the energy levels as:

| Level | Energy |
|-------|--------|
| (4) | $+\frac{1}{2}(\nu_A + \nu_B) + \frac{1}{4}J_{AB}$ |
| (3) | $-D - \frac{1}{4}J_{AB}$ |
| (2) | $+D - \frac{1}{4}J_{AB}$ |
| (1) | $-\frac{1}{2}(\nu_A + \nu_B) + \frac{1}{4}J_{AB}$ |

$$\qquad \textbf{1.42}$$

where we have defined the positive quantity:

$$2D = \{(\nu_A - \nu_B)^2 + J_{AB}^2\}^{\frac{1}{2}} \qquad \textbf{1.43}$$

There are four allowed transitions:

| Spin | Levels | Frequency | Intensity |
|------|--------|-----------|-----------|
| A | (1)–(3) | $+\frac{1}{2}(\nu_A + \nu_B) - \frac{1}{2}J_{AB} - D$ | $1 - \sin 2\theta$ |
| A | (2)–(4) | $+\frac{1}{2}(\nu_A + \nu_B) + \frac{1}{2}J_{AB} - D$ | $1 + \sin 2\theta$ |
| B | (1)–(2) | $+\frac{1}{2}(\nu_A + \nu_B) - \frac{1}{2}J_{AB} + D$ | $1 + \sin 2\theta$ |
| B | (3)–(4) | $+\frac{1}{2}(\nu_A + \nu_B) + \frac{1}{2}J_{AB} + D$ | $1 - \sin 2\theta$ |

$$\qquad \textbf{1.44}$$

Note that $J_{AB}$ may be positive or negative, but this has no effect on the form of the spectrum, merely reassigning the transitions to different energy levels. These expressions allow us to picture the way the two-spin spectrum changes as we go from weak coupling ($AX$) through strong coupling ($AB$) to the degenerate case ($A_2$).

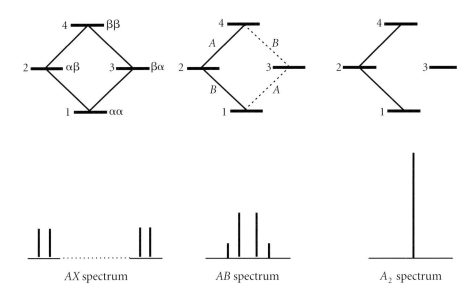

AX spectrum                    AB spectrum                    $A_2$ spectrum

Although the outer splittings are always equal to $J_{AB}$, the centres of these doublets do not correspond to the exact chemical shifts, since there is a difference between $2D$ and $(\nu_A - \nu_B)$. Care must always be exercised when extracting chemical shift values from strongly coupled spectra.

Note that the inner lines gradually gain intensity at the expense of the outer lines as the coupling gets stronger. In the limit where the chemical shift difference vanishes, level (2) is exactly midway between levels (1) and (4) and the two strong lines both fall at the same frequency and acquire all the intensity ($\sin 2\theta = 1$). The outer lines vanish.

This $A_2$ spin system has one energy level with a purely antisymmetrical wavefunction, $\Psi_3 = 1/(2^{\frac{1}{2}})\{|\beta\alpha> - |\alpha\beta>\}$, because $\sin\theta = \cos\theta = 1/(2^{\frac{1}{2}})$. Transitions to this level are strictly forbidden by a symmetry-based selection rule. This illustrates the general rule that we cannot detect the spin–spin coupling between equivalent spins (such as the protons within a methyl group).

## 1.5.2 Three-spin system $AX_2$

If we introduce another spin into an $AX$ system, one possibility is that the new spin has a different chemical shift from the other two, so this would be an $AMX$ system. If the molecular symmetry is such that the new spin has exactly the same shift as $X$, the two spins are said to be chemically equivalent or isochronous. This is called an $AX_2$ spin system, and it has the energy level diagram shown below.

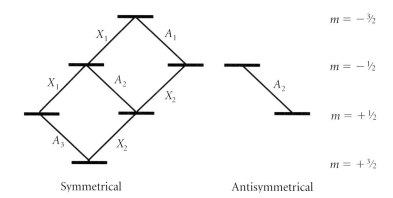

Symmetrical          Antisymmetrical

The interesting new feature is that the energy level diagram consists of two quite separate manifolds and there are no transitions allowed between the two. This arises because of the symmetry of the problem. One manifold consists entirely of symmetrical states, the other of antisymmetrical states. There is another remarkable feature. One of the $A$ transitions ($A_2$ in the antisymmetrical manifold) has no energy levels in common with the $X$ spins. It is thus completely unaffected by double irradiation experiments involving the $X$ spins, or by coherence transfer from the $X$ spins. In some molecules, relaxation through the spin–spin coupling to the $X$ nuclei broadens the other $A$ lines, but leaves the antisymmetrical $A_2$ line unusually sharp.[32]

### 1.5.3 Magnetic equivalence

There is another type of three-spin system which must be carefully distinguished from the $AX_2$ case. Two $X$ spins that are chemically equivalent may nevertheless have different couplings to the $A$ spin, $J_{AX}$ and $J_{AX'}$. These spins are said to be magnetically nonequivalent and the spin system is designated $AXX'$. Magnetic nonequivalence is more easily visualized in the much more common $AA'XX'$ case which is characteristic of *para*-disubstituted benzenes. The two $A$ spins (and the two $X$ spins) are obviously chemically equivalent by symmetry, but there are two distinct couplings $J_{AX}$ and $J_{AX'}$ corresponding to the *ortho* and *para* proton pairs.

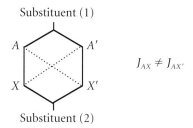

Substituent (1)

$A$     $A'$

$X$     $X'$

$J_{AX} \neq J_{AX'}$

Substituent (2)

The corresponding strongly coupled cases $ABB'$ and $AA'BB'$ are treated in the standard texts on spectral analysis; however, most of the important principles may be illustrated by reference to the simpler spin system $ABX$.

### 1.5.4 Three-spin system *ABX*

Although few spectroscopists bother with analytical solutions now that computer programs are available for most spin systems, it is worth working through the *ABX* case in order to illustrate some important general features. The *ABX* spin system has *A* and *B* strongly coupled, while *X* is weakly coupled to both *A* and *B*:

$$\nu_A - \nu_B \approx J_{AB}$$
$$|\nu_A - \nu_X| \gg |J_{AX}|$$
$$|\nu_B - \nu_X| \gg |J_{BX}| \qquad \textbf{1.45}$$

The key to *ABX* spectra is the factorization of the *AB* region into two separate *AB* subspectra, one with the effective chemical shifts

$$\nu_A + \tfrac{1}{2}J_{AX} \text{ and } \nu_B + \tfrac{1}{2}J_{BX} \qquad \textbf{1.46}$$

and the other with the effective chemical shifts:

$$\nu_A - \tfrac{1}{2}J_{AX} \text{ and } \nu_B - \tfrac{1}{2}J_{BX} \qquad \textbf{1.47}$$

This is the *X* approximation referred to earlier in § 1.3.6. The expressions for the energy levels may be written:

| ABX | Energy |
|-----|--------|
| βββ | $E_8 = +\tfrac{1}{2}\nu_A + \tfrac{1}{2}\nu_B + \tfrac{1}{2}\nu_X + \tfrac{1}{4}J_{AB} + \tfrac{1}{4}J_{AX} + \tfrac{1}{4}J_{BX}$ |
| βαβ | $E_7 = +\tfrac{1}{2}\nu_X - \tfrac{1}{4}J_{AB} + D_+$ |
| αββ | $E_6 = +\tfrac{1}{2}\nu_X - \tfrac{1}{4}J_{AB} - D_+$ |
| ββα | $E_5 = +\tfrac{1}{2}\nu_A + \tfrac{1}{2}\nu_B - \tfrac{1}{2}\nu_X + \tfrac{1}{4}J_{AB} - \tfrac{1}{4}J_{AX} - \tfrac{1}{4}J_{BX}$ |
| ααβ | $E_4 = -\tfrac{1}{2}\nu_A - \tfrac{1}{2}\nu_B + \tfrac{1}{2}\nu_X + \tfrac{1}{4}J_{AB} - \tfrac{1}{4}J_{AX} - \tfrac{1}{4}J_{BX}$ |
| βαα | $E_3 = -\tfrac{1}{2}\nu_X - \tfrac{1}{4}J_{AB} + D_-$ |
| αβα | $E_2 = -\tfrac{1}{2}\nu_X - \tfrac{1}{4}J_{AB} - D_-$ |
| ααα | $E_1 = -\tfrac{1}{2}\nu_A - \tfrac{1}{2}\nu_B - \tfrac{1}{2}\nu_X + \tfrac{1}{4}J_{AB} + \tfrac{1}{4}J_{AX} + \tfrac{1}{4}J_{BX}$ |

$$\textbf{1.48}$$

where the new parameters $D_+$ and $D_-$ (always positive) are defined by:

$$2D_+ = \{[(\nu_A - \nu_B) + \tfrac{1}{2}(J_{AX} - J_{BX})]^2 + J_{AB}^2\}^{\frac{1}{2}} \qquad \textbf{1.49}$$

$$2D_- = \{[(\nu_A - \nu_B) - \tfrac{1}{2}(J_{AX} - J_{BX})]^2 + J_{AB}^2\}^{\frac{1}{2}} \qquad \textbf{1.50}$$

We see that these are derived from the *D* parameter defined above for the *AB* system, but with the two effective *AB* shifts in place of $(\nu_A - \nu_B)$:

$$\Delta_{AB^+} = (\nu_A - \nu_B) + \tfrac{1}{2}(J_{AX} - J_{BX}) \qquad \textbf{1.51}$$

$$\Delta_{AB^-} = (\nu_A - \nu_B) - \tfrac{1}{2}(J_{AX} - J_{BX}) \qquad \textbf{1.52}$$

These energy levels can be assigned to two separate manifolds $X(\alpha)$ and $X(\beta)$ with the effective *AB* shifts $\Delta_{AB^+}$ and $\Delta_{AB^-}$. Certain energy levels, (2), (3), (6) and (7), have mixed wavefunctions and are shifted if the mixing is changed by altering

$(\nu_A - \nu_B)$; these are shown shaded. The remaining levels (1), (4), (5) and (8) can still be represented by the basis functions ($\alpha\alpha\alpha$ etc.) and are unaffected by a change in $(\nu_A - \nu_B)$; they are shown as solid lines:

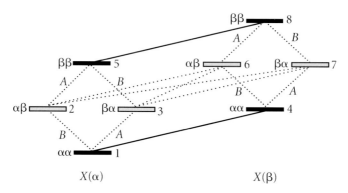

$$X(\alpha) \hspace{6cm} X(\beta)$$

Consequently there are two $X$ transitions (solid lines) that always remain at the same frequencies and retain the same intensities (unity) whatever the chemical shift difference $(\nu_A - \nu_B)$. They appear at the frequencies

$$\nu_X \pm \tfrac{1}{2}|J_{AX} + J_{BX}|$$

There are eight transition frequencies in the $AB$ region; it is convenient to measure them from the mean $AB$ chemical shift $\tfrac{1}{2}(\nu_A + \nu_B)$:

| Spin | Levels | Frequency | Intensity |
|------|--------|-----------|-----------|
| $A$ | (6)–(8) | $\tfrac{1}{2}(\nu_A + \nu_B) + \tfrac{1}{2}J_{AB} + \tfrac{1}{4}J_{AX} + \tfrac{1}{4}J_{BX} + D_+$ | $1 - \sin2\theta_+$ |
| $A$ | (2)–(5) | $\tfrac{1}{2}(\nu_A + \nu_B) + \tfrac{1}{2}J_{AB} - \tfrac{1}{4}J_{AX} - \tfrac{1}{4}J_{BX} + D_-$ | $1 - \sin2\theta_-$ |
| $A$ | (4)–(7) | $\tfrac{1}{2}(\nu_A + \nu_B) - \tfrac{1}{2}J_{AB} + \tfrac{1}{4}J_{AX} + \tfrac{1}{4}J_{BX} + D_+$ | $1 + \sin2\theta_+$ |
| $A$ | (1)–(3) | $\tfrac{1}{2}(\nu_A + \nu_B) - \tfrac{1}{2}J_{AB} - \tfrac{1}{4}J_{AX} - \tfrac{1}{4}J_{BX} + D_-$ | $1 + \sin2\theta_-$ |
| $B$ | (7)–(8) | $\tfrac{1}{2}(\nu_A + \nu_B) + \tfrac{1}{2}J_{AB} + \tfrac{1}{4}J_{AX} + \tfrac{1}{4}J_{BX} - D_+$ | $1 + \sin2\theta_+$ |
| $B$ | (3)–(5) | $\tfrac{1}{2}(\nu_A + \nu_B) + \tfrac{1}{2}J_{AB} - \tfrac{1}{4}J_{AX} - \tfrac{1}{4}J_{BX} - D_-$ | $1 + \sin2\theta_-$ |
| $B$ | (4)–(6) | $\tfrac{1}{2}(\nu_A + \nu_B) - \tfrac{1}{2}J_{AB} + \tfrac{1}{4}J_{AX} + \tfrac{1}{4}J_{BX} - D_+$ | $1 - \sin2\theta_+$ |
| $B$ | (1)–(2) | $\tfrac{1}{2}(\nu_A + \nu_B) - \tfrac{1}{2}J_{AB} - \tfrac{1}{4}J_{AX} - \tfrac{1}{4}J_{BX} - D_-$ | $1 - \sin2\theta_-$   **1.53** |

and there are six transitions in the $X$ region symmetrically arranged about $\nu_X$:

| Spin | Levels | Frequencies | Intensity |
|------|--------|-------------|-----------|
| $X$ | (5)–(8) | $\nu_X + \tfrac{1}{2}(J_{AX} + J_{BX})$ | $1$ |
| $X$ | (2)–(7) | $\nu_X + D_+ + D_-$ | $\sin^2(\theta_+ - \theta_-)$ |
| $X$ | (3)–(7) | $\nu_X + D_+ - D_-$ | $\cos^2(\theta_+ - \theta_-)$ |
| $X$ | (2)–(6) | $\nu_X - D_+ + D_-$ | $\cos^2(\theta_+ - \theta_-)$ |
| $X$ | (3)–(6) | $\nu_X - D_+ - D_-$ | $\sin^2(\theta_+ - \theta_-)$ |
| $X$ | (1)–(4) | $\nu_X - \tfrac{1}{2}(J_{AX} + J_{BX})$ | $1$   **1.54** |

The angles $\theta_+$ and $\theta_-$ have no physical significance but are convenient for expressing the intensities, based on the definitions

$$\sin 2\theta_+ = \tfrac{1}{2} J_{AB}/D_+ \qquad\qquad \textbf{1.55}$$

$$\sin 2\theta_- = \tfrac{1}{2} J_{AB}/D_- \qquad\qquad \textbf{1.56}$$

$$\cos 2\theta_+ = \tfrac{1}{2}\left[(\nu_A - \nu_B) + \tfrac{1}{2}(J_{AX} - J_{BX})\right]/D_+ \qquad\qquad \textbf{1.57}$$

$$\cos 2\theta_- = \tfrac{1}{2}\left[(\nu_A - \nu_B) - \tfrac{1}{2}(J_{AX} - J_{BX})\right]/D_- \qquad\qquad \textbf{1.58}$$

where $\theta_+$ and $\theta_-$ are confined to the range from 0 to $\pi$ radians. We find that the sign of $J_{AB}$ has no influence on the form of the spectrum, whereas the relative signs of $J_{AX}$ and $J_{BX}$ are important.

A vivid illustration of the effect of varying the chemical shift difference $(\nu_A - \nu_B)$ is provided by the simulation in Figure 1.4, where the relevant fixed parameters are $J_{AB}$ = 16 Hz, $J_{AX}$ = 2 Hz, $J_{BX}$ = 9 Hz. These were chosen to correspond approximately with those found experimentally in cinnamalazine, where the $AB$ shift can be manipulated by changing the composition of the solvent.[33] At the extremes of the $AB$ shift range, the spectrum tends towards the weakly coupled case; we can easily recognize two separate $AB$ patterns and the intensities become much more uniform.

As the $AB$ shift is reduced, it is still possible to pick out the two interpenetrating $AB$ patterns. Each of these $AB$ patterns becomes degenerate (a singlet) when the respective effective chemical shift difference ($\Delta_{AB^+}$ or $\Delta_{AB^-}$) goes to zero. For this case the combination lines are always weak, but they can just be discerned in the $X$ spectrum when $\nu_A$ is close to $\nu_B$. In some $ABX$ spectra combination lines can be quite intense.

### 1.5.5 Deceptive simplicity

Figure 1.4 illustrates one of the problems that can arise if we make a cursory assign-

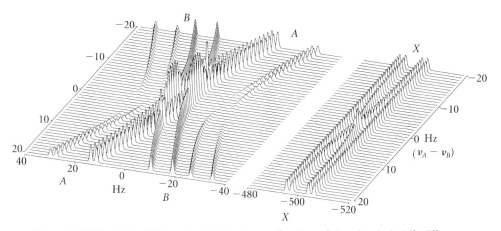

**Figure 1.4** Simulated $ABX$ spectra displayed as a function of the chemical shift difference $(\nu_A - \nu_B)$. The fixed parameters are $J_{AB}$ = 16 Hz, $J_{AX}$ = 2 Hz, $J_{BX}$ = 9 Hz. Note the deceptively simple $X$ spectrum (a 1:2:1 triplet) when $(\nu_A - \nu_B)$ is very small. There are two very weak combination lines that are just visible in the $X$ region at small values of $(\nu_A - \nu_B)$.

ment of an NMR spectrum assuming weak coupling, when in fact two or more spins are strongly coupled. If the $A$ and $B$ spins are sufficiently strongly coupled that both the effective $AB$ shifts ($\Delta_{AB^+}$ and $\Delta_{AB^-}$) are small with respect to $|J_{AB}|$,

$$\left| (\nu_A + \tfrac{1}{2}J_{AX}) - (\nu_B + \tfrac{1}{2}J_{BX}) \right| << |J_{AB}| \qquad \textbf{1.59}$$

$$\left| (\nu_A - \tfrac{1}{2}J_{AX}) - (\nu_B - \tfrac{1}{2}J_{BX}) \right| << |J_{AB}| \qquad \textbf{1.60}$$

Then $D_+$ is very nearly equal to $D_-$, and instead of the expected doublet of doublets in the X spectrum, we find an approximate 1:2:1 triplet with splitting $\tfrac{1}{2}|J_{AX} + J_{BX}|$. One could perhaps be forgiven for concluding that $J_{AX}$ and $J_{BX}$ were equal. This is a classic case of deceptive simplicity caused by the strong coupling. In general, any spins that are tightly coupled among themselves behave as if they were all coupled equally to other spins with the algebraic mean of the true coupling constants.

### 1.5.6 Virtual coupling

The example above also warns us to make a clear distinction between the splitting, which is what is observed in the spectrum, and the corresponding coupling constant, which is a molecular parameter determined by the electronic structure. The two are usually represented by $J$ and are measured in Hz, but they may or may not be equal. Indeed, we can see clearly from Figure 1.4 that splittings observed in the $X$ region change markedly as we alter the $AB$ chemical shift although the relevant coupling constants are indeed constant. Even if $J_{AX}$ had been zero we would have observed a nonzero splitting in the central range of Figure 1.4, where ($\nu_A - \nu_B$) is small. A spurious splitting of this kind is known as virtual coupling and it could prove very misleading to the structural chemist. Virtual coupling disappears from the X spectrum if the $AB$ shift becomes very large compared with $|J_{AB}|$, that is to say when the $ABX$ system approaches the first-order $AMX$ case.

### 1.5.7 Combination lines

The two X transitions (2)–(7) and (3)–(6) are usually called combination lines by analogy with certain forbidden transitions in infrared spectroscopy. Although, under certain circumstances, they can be just as intense as the formally allowed transitions, they are more often weak and eventually disappear as the $AB$ coupling becomes weak. They are the weak outer lines just visible in the X region of Figure 1.4. One other combination line (4)–(5) is strictly forbidden in the $ABX$ case but becomes allowed in the $ABC$ case where the chemical shift of the third spin approaches those of $A$ and $B$. There is no analytical solution for the $ABC$ case; the spectra are analysed numerically by computer.

We may conclude that strong coupling can cause several disconcerting effects in high resolution NMR spectra—the appearance of more resonance lines than expected, distortion of the relative intensities, and very complex patterns of resonance lines. In a system of three or more coupled spins, two sites $I$ and $S$ may exhibit a virtual coupling even though $J_{IS}$ is identically zero. A strongly coupled

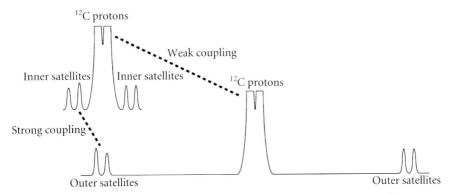

**Figure 1.5** Schematic representation (not to scale) of the case where one set of $^{13}C$ satellites in a proton spectrum is strongly coupled (*AB* system) whereas the protons attached to $^{12}C$ nuclei are only weakly coupled (*AX* system). The outer satellies arise from directly bound $^{13}C$–H, whereas the inner satellites arise from long-range coupling. The two $^{13}C$–H couplings are assumed to have the same sign.

spectrum may appear to contain fewer splittings than anticipated (deceptive simplicity). Finally, it is not always feasible to measure the chemical shifts and coupling constants directly from such spectra.

Fortunately there are several computer programs available to analyse strongly coupled spectra. Although strong coupling effects become less serious as NMR spectrometers are operated at higher and higher magnetic fields, they can still surprise the unwary spectroscopist on occasion. For example, a proton spectrum may be quite clearly weakly coupled, but the $^{13}C$ satellites can still show strong coupling effects because the difference between the effective chemical shifts is comparable with the relevant proton–proton coupling (Figure 1.5). This can have unfortunate ramifications[34] in $^{13}C$—H correlation spectroscopy (§ 8.3).

## 1.6 **Discussion**

For those who use high resolution NMR simply as a tool for structural studies, there is a temptation to concentrate on the spectrum and ignore the fact that one is observing transitions between energy levels. This chapter has set out to show that this approach may overlook several important features such as spin sorting, the Overhauser effect and polarization transfer. Furthermore, there is a danger of excluding the rich field of forbidden transitions and double resonance phenomena, not to mention many aspects of two-dimensional spectroscopy.

## **References**

1. S. Forsén and R. A. Hoffman, *J. Chem. Phys.* **39**, 2892 (1963).
2. J. H. Noggle and R. E. Schirmer, *The Nuclear Overhauser Effect, Chemical Applications*, Academic Press, New York, 1971.

3. D. Neuhaus and M. Williamson, *The Nuclear Overhauser Effect in Structural and Conformational Analysis*, VCH Publishers, Weinheim, 1989.

4. K. F. Kuhlman and D. M. Grant, *J. Am. Chem. Soc.* **90**, 7355 (1968).

5. G. A. Morris and R. Freeman, *J. Am. Chem. Soc.* **101**, 760 (1979).

6. A. A. Maudsley and R. R. Ernst, *Chem. Phys. Lett.* **50**, 368 (1977).

7. R. Kaptein and L. J. Oosterhoff, *Chem. Phys. Lett.* **4**, 195, 214 (1969).

8. G. L. Closs, *J. Am. Chem. Soc.* **91**, 4552 (1969).

9. C. R. Bowers and D. P. Weitekamp, *Phys. Rev. Lett.* **57**, 2645 (1986).

10. C. R. Bowers and D. P. Weitekamp, *J. Am. Chem. Soc.* **109**, 5541 (1987).

11. S. I. Hommeltoft, D. H. Berry and R. Eisenberg, *J. Am. Chem. Soc.* **108**, 5346 (1986).

12. N. D. Bhaskar, W. Happer and T. McClelland, *Phys. Rev. Lett.* **49**, 25 (1982).

13. H. W. Long, H. C. Gaede, J. Shore, L. Reven, C. R. Bowers, J. Kritzenberger, T. Pietrass, A. Pines, P. Tang and J. A. Reimer, *J. Am. Chem. Soc.* **115**, 8491 (1993).

14. J. Jeener, Ampere International Summer School, Basko Polje, Yugoslavia, 1971, published in *NMR and More. In Honour of Anatole Abragam*, Eds: M. Goldman and M. Porneuf, Les Editions de Physique, Les Ulis, France, 1994.

15. W. P. Aue, E. Bartholdi and R. R. Ernst, *J. Chem. Phys.* **64**, 2229 (1976).

16. W. A. Anderson, R. Freeman and C. A. Reilly, *J. Chem. Phys.* **39**, 1518 (1963).

17. K. A. McLauchlan and D. H. Whiffen, *Proc. Chem. Soc. (London)*, 144 (1962).

18. X. L. Wu, P. Xu and R. Freeman, *J. Magn. Reson.* **88**, 417 (1990).

19. J. A. Pople and T. Schaefer, *Mol. Phys.* **3**, 547 (1962).

20. P. Diehl and J. A. Pople, *Mol. Phys.* **3**, 557 (1960)

21. R. Freeman and W. A. Anderson, *J. Chem. Phys.* **37**, 2053 (1962).

22. J. A. Ferretti and R. Freeman, *J. Chem. Phys.* **44**, 2054 (1966).

23. F. Bloch and A. Siegert, *Phys. Rev.* **57**, 522 (1940).

24. N. F. Ramsey, *Phys. Rev.* **100**, 1191 (1955).

25. W. A. Anderson and R. Freeman, *J. Chem. Phys.* **37**, 85 (1962).

26. R. Freeman, J. B. Grutzner, G. A. Morris, and D. L. Turner, *J. Am. Chem. Soc.* **100**, 5637 (1978).

27. S. L. Hartmann and E. L. Hahn, *Phys. Rev.* **128**, 2042 (1962).

28. M. H. Levitt, R. Freeman and T. Frenkiel, *J. Magn. Reson.* **47**, 328 (1972).

29. A. J. Shaka, J. Keeler and R. Freeman, *J. Magn. Reson.* **53**, 313 (1983).

30. M. Kadkhodaie, O. Rivas, M. Tan, A. Mohebbi and A. J. Shaka, *J. Magn. Reson.* **91**, 437 (1991).

31. J. A. Pople, W. G. Schneider and H. J. Bernstein, *High-Resolution Nuclear Magnetic Resonance*, McGraw-Hill, New York, 1959.

32. W. A. Anderson, *Phys. Rev.* **102**, 151 (1956).

33. R. Freeman and N. S. Bhacca, *J. Chem. Phys.* **45**, 3795 (1966).

34. G. A. Morris and K. I. Smith, *J. Magn. Reson.* **65**, 506 (1985).

# 2

# Vector model

## 2.1 Introduction

NMR phenomena can be described by several different theoretical models, some of them quite sophisticated in their mathematical formalism. However, for a chemist who wishes to use NMR spectroscopy as a tool to solve structural problems, simplicity is probably the most attractive feature of any proposed model, provided that the treatment is valid for the most common experiments. Density operator theory is elegant and comprehensive, but the resulting algebra becomes unwieldy for systems of three or more coupled spins, and it is then difficult to retain a clear picture of the physical processes involved. Physical pictures are important to chemists. Much of their work involves the visualization of the complex three-dimensional relationships characteristic of molecular interactions, and hence geometry is probably preferable to algebra. The chemist is, above all, a pragmatist, and does not normally seek more complicated solutions than are absolutely necessary for the task in hand. This is why the vector model is so important for the understanding of modern pulsed Fourier transform NMR and the myriad variations on this theme.

Bloch[1] emphasized that quantum mechanical systems, such as nuclear spins, may be treated by classical mechanics provided we concentrate on the ensemble average over a large number of spins. We avoid unproductive questions about one-spin flipping, just as we might side-step the philosophical concept of one-hand clapping. Any realistic NMR sample contains upwards of $10^{17}$ spins within the active volume of the receiver coil. Each spin is associated with a tiny magnetic moment that may be represented by a vector $m$; we simply focus our attention on the resultant vector $M$, summed over all the active spins. This macroscopic magnetization $M$ behaves classically. It may rotate about a magnetic field, and decay or recover through the influence of relaxation processes. Strictly the model applies to an ensemble of isolated, noninteracting spins, but it is usually generalized to treat the more complex spin systems appropriate to high-resolution NMR spectra.

## 2.2 Precession

Precession is the motion executed by a spinning top or gyroscope. The nuclear magnetization vector sweeps out a cone of precession, keeping a fixed angle with respect to the magnetic field direction ($+z$):

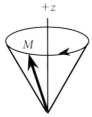

For electron or nuclear spins the corresponding motion is known as Larmor precession and, neglecting chemical shifts for the moment, is governed by the equation:

$$\omega_L = \gamma B_0. \tag{2.1}$$

It is common practice to take this universal proportionality of frequency and field for granted, expressing magnetic fields in frequency units without further comment and describing NMR spectrometers in terms of MHz (for protons) rather than tesla, the unit of magnetic field.

When the spin system is at Boltzmann equilibrium and has not been exposed to radiofrequency fields, the individual spins precess about the applied magnetic field with no coherence between their phases. Consequently, the resultant transverse magnetization components $M_x$ and $M_y$ are zero. There is nevertheless a net component ($M_0$) along the magnetic field direction because there is a slight excess of spins in the lower spin state. This is the initial condition for most NMR experiments. In the general case, where the system is not necessarily at equilibrium, the component $M_z$ is proportional to the population difference.

### 2.2.1 Rotating reference frame

On earth, we perform experiments in mechanics in a rotating frame—the motion of a falling stone is in fact much more complicated than it appears to observers on earth because, to them, the earth's rotation appears to have been neutralized. A similar simplification can be achieved for the motion of nuclear spins. The interaction of the radiofrequency field with the nuclear magnetization vector is rather complex when considered in the usual laboratory coordinate frame. The radiofrequency field is linearly polarized, but we may decompose it into two counter-rotating circularly polarized fields, only one of which has the right sense of rotation to interact constructively with the nuclear spins (§ 1.4.3). This is the field $B_1$. If the problem is reformulated in a coordinate system that rotates about the $z$-axis at the transmitter frequency ($\omega_0$), $B_1$ becomes a static vector and the motion of nuclear magnetization is much easier to follow. The normal convention is to set $B_1$ along the $+x$ axis of the rotating frame. It may of course be phase-shifted when necessary, and the classic example is the Meiboom–Gill spin echo method, where $B_1$ is aligned along the $+y$ axis (§ 4.1.6).

There is another important consequence of transformation into the rotating frame. In the laboratory frame, magnetization precesses about the static field at the Larmor frequency, but in the rotating frame it appears to precess much more slowly (at the difference between the Larmor frequency $\omega_L$ and the rotating frame frequency $\omega_0$). Since the precession equation must be obeyed in both reference frames, we have to accept that the apparent field intensity is proportionately reduced in the rotating frame, leaving a residual field

$$\Delta B = (\omega_L - \omega_0)/\gamma \qquad\qquad \textbf{2.2}$$

Whereas the applied field is very intense ($|B_0| \gg |\Delta B|$), the residual field $\Delta B$ is of the same order as the radiofrequency field $B_1$. For exact synchronization of the frame frequency with the Larmor frequency, $\Delta B = 0$.

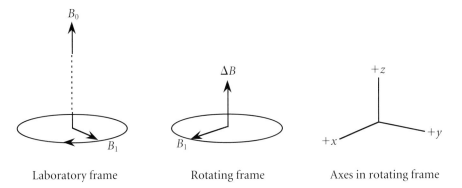

| Laboratory frame | Rotating frame | Axes in rotating frame |

In describing pulsed NMR experiments, transformation into the rotating frame is normally taken for granted; in what follows, all vector diagrams involving radiofrequency pulses will be considered in the rotating frame, with the axes $x$, $y$ and $z$. The radiofrequency field $B_1$ will normally be aligned along $+x$.

## 2.2.2 **Bloch equations**

The motion of nuclear magnetization vectors is governed by a set of phenomeno-logical equations called the Bloch equations. Formally, the magnetization vector $M$ obeys the (classical) torque equation

$$\frac{\mathrm{d}M}{\mathrm{d}t} = \gamma(M \wedge B) \qquad\qquad \textbf{2.3}$$

where $B$ is the magnetic field vector. This means that any magnetization vector $M$ precesses about the direction of an applied field $B$ with an angular frequency $\gamma|B|$ rad s$^{-1}$. It is convenient to rewrite this as three separate equations that tell us how the three components of magnetization $M_x$, $M_y$ and $M_z$ behave in the presence of fields $B_z$, $B_x$ and $B_y$. For the moment we neglect relaxation processes.

$$\mathrm{d}M_x/\mathrm{d}t = \gamma(M_y\,B_z - M_z\,B_y)$$
$$\mathrm{d}M_y/\mathrm{d}t = \gamma(M_z\,B_x - M_x\,B_z)$$
$$\mathrm{d}M_z/\mathrm{d}t = \gamma(M_x\,B_y - M_y\,B_x). \qquad\qquad \textbf{2.4}$$

In the case where no radiofrequency fields are present ($B_x = B_y = 0$), we simply have free precession about the residual field $B_z = \Delta B$ at a frequency.

$$\omega_L - \omega_0 = \gamma\Delta B \qquad\qquad \textbf{2.5}$$

If a radiofrequency field $B_1$ is applied, with the transmitter frequency set at exact resonance, $B_z = B_y = 0$, and the magnetization rotates about the $B_1$ field at an angular frequency $\gamma B_1$ rad s$^{-1}$. This particular type of motion is sometimes referred to as nutation. If we neglect relaxation, it continues as long as the $B_1$ field is applied, $M$ continuing to turn about the $+x$ axis, giving rise to the transient oscillatory signal first described by Torrey.[2] More commonly we use pulse excitation, where $B_1$ is applied for a short period $t_p$ (a few microseconds) giving a rotation through an angle $\gamma B_1 t_p$ radians.

## 2.2.3 **Sign conventions for rotations**

Vector diagrams are simpler if we adopt the convention that magnetization vectors rotate counter-clockwise about magnetic fields (looking along the field vector towards the tip). Thus a field along $+x$ rotates a vector from $+z$ to $+y$, while precession about a field along $+z$ carries magnetization from $+y$ towards $+x$. (As we shall see in § 3.3.1, this is consistent with the convention used for product operators by Sørensen et al.[3]) This makes it possible to keep most of the diagrams of magnetization trajectories in the front part of the unit sphere.

Now consider the action of a radiofrequency field $B_1$ applied in the form of a pulse of length $t_p$. The nuclear spins see the resultant of a field $\Delta B$ aligned along $+z$ and another field $B_1$ along $+x$. This is called the effective field in the rotating frame ($B_{\mathrm{eff}}$) and has a magnitude

$$B_{\mathrm{eff}} = (\Delta B^2 + B_1{}^2)^{1/2}. \qquad\qquad \textbf{2.6}$$

It is inclined at an angle $\theta$ with respect to the $x$ axis, where

$$\tan\theta = \Delta B/B_1. \qquad\qquad \textbf{2.7}$$

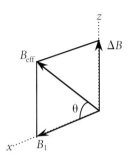

A magnetization vector $M$ is rotated about the effective field through an angle $\alpha = \gamma B_{\text{eff}} t_p$, moving on a cone whose axis is $B_{\text{eff}}$. In many practical cases we would ensure that $B_1 \gg \Delta B$, for all offsets $\Delta B$ in the spectrum. This is known as a hard pulse; in the other limit, where $B_1 \ll \Delta B$, we speak of a soft radiofrequency pulse (§ 5). With a hard pulse the rotation axis is essentially the $x$ axis, and $B_{\text{eff}}$ is essentially equal to $B_1$. For the maximum detected signal, the pulse length $t_p$ is set to give a flip angle $\alpha = \pi/2$ radians, carrying the vector $M$ from the $+z$ axis to the $+y$ axis. In what follows, a radiofrequency pulse will be represented by its flip angle and the rotation axis, for example $(\pi/2)_x$ would be the usual excitation pulse, $(\pi)_x$ the usual spin inversion pulse.

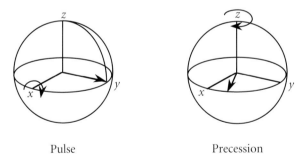

Pulse                                    Precession

When the pulse is extinguished, this vector is free to precess about the $+z$ axis at a frequency $\gamma\Delta B$ rad s$^{-1}$. The field from this rotating magnetization vector intersects the turns of the receiver coil and induces a small radiofrequency voltage at the Larmor frequency which we call the NMR signal. The signal is strongest if the vector is in the $xy$ plane.

## 2.2.4 Free induction decay

In a spectrum with many lines, each line can be associated with a separate magnetization vector with an amplitude proportional to the number of spins giving rise to that line. Even the constituent lines of a spin multiplet can be represented by vectors of the appropriate frequencies and intensities. This is an extension of the vector model proposed by Bloch and must be used with care. For example, it is not applicable to experiments that involve the excitation of multiple-quantum coherences (§ 1.3.2). In a $\pi/2$ pulse experiment the radiofrequency field applied during the pulse is so intense that each magnetization vector, although associated with a line at a different offset, simply rotates from $+z$ to $+y$ and then precesses in the $xy$ plane at its offset frequency $\Omega$ (representing shifts and coupling constants). The detected signal is the resultant of all of these vectors, for example if there are three lines,

$$M_y(t) = M_1 \cos\Omega_1 t + M_2 \cos\Omega_2 t + M_3 \cos\Omega_3 t \qquad \textbf{2.8}$$

This superposition of oscillating terms generates the familiar form of a free induction signal. We shall see later that these signal components decay with time, so it is more usually called the free induction decay. In general this interferogram or beat pattern is too complicated to be readily analysed by inspection and has to be converted into a spectrum by Fourier transformation. This picture of the free induction decay as the resultant of many different precessing vectors, one for each distinct resonance in the high resolution spectrum, helps us appreciate the Fourier transform relationship between the oscillating time-domain signal and the frequency-domain spectrum.

The free precession signal does not continue indefinitely because the individual spins gradually lose phase coherence, the process of spin–spin relaxation (§ 2.3.2). This is an irreversible effect arising from the perturbation of the time-keeping of a particular spin due to the magnetic fields generated by adjacent spins. The resultant signal decays exponentially with time constant $T_2$. There is an additional damping effect that arises for purely instrumental reasons. The NMR sample is immersed in a magnetic field that is not perfectly homogeneous in space so there is interference between signals of isochromats[4] from different sample regions that precess about the $z$ axis at different rates, losing phase coherence in the process (§ 4.1). Although the form of this decay curve is determined by the details of the static field distribution, the combined effects of spin–spin relaxation and field inhomogeneity are usually represented by a single decay constant $T_2^*$ (not to be confused with $T_2$). After Fourier transformation, this damping gives rise to the instrumental line width.

## 2.3 Relaxation

### 2.3.1 Spin–lattice relaxation

The vector model can be used to represent relative intensities of NMR signals and a particularly useful illustration is the measurement of spin–lattice relaxation times. At Boltzmann equilibrium the magnetization is represented by the vector $M_0$ aligned along the $+z$ axis of the rotating frame. Any deviation from equilibrium spin populations results in a change in the length of the $z$ component of magnetization, $M_z$. For example, when we make up an NMR sample we do it in the earth's field where there is essentially zero polarization (equal spin populations), so $M_z = 0$. If we suddenly insert this sample into the magnetic field, no signal can be excited until relaxation has created a population difference. Equal populations may also be achieved by prolonged irradiation with a continuous radiofrequency field, or by a suitable repetitive sequence of pulses. This is called saturation.

The return to equilibrium is assumed to be first order, and is characterized by the spin–lattice relaxation time $T_1$:

$$\frac{dM_z}{dt} = \frac{M_0 - M_z}{T_1}$$

**2.9**

This implies that if the $z$ magnetization is perturbed from its equilibrium value $M_0$, the recovery curve is exponential:

$$M_0 - M_z(t) = [M_0 - M_z(0)] \exp(-t/T_1)$$

**2.10**

where $M_z(0)$ is the initial value of the longitudinal magnetization. Unlike the situation in most other forms of spectroscopy, nuclear spin relaxation is very slow, reflecting the fact that the interactions between the spins and their environment are very weak, provided that $I = \frac{1}{2}$. Although the nucleus experiences all the same violent motions endured by the molecules of a liquid, the nuclear spin states remain unaffected unless a fluctuating magnetic field is involved. Even then, the interaction is only effective if this magnetic field is modulated at the correct frequency (the Larmor frequency). For more on the mechanism of spin–lattice relaxation, see § 9.1.

There are several ways of disturbing the equilibrium populations in order to study spin–lattice relaxation. In the common inversion recovery experiment we adjust the width of an initial excitation pulse ($t_p$) so that

$$\gamma B_1 t_p = \pi$$

**2.11**

This population inversion pulse converts equilibrium magnetization $+M_0$ into $-M_0$, that is to say, a vector aligned along the negative $z$ axis. There is no coherence between the precession phases of individual spins ($M_x = M_y = 0$) but the spin populations of the energy levels are interchanged, with the upper level now more highly populated than the lower level. We might represent this schematically in terms of the small excess population which eventually falls back to the lower level:

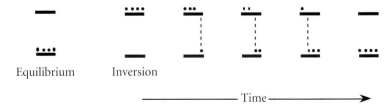

Equilibrium          Inversion

───────────────── Time ─────────────────→

It is not strictly correct to speak of a spin temperature under these circumstances but, in a very general sense, we may think of the spins as being hotter than their surroundings, and consider the interactions with the environment as cooling the spins down to the temperature of the rest of the sample. Because early theories of relaxation dealt with crystals, the environment was always called the lattice and we still speak of spin–lattice relaxation for all three states of matter.

If the initial population inversion is complete ($M_z(0) = -M_0$), the deviation from equilibrium magnetization after an interval $t$ follows an exponential curve:

$$M_0 - M_z(t) = 2 M_0 \exp(-t/T_1) \qquad \textbf{2.12}$$

The magnetization $M_z$ is initially inverted, but as the excess spins fall back to the ground state, $M_z$ decreases in amplitude, passes through a null when $t = T_1\ln2$, becomes positive and grows asymptotically towards the equilibrium value $M_0$.

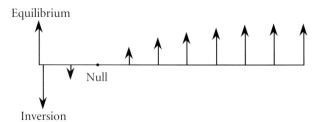

The entire recovery curve can be monitored by performing a series of experiments at different values of $t$, applying a $(\pi/2)_x$ read pulse to create observable $M_y$ magnetization. The arrows in this diagram may be taken to represent either magnetization vectors ($M_z$) or detected NMR signals.

Precautions must be taken to ensure that the experiment starts with the true equilibrium magnetization $M_0$, and that when the experiment is repeated, a suitably long waiting interval is included between measurements. This interval should be about three or four times the longest $T_1$ in the molecule, but since this is not known initially, we have to use a little common sense, and perhaps rerun the whole series of measurements if the apparent relaxation times seem to be comparable with the waiting times initially used.

In a modern Fourier transform spectrometer, this inversion–recovery experiment can be used to measure the spin–lattice relaxation times of all the different chemical sites in the molecule, and the results can be displayed in the form of a stacked-trace plot of spectra where the recovery curves of the different chemical sites can be followed by eye.[5] The stacked-trace plot of Figure 2.1 shows the spin–lattice relaxation of the carbon-13 sites in a sample of *m*-xylene. Note how well this diagram corresponds with the vector picture. Standard computer routines are available to fit the recovery curves and extract the spin–lattice relaxation times.

Alternatively, a selective inversion pulse may be employed to monitor the relaxation of a chosen site, but in systems of many spins the relaxation rate is not necessarily the same as in the nonselective experiment. Cross-relaxation effects[6] can cause the recovery curve to be nonexponential, a problem that is usually handled by measuring initial rates of recovery. These initial rates depend on how many sites are

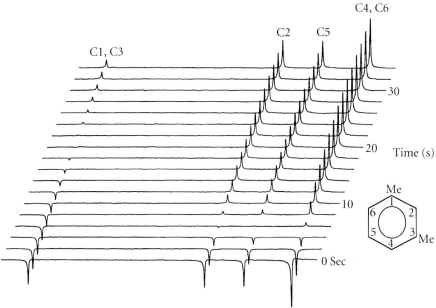

C4, C6

C2  C5

C1, C3

30

20  Time (s)

Me

10

Me

0 Sec

**Figure 2.1** A stacked trace plot of spin–lattice relaxation of $^{13}$C of the aromatic ring in *m*-xylene. At $t=0$ all the resonance lines are inverted by a $\pi$ pulse. Each resonance recovers exponentially by spin–lattice relaxation, passing through a null condition at $t = T_1 \ln 2$. Note the very slow recovery of sites $C_1$ and $C_3$ which have no directly attached protons. There is a perceptible difference in relaxation times of $C_4$ and $C_6$ compared with those of $C_2$ and $C_5$, attributable to a slight anisotropy of the reorientational motion.

inverted by the $\pi$ pulse and on the strengths of the various cross-relaxation terms, which are responsible for the nuclear Overhauser effect (§ 5.4.7 and § 9.2.3).

Many other NMR experiments which involve changing spin populations (or even changing concentrations of reagents) lend themselves naturally to treatment by the vector model. For example in chemical exchange, when an atom moves physically to another molecule or, through a conformational change, to a different chemical site in the same molecule, it carries with it the nucleus in a well-defined spin state ($\alpha$ or $\beta$). It is possible to label that atom by perturbing the populations of the spin states corresponding to a particular resonance line.[7] We might, for example, selectively invert these populations, or simply equalize them (saturation). The actual rate of chemical exchange can then be measured by monitoring the transfer of spin inversion or spin saturation to the other site, provided that this occurs at a rate at least comparable with the rate of spin–lattice relaxation. This has the advantage over most other methods for the study of kinetics in that it is applicable to systems in thermodynamic chemical equilibrium. Furthermore, the labelling process itself is such a mild and innocuous change that it is extremely unlikely that it could perturb the exchange process under investigation. Nuclear spin states can in fact influence chemical reactions, but only in the very special circumstances prevailing during free-radical recombination reactions (see § 1.2.3).

### 2.3.2 **Spin–spin relaxation**

If a transverse component of nuclear magnetization ($M_x$ or $M_y$) is excited by a radiofrequency pulse, it decays exponentially with a time constant $T_2$ the spin–spin (or transverse) relaxation time:

$$dM_x/dt = - M_x/T_2$$
$$dM_y/dt = - M_y/T_2 \hspace{3cm} \textbf{2.13}$$

The mechanism involves slight perturbations of the Larmor frequency by the fluctuating local magnetic fields, for example those from nearby magnetic nuclei. This leads to an irreversible loss of phase coherence and a decay of the resultant magnetization. In the absence of instrumental broadening effects, spin–spin relaxation determines the natural width of an individual high resolution line. In this approximation the line shapes are Lorentzian (the Fourier transform of an exponential decay). In practice, for most high-resolution spectra of mobile liquids, imperfections such as the spatial inhomogeneity of the applied field usually obscure the milder effects of spin–spin relaxation. Larger molecules and viscous liquids have shorter spin–spin relaxation times (§ 9.3.1). The measurement of $T_2$ normally relies on monitoring the decay of spin-echo amplitude in a multiple-echo train (§ 4.1.7).

Many modern NMR experiments involve the application of a sequence of radiofrequency pulses and periods of free precession when spin–spin relaxation occurs. There is a consequent loss of signal intensity if the length of the sequence becomes comparable with the spin–spin relaxation time. Related considerations apply to the use of selective (soft) radiofrequency pulses which may have durations of tens or even hundreds of milliseconds. Relaxation losses during a pulse involve both spin–spin and spin–lattice relaxation[2] and cannot be neglected on these long time-scales.

## 2.4 **Manipulation of vectors**

### 2.4.1 **Radiation damping**

We see from the above that there are four principal kinds of motion associated with the macroscopic nuclear magnetization vector—nutation by a pulse, free precession, transverse and longitudinal relaxation. It would be easy to fall into the trap of thinking that these are the only torques and damping terms that need to be considered. There is, however, a less-common phenomenon called radiation damping.[8] This is only important when the NMR signal itself is very intense (from a concentrated sample of a nucleus with high gyromagnetic ratio). The NMR response induces a small radiofrequency current in the receiver coil which creates a radiofrequency field $B_{rad}$. If the magnetization vector lies along the $+y$ axis, $B_{rad}$ lies along $-x$ and rotates $M_y$ back towards the $+z$ axis. It therefore accelerates the decay of the free induction signal and thus broadens the resonance line.

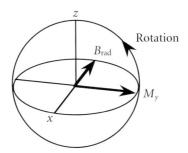

The time constant of the decay due to radiation damping[8] is given by

$$\tau_{rad} = (2\pi\eta Q\gamma M)^{-1} \qquad \textbf{2.14}$$

where $M$ is the magnetization in the transverse plane, $Q$ is the quality factor of the coil and $\eta$ is the filling factor. For the protons of water the radiation damping time constant can be significantly shorter than the spin–spin relaxation time ($T_2 \approx 3$ s). The radiation damping field $B_{rad}$ is nevertheless quite weak in comparison with most resonance offsets. Consequently, in a spectrum of many lines where we are interested in the broadening of a given resonance line $A$, the radiation damping term is mainly determined by $M_A$ and the magnetization of any neighbouring resonances can be largely neglected. This greatly dilutes the radiation damping effect. The coil quality factor $Q$ enters the formula for radiation damping because the current induced in the receiver coil by the precessing magnetization is increased by a factor $Q$ when the coil is tuned to the Larmor frequency. Consequently, radiation damping is easily reduced in practice by detuning the receiver coil, moving down the response curve of the coil; this is often used as a practical test.

Radiation damping is most commonly encountered as a spurious broadening of an intense resonance line, such as that from the protons in water (§ 11.3.1), but it may make itself evident in other, more subtle, ways. It can distort the relative peak heights of the lines of an intense spin multiplet; for example, a 1:3:3:1 quartet may have the inner lines preferentially broadened with respect to the outer lines. The effect on a neighbouring resonance line can be quite complicated, involving a phase shift and distortion of the line shape. Radiation damping also has a dramatic effect on the signal following a $\pi$ pulse.[9] For this reason it seriously interferes with the measurement of spin–lattice relaxation times by the inversion–recovery technique (§ 11.3.2).

Although it is easily overlooked, radiation damping may give rise to serious problems. In a typical pulse sequence, what is regarded as period of free precession may turn out to be far from free if there is a spurious radiofrequency field $B_{rad}$ influencing the motion. For this reason the testing of new pulse sequences is best done with dilute proton samples (for example residual HDO in heavy water) rather than with neat $H_2O$. In some situations a field gradient pulse may be used to purge the transverse magnetization that gives rise to the radiation damping. Radiation damping in the context of NMR spectroscopy of aqueous solutions is treated in more detail in § 11.3.

## 2.4.2 **Adiabatic rapid passage**

Most modern NMR experiments employ radiofrequency pulses for excitation or spin inversion, but there is another mode that is occasionally useful—adiabatic rapid passage. In its simplest form, this involves sweeping a continuous radiofrequency field $B_1$ through resonance, starting at a large positive offset and terminating at a large negative offset. The effective field $B_{eff}$ describes an arc in the $xz$ plane from the $+z$ axis to the $-z$ axis.

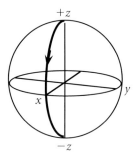

Since the only field seen by the spins is $B_{eff}$, it follows that at equilibrium the magnetization vector $M$ is aligned along $B_{eff}$. What is more surprising, is the fact that $M$ remains aligned with $B_{eff}$ when the latter is swept, but only if the sweep rate is sufficiently slow. This is called the adiabatic condition. If the inclination of $B_{eff}$ with respect to $+x$ is $\theta$, then this restriction may be written as

$$|d\theta/dt| << \omega_{eff} = \gamma B_{eff} \qquad \textbf{2.15}$$

If an adiabatic sweep (from $+z$ to $-z$) is executed in a time short compared with the spin–spin and spin–lattice relaxation times (rapid as opposed to slow passage), a complete population inversion is achieved. A typical sweep rate would be 10–100 MHz s$^{-1}$ for a pulse duration of 1 ms.

The adiabatic condition may be expressed as an adiabaticity factor:

$$Q = \frac{\omega_{eff}}{|d\theta/dt|} \qquad \textbf{2.16}$$

This $Q$ factor (not to be confused with the quality factor of a radiofrequency coil) should normally be large compared with unity. Since $\omega_{eff} \geq \omega_1$, the adiabatic condition is most critical near resonance and is easily satisfied at large offsets. This $Q$ factor may be expressed in a different form:

$$Q = \frac{(\omega_1^2 + \Delta\omega^2)^{3/2}}{\left(\omega_1 \dfrac{d\Delta\omega}{dt} - \Delta\omega \dfrac{d\omega_1}{dt}\right)} \qquad \textbf{2.17}$$

where $\omega_1$ is the radiofrequency field intensity expressed in frequency units and $\Delta\omega$ is the resonance offset. This expression indicates that the adiabatic sweep may be implemented by sweeping either the amplitude $\omega_1$ or the offset frequency $\Delta\omega$, or both simultaneously.

We could be forgiven for thinking that the adiabatic condition is a new restriction, not anticipated by the Bloch equations, which simply describe precession about magnetic fields, and say nothing about the rates at which these fields may be changed. In fact, it is a direct consequence of these equations of motion. Consider the simple case of a linear frequency sweep where $B_1$ is suddenly switched on at the beginning of the sweep. Since we start with a finite offset $\Delta B$, there is a slight misalignment of the effective field with respect to the magnetization vector $M$ (which is initially along $+z$). Consequently, $M$ precesses about $B_{eff}$ at an angular rate $\gamma B_{eff}$. If $\Delta B$ is swept sufficiently slowly to satisfy the adiabatic condition ($Q = 10$), $M$ precesses many times around $B_{eff}$ for a small change in $\theta$ and the trajectory takes the form of a prolate cycloid (Figure 2.2(a)), alternately accelerating and decelerating as $M$ is carried towards the $-z$ axis. The magnetization vector can be said to follow the effective field. As the sweep rate is increased, $Q$ becomes critical near exact resonance, and the precession loop cannot be completed before $B_{eff}$ has moved on; at this point the trajectory degenerates into a curtate cycloid (Figure 2.2(c)). With further increases in sweep rate, the trajectories lose their simple cycloidal character and execute very large excursions (Figure 2.2(e)). When the sweep is so fast that the adiabatic condition is seriously violated ($Q = 0.5$), the vector $M$ lags behind the effective field by an ever-increasing amount and terminates far short of the $-z$ axis (Figure 2.2(f)). Clearly this would be a very poor inversion pulse.

An adiabatic sweep achieves wideband spin inversion very effectively without the need for a very intense radiofrequency field $B_1$. This makes it preferable to a hard radiofrequency pulse when a wide frequency range must be covered, or when radiofrequency heating of the sample must be avoided. Note that high resolution spectrometers are expected to be operating soon at magnetic fields where protons resonate at 1000 MHz and have a chemical shift range of 10 kHz, while the $^{13}$C chemicals shifts cover more than 50 kHz. Broadband heteronuclear decoupling becomes particularly difficult at such high fields (§ 7.4.3)

Adiabatic pulses tolerate a wide range of $B_1$ intensities above the minimum level set by the adiabatic condition; the performance is then quite insensitive to spatial inhomogeneities in $B_1$ or $B_0$. If the $B_1$ field is switched off at the extremes of the sweep it is useful to round off these amplitude discontinuities so as to retain adiabaticity. The important application of adiabatic pulses to broadband heteronuclear decoupling is discussed in § 7.4.4.

Excitation rather than spin inversion is achieved by terminating the adiabatic sweep at the exact resonance condition, leaving the magnetization vector along the $+x$ axis. This type of excitation is really only suitable for a single resonance line or an isolated group of lines spanning a range $\Delta f << |\gamma B_1/2\pi|$. If $B_1$ is then extinguished, the excited spins are left free to precess in $xy$ plane in the normal manner.

## 2.4.3  Spin locking

We may of course satisfy the adiabatic condition (Equation 2.15) by not sweeping the applied field at all ($d\theta/dt = 0$). This is the situation if an adiabatic fast passage is interrupted at exact resonance and the $B_1$ field is not extinguished, leaving the

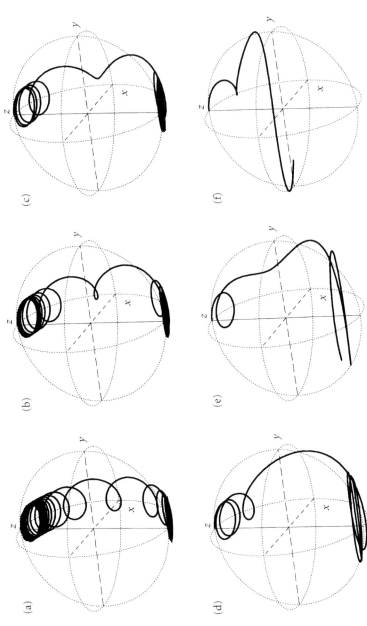

**Figure 2.2** Magnetization trajectories calculated according to the Bloch equations for a constant-amplitude, linear-sweep adiabatic pulse with different adiabaticity factors at resonance: (a) $Q_0 = 10$, (b) $Q_0 = 5$, (c) $Q_0 = 3$, (d) $Q_0 = 2$, (e) $Q_0 = 1$, (f) $Q_0 = 0.5$. For a sufficiently slow sweep rate (a), the trajectory approximates a prolate cycloid and it terminates close to the $-z$ axis. When the adiabaticity condition is violated near resonance during a more rapid sweep (c), the magnetization vector cannot complete a full cycle of precession about $B_{eff}$ before $B_{eff}$ has moved on. At even faster sweep rates, (e) and (f), the trajectory loses its cycloidal character, executing wild excursions and terminating far from the $-z$ axis.

magnetization aligned along the effective field $B_{\text{eff}} = B_1$. The same result can be more simply achieved by a $(\pi/2)_x$ pulse followed by a $\pi/2$ phase shift of the $B_1$ field so that it is aligned with the magnetization vector along $+y$.

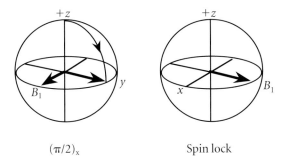

$(\pi/2)_x$                                              Spin lock

This is known as spin locking.[10] The presence of $B_1$ inhibits the usual free precession, and the magnetization simply decays with a time constant $T_{1\rho}$, the spin–lattice relaxation time in the rotating frame. The parameter $T_{1\rho}$ is sensitive to molecular motions in the frequency range around $\gamma B_1$, whereas the spin–spin relaxation time $T_2$ is sensitive to those near zero frequency. Since molecular motion in liquids is invariably much faster than $\gamma B_1$ rad s$^{-1}$, $T_{1\rho}$ is essentially equal to $T_2$.

Spin-locking is a useful method for temporarily inhibiting precession due to chemical shifts or spin–spin coupling. For example, in certain heteronuclear multiple-quantum coherence (HMQC) experiments (§ 8.3) involving the double transfer H $\rightarrow$ $^{13}$C $\rightarrow$ H, the proton signal is prepared in an antiphase state and a $(\pi/2)$ pulse is applied to $^{13}$C to create heteronuclear multiple-quantum coherence. Usually a hard $(\pi/2)$ pulse is employed, but if a selective pulse is required, it is ineffective if its duration is long enough to allow appreciable precession due to chemical shifts or divergence due to $^{13}$C–H coupling. The problem can be solved by applying an intense proton spin-lock field to prevent chemical shift evolution and to freeze the antiphase proton magnetization for the duration of the soft pulse.[11]

At some point in a complex pulse sequence, some components of magnetization may be aligned along the $x$ axis and others along the $y$ axis. If a spin-lock field $B_1$ is applied along the $y$ axis, the $x$ components rotate about $B_1$ at an angular frequency $\gamma B_1$, and if $B_1$ is spatially inhomogeneous they are soon dispersed around a circle and have a vanishing resultant. The $y$ components of magnetization are retained. This is often exploited as a method of purging undesirable magnetization components.[12]

Spin locking is the mode of preparation used in the Hartmann–Hahn experiment.[13] In a two-spin ($IS$) system, the $I$ spins are locked along a radiofrequency field $B_1$ and a second radiofrequency field $B_2$ is applied to the $S$ spins. If the Hartmann–Hahn matching condition

$$\gamma_I B_1 = \gamma_S B_2 \qquad\qquad \textbf{2.19}$$

is satisfied in the respective rotating reference frames, then coherence can be transferred between the two spin species (§ 1.4.5 and § 5.5.6).

In the solid state, the $I$ spins are tightly coupled by strong dipolar interactions $(T_2 \ll T_1)$ and they can be described in terms of a spin temperature that differs from the sample or lattice temperature. The $I$ spins (which are far more abundant than the $S$ spins) serve as a thermal reservoir. They are cold because they were polarized in the intense $B_0$ field of the magnet, but now find themselves in a much weaker (radiofrequency) $B_1$ field; their spin temperature is reduced by the factor $B_1/B_0$. By providing a common communication frequency, the Hartmann–Hahn condition permits thermal contact between the two spin systems so that the $I$ spins cool (polarize) the hotter $S$ spins.

In the liquid phase the thermodynamic considerations invoked above are not appropriate $(T_2 \approx T_1)$. The $I$ and $S$ spins interact through spin–spin coupling, exhibiting a cyclic interchange of coherence between the two sites (§ 5.5.6), rather like a pair of coupled pendulums.

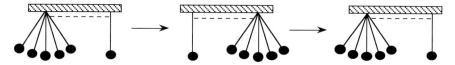

This can be used as a method for transmitting coherence in a step-wise manner along a chain of coupled spins—multistage Hartmann–Hahn transfer (§ 5.5.7).

### 2.4.4 Spin echoes

The vector picture is essentially indispensable to the understanding of the spin echo phenomenon (which is treated in detail in § 4.1). Magnetization from a small volume element of the sample is defined as an **isochromat**. A distribution of isochromatic vectors may be represented by (for example) three typical vectors, fast $f$, medium $m$, and slow $s$, as they diverge in the $xy$ plane during the first period $\tau$ of free precession (Figure 2.3). By rotating these vectors through $\pi$ radians about the $+x$ axis, we may set them up in reverse order with the $f$ vector behind $m$, which in turn is behind $s$, so that after a further period ($\tau$) of free precession all three come to a focus along the $-y$ axis. An alternative way to visualize the focusing effect is to plot a phase evolution diagram:

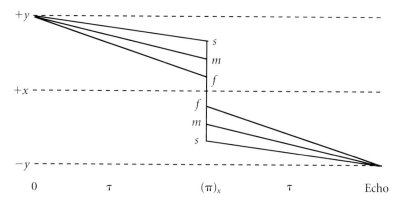

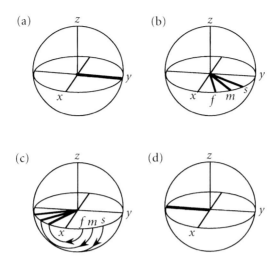

**Figure 2.3** Spin-echo formation by the Carr–Purcell sequence. (a) A $(\pi/2)_x$ pulse rotates the magnetization vectors to the $+y$ axis. (b) Representative fast ($f$), medium ($m$) and slow ($s$) vectors fan out in the $xy$ plane. (c) At time $\tau$ a $(\pi)_x$ pulse rotates them into mirror-image positions with respect to the $xz$ plane. (d) Free precession brings the fast, medium and slow vectors to a focus along the $-y$ axis at time $2\tau$.

This emphasizes the analogy with the focusing of light rays by a lens. It is difficult to imagine a description of Hahn[14] or Carr-Purcell[15] spin echoes without the vector representation. Indeed, the analysis of magnetization trajectories is implicit in most pulse experiments. We could argue that Bloch and Hahn set the stage for all of modern spin choreography.

## 2.4.5 **Echo modulation**

If the vector picture is to be extended to cover present-day pulse experiments, the effects of spin–spin coupling must also be considered. For this purpose it is useful to introduce one further concept—the labelling of spin multiplet components according to the spin states of the coupled partner. For example, in a two-spin ($IS$) system the two resonances of the $S$ doublet would be labelled $\alpha$ and $\beta$, reflecting the two quantum states of the $I$ spin. The $I$ spin exerts a tiny positive or negative magnetic field at the site of the $S$ spin, the sign depending on the sign of $J_{IS}$ and on whether $I$ is aligned along or against the applied magnetic field $B_0$. If a $\pi$ pulse is applied to the $I$ spins, these two alignments are reversed, interchanging the two $S$-spin labels. Similarly a $\pi$ pulse on the $S$ spins interchanges the $\alpha$ and $\beta$ labels of the $I$ spins. In this manner the precession frequencies can be changed simply by the application of a $\pi$ pulse to an adjacent spin.

Perhaps the most important manifestation of this effect is spin echo modulation.[16] Consider the case of echo formation in a weakly coupled $IS$ spin system where the $\pi$ refocusing pulse is applied to the $S$ spins but does not affect the $I$ spins.

The two $S$-spin vectors, labelled $\alpha$ and $\beta$, diverge during the first $\tau$ interval but reconverge to an echo during the second $\tau$ interval. The $\alpha$ vector always precesses faster than the $\beta$ vector.

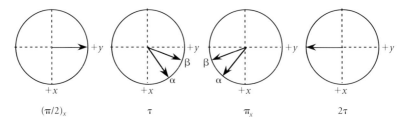

In other words, spin–spin coupling is refocused in the same manner as chemical shifts and field inhomogeneity. On the other hand, if the $I$ spins do experience the effect of the $\pi$ pulse, the $\alpha$ and $\beta$ labels of the $S$ spins are interchanged at time $\tau$ and the divergence due to spin–spin coupling persists throughout the second $\tau$ period.

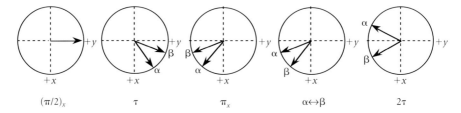

At time $2\tau$ the vectors are left at inclinations $\phi = \pm 2\pi J_{IS}\tau$ with respect to the $-y$ axis. If the spin echoes are monitored as a function of a variable period $2\tau$, they exhibit an amplitude modulation $\cos(2\pi J_{IS}\tau)$.

This phenomenon allows us to retain or suppress echo modulation in homonuclear systems by using hard or soft $\pi$ pulses respectively.[17] In heteronuclear systems, echo modulation of the $S$ spins can be switched on by deliberately applying a $\pi$ pulse to the $I$ spins to interchange the $\alpha$ and $\beta$ states. Such experiments are the basis of many techniques for editing high resolution spectra or for filtering the responses of coupled and noncoupled spins (§ 6.4). Similar concepts are used to explain broadband decoupling and $J$ scaling experiments, which use a repetitive train of $\pi$ pulses applied to a heteronuclear spin species (§ 7.3.2).

## 2.5  Pulse excitation

### 2.5.1  DANTE sequences

The vector picture has proved to be the key to several useful inventions in high resolution NMR. For example, suppose we wish to excite a nuclear spin system in a frequency-selective fashion. Some spectrometers have only hard radiofrequency pulses in the sense that $|B_1| \gg |\Delta B|$, for all offsets $\Delta B$. The effect of a soft or selective pulse can be obtained by excitation with a regular train of $N$ hard pulses, each with a small nutation angle $\alpha$, giving a total flip angle $N\alpha$. Free precession occurs in the short intervals $\tau$ between the pulses. At exact resonance, where there is no precession, the pulses have a cumulative effect, taking $M$ from $+z$ to $+y$ if $N\alpha = \pi/2$. We may call this the centreband response. The same overall result is obtained whenever the spins execute a whole number of complete revolutions about the $z$ axis between pulses; these are called sideband responses. We may illustrate this by considering only a small number of pulses ($N=4$):

Centreband response

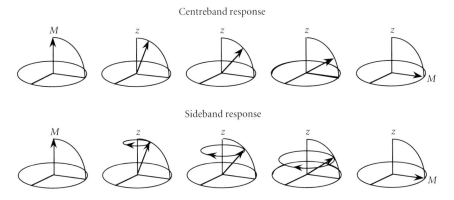

Sideband response

The centreband response occurs when the transmitter frequency is set at exact resonance. Sidebands occur at offsets equal to $\pm n/\tau$ Hz where $n$ is an integer and $\tau$ is the interval between pulses. Excitation at a sideband condition can be useful for fine-tuning the resonance condition by varying the pulse interval $\tau$.

In the general case, the magnetization trajectory is made up of a zig-zag of alternate rotations about $+x$ and precessions about $+z$. Instead of following the arc from $+z$ to $+y$, the trajectory moves out of the $zy$ plane towards the $+x$ axis.

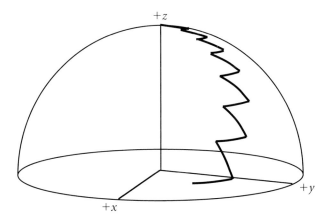

When there is a large number of pulses in the sequence, the resulting trajectory approaches that for a single weak pulse with the same overall duration $N\tau$. At large offsets from resonance, the nuclear magnetization vector executes a series of tight loops near to the $+z$ axis and the net excitation is very small. Shaped DANTE sequences can be implemented by varying the individual pulse widths (or amplitudes) and they have essentially the same frequency-domain excitation profile as a soft pulse with the same time-domain envelope.

It would be difficult to visualize the overall effect of such a sequence of narrow intense pulses without the vector picture. Morris[18] noted that the trajectory for the first sideband response corresponds closely with that prescribed for the souls in Dante's 'Purgatory' (one circumnavigation of each ledge on the mountain before being permitted to climb to the next ledge), hence the name for the sequence.

A DANTE sequence is used in situations where the corresponding soft radio-frequency pulse is not easily implemented, or where suitable pulse shaping equipment is not available (§ 5.3.1). DANTE does not require any change in the radiofrequency transmitter level, and the resonance condition can be finely tuned by setting the interval $\tau$. It also has the useful property that the NMR response can be acquired (one point at a time) in the intervals between the hard pulses; this makes it possible to monitor the behaviour of the nuclear magnetization vector *during* selective excitation.

### 2.5.2 **Magnetization trajectories**

As magnetization vectors move in response to radiofrequency pulses, free precession or the effect of field gradients, they trace out trajectories on the surface of the unit sphere that can provide valuable insight into the mechanism of the process under investigation. If relaxation is taken into account, the vectors also change their amplitudes as they evolve. The classic illustration records the recovery of a single spin species after a $(\pi/2)_x$ excitation pulse (Figure 2.4). The magnetization is slightly off-resonance in this frame and is subject to both spin–spin and spin–lattice relaxation. After the excitation has placed it along $+y$, the magnetization vector spirals back to the $+z$ axis as $M_x$ and $M_y$ contract due to spin–spin relaxation, while $M_z$ simultaneously grows as a result of spin–lattice interactions.

Calculations of trajectories are based on the Bloch equations and would typically involve rotation matrices used in an iterative fashion. Relaxation effects may be included where necessary. Instead of a display of trajectories, it is sometimes advantageous to plot the locus of the termini of the trajectories at suitable intervals during the soft pulse (or hard pulse sequence). A very early example of this mode of display is Hahn's 8-ball echo[14] (§ 4.1.1).

### 2.5.3 **Pulse imperfections**

The visualization of magnetization trajectories played an important part in the discovery of composite pulses. This provides considerable insight into the origin of the pulse imperfections and may suggest practical remedies. Whereas an ideal $(\pi/2)_x$

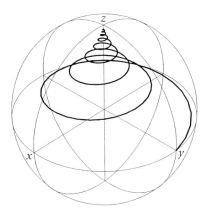

**Figure 2.4** Recovery of magnetization after excitation by a $(\pi/2)_x$ pulse. The trajectory spirals back from $+y$ to $+z$, losing intensity due to spin–spin relaxation but also gaining intensity due to spin–lattice relaxation. This is typical of the motion of a single-frequency component in a free induction decay.

pulse would take magnetization from $+z$ to $+y$, pulse imperfections tend to cause a divergence of trajectories away from the ideal path. Fortunately there are tricks that allow this divergence to be refocused, and thus compensate the inherent shortcomings of the pulse. These schemes are all related in some way to Hahn's concept of the spin echo; we allow some divergence due to resonance offset, and then transform the geometry in such a manner that the offset works in the opposite sense, bringing the trajectories to a focus. Suppose an amplitude-modulated pulse starts with a negative lobe, taking magnetization vectors into the back hemisphere (where $y$ is negative) so that the trajectories diverge. If this is followed by a larger positive lobe of the pulse envelope, the vectors are brought into the front hemisphere where the effect of the free precession is reversed, leading to reconvergence to a focus. This is the basis of the design of pure-phase pulses described below.[19]

### 2.5.4 Composite pulses

High resolution NMR experiments often employ an appreciable number of radiofrequency pulses that must be calibrated to the $\pi/2$ or $\pi$ condition. Even if the calibration is properly carried out, the spatial inhomogeneity of the radiofrequency field causes local deviations from the nominal flip angle. There is also the problem of resonance offset. Levitt[20] has shown that it is feasible to design composite radiofrequency pulses that are self-compensating, in a manner analogous to the compensated pendulum which did so much to improve the time-keeping of the early clocks.

Consider first of all the problem of an incorrectly calibrated $\pi$ pulse (Figure 2.5). In a loose analogy with the spin echo experiment, we can see that a slight undershoot of the $+y$ axis caused by a short $(\pi/2)_x$ pulse may be compensated by an equally short second $(\pi/2)_x$ pulse if there is a $(\pi)_y$ pulse sandwiched between them.

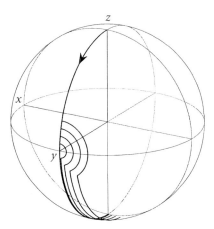

**Figure 2.5** Family of magnetization trajectories for the composite $(\pi/2)_x$ $(\pi)_y$ $(\pi/2)_x$ pulse with pulse lengths 80%, 85%, 90%, 95% and 100% of the nominal value. The shortfall of the first pulse is compensated by an equal shortfall of the third pulse through the action of the $(\pi)_y$ pulse.

The fact that the $(\pi)_y$ pulse is itself slightly short is only a second-order correction and scarcely affects the compensation. The shortfall on the rotation in the $yz$ plane is changed in sign by the $(\pi)_y$ pulse so that the final $(\pi/2)_x$ pulse carries the magnetization vector to the $-z$ axis. A long pulse would be compensated in an analogous fashion. Figure 2.5 illustrates this effect for a family of pulse length missets representing 80%, 85%, 90%, 95% and 100% of the nominal flip angle; in each case the $(\pi)_y$ pulse is exactly twice as long as the $(\pi/2)_x$ pulse.

The composite pulse $(\pi/2)_x$ $(\pi)_y$ $(\pi/2)_x$ also compensates for errors due to resonance offset, where the rotation takes place about a tilted effective field $(B_{\text{eff}})$ and where the actual rotation angle $\gamma B_{\text{eff}} t_{\text{p}}$ exceeds the nominal flip angle $\gamma B_1 t_{\text{p}}$. If we study a representative family of trajectories during the initial $(\pi/2)_x$ pulse, they terminate along a curve that lies close to the $xy$ plane (Figure 2.6). We can see from symmetry considerations that a subsequent ideal $(\pi)_y$ pulse would carry these vectors to positions where the $x$ and $z$ coordinates are exactly reversed while the $y$ components are unchanged. Consequently, the final tilted $(\pi/2)_x$ pulse would rotate these vectors to a focus along the $-z$ axis. The action is thus closely related to the formation of a Carr–Purcell spin echo. In the event, the $(\pi)_y$ pulse is not perfect but is tilted in the $zy$ plane, and the resulting focus is satisfactory but not exact. Figure 2.6(a) shows magnetization trajectories for the composite pulse $(\pi/2)_x$ $(\pi)_y$ $(\pi/2)_x$ for resonance offsets $\Delta B/B_1 = 0.0, 0.1, 0.2, 0.3$ and $0.4$. Interestingly, the related composite pulse $(\pi/2)_x$ $(3\pi/2)_y$ $(\pi/2)_x$ also has a similar compensatory effect, as can be appreciated from the magnetization trajectories of Figure 2.6(b). The increase in the flip angle of the central pulse has the effect of compensating the tilt of the effective field (which increases with increasing offset).

Unfortunately the two types of compensation (for amplitude misset and resonance offset) do not combine at all well for either of these composite pulses. How-

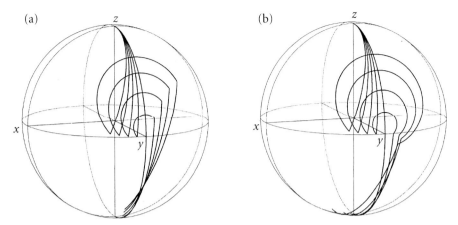

**Figure 2.6** Families of magnetization trajectories for (a) the composite $(\pi/2)_x\,(\pi)_y\,(\pi/2)_x$ pulse and (b) the composite $(\pi/2)_x\,(3\pi/2)_y\,(\pi/2)_x$ pulse for resonance offsets $\Delta B/B_1 = 0.0, 0.1, 0.2,$ 0.3 and 0.4. The effect of the tilt of the effective field during the first and third pulses is largely compensated by the action of the central pulse.

ever, it is possible to design more complicated pulse sandwiches that do possess this property of dual compensation. Many different types of composite pulses have now been invented.[21] Most are designed to have an increased tolerance to some kind of pulse imperfection, although others are deliberately made more sensitive to a particular pulse parameter. This would be useful, for example, for in vivo spectroscopy where spatial localization is required; the enhanced sensitivity to the spatial inhomogeneity of the $B_1$ field would decrease the volume of sample excited.

### 2.5.5 Pulse shaping

A rectangular soft $(\pi/2)$ pulse has definite shortcomings. Its frequency-domain excitation profile has a central response flanked by a symmetrical set of sidelobes similar to those of a sinc (i.e. $\sin x/x$) function, which can excite undesirable responses from neighbouring resonances at surprisingly large offsets (§ 5.1.1). The effect may be illustrated by calculating the appropriate magnetization trajectories. At appreciable offsets, a rectangular $(\pi/2)_x$ pulse induces circular excursions near the $+z$ axis of the unit sphere; the radii decrease with increasing offset but only quite slowly. Because the end of the rectangular pulse may occur at any point on this circle, this can leave appreciable absorption or dispersion components.

A much better excitation profile is afforded by a soft Gaussian pulse.[22] At large offsets, where the rectangular pulse has appreciable sidelobes, the response after a Gaussian pulse is essentially negligible. This does not mean that the magnetization vector stays aligned along the $+z$ axis; instead it follows a 'tear-drop' trajectory that always carries it right back to the $+z$ axis at the end of the pulse, leaving a negligible signal. The behaviour differs from that of a rectangular pulse because, as the amplitude $B_1$ varies, the effective field ($B_{\text{eff}}$) changes its tilt angle during the pulse, starting along the $+z$ axis (vanishing $B_1$ field), swinging towards the $+x$ axis (maximum $B_1$ field) and then returning to the $+z$ axis. At appreciable offsets, although the maximum excursion generated by the Gaussian pulse exceeds that after an equivalent rectangular pulse, the residual signal is essentially zero. The lack of sidelobe responses is a general property of pulses that have no sharp leading and trailing edges.

### 2.5.6 **Pure-phase pulses**

Unfortunately most soft pulses (including the Gaussian) have the drawback that they induce a phase gradient into the observed spectrum. Adjustment for pure absorption across the entire spectral width requires a frequency-dependent phase correction. There is an appreciable divergence of magnetization vectors during the pulse, just as for the case of a rectangular pulse. A certain degree of compensation can be achieved[23] by using a Gaussian with a $3\pi/2$ flip angle rather than $\pi/2$. A better solution is a pure-phase pulse designed specifically to bring magnetization vectors to a focus along the $+y$ axis for a specified range of offsets. The E-BURP soft pulses,[19] designed by simulated annealing[24] achieve this by carrying the magnetization vectors backwards and forwards several times between the $+y$ and $-y$ hemispheres of the unit sphere, balancing the divergence and convergence so as to attain a sharp focus condition at the end of the pulse. Pure-phase pulses are described in more detail in § 5.2.6.

Ramsey[25] extended the concept of the Bloch–Siegert shift[26] to cover the interaction that occurs when two radiofrequencies are applied simultaneously during a double resonance experiment. When two soft pulses are applied at the same time, there is an analogous interaction which increases with the radiofrequency intensities ($B_1$ and $B_2$) and as the inverse of their separation ($\Delta F$). In a frame synchronous with the $B_1$ field, the $B_2$ field rotates about the $z$ axis at $2\pi\Delta F$ rad s$^{-1}$. The resultant radiofrequency field ($B_1 + B_2$) is modulated in amplitude and phase, and traces out a circular locus in the $xy$ plane:

$B_1$ $\qquad\qquad\qquad\qquad$ $B_2$ $\qquad\qquad\qquad\qquad$ Resultant

The corresponding trajectories[27] have some of the character of a cycloid, reflecting the fact that the two fields act part of the time in concert, and part of the time in opposition (Figure 2.7). Instead of moving in a smooth arc from $+z$ to $+y$, the magnetization vector accelerates and decelerates, pausing at the cusps of the cycloid where the two radiofrequency fields cancel exactly. The final net effect is a slight phase shift of the signal excited by one radiofrequency field as a result of perturba-

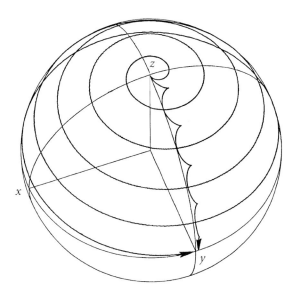

**Figure 2.7** Interaction between two equal soft rectangular $(\pi/2)_x$ pulses spaced 5 Hz apart. The resulting magnetization trajectories, viewed in a frame synchronous with one of the radiofrequencies, follow cycloidal paths as the two fields alternately reinforce and cancel. This distortion of the usual smooth arc is clearly visible for the trajectory at resonance but is disguised on the off-resonance trajectory.

tion by the other (off-resonance) field. As the separation $\Delta F$ decreases, the cycloidal contributions become more prominent and the phase shift more serious. It also depends in a cyclic manner on $T\Delta F$, where $T$ is the soft pulse duration. Interaction between two soft pulses is illustrated in § 5.3.7.

## 2.5.7 **Antiphase magnetization**

The first step in many pulse experiments is the creation of antiphase magnetization, with the two vectors ($\alpha$ and $\beta$) of a $J$ doublet in opposition along the $\pm x$ axes of the rotating frame. One method of preparation uses the spin-echo modulation scheme (§ 2.4.5) with the delay $\tau$ set equal to $1/(4J_{IS})$. Consider the evolution of the $S$-spin vectors. Chemical shift effects are refocused by the $(\pi)_x$ pulse, but the divergence due to $J$ coupling persists for the period $2\tau$, leaving the $\alpha$ and $\beta$ vectors aligned along the $\pm x$ axes.

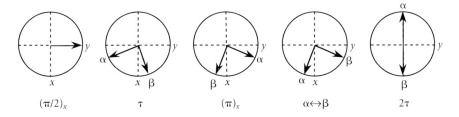

The interchange of spin labels ($\alpha \leftrightarrow \beta$) is brought about by the $(\pi)_x$ pulse on the coupled $I$ spins. If a $(\pi/2)_y$ pulse is now applied to the $S$ spins (note the 90° phase shift), it rotates the antiphase vectors so that they lie along the $\pm z$ axes, creating antiphase $z$ magnetization. This is equivalent to a selective population inversion of one of the $S$-spin transitions, a situation that goes by the name of longitudinal two-spin order (§ 3.2.1). Such a preparation of antiphase $z$ magnetization is the key step in the INEPT sequence,[28] and typical of an entire family of heteronuclear polariza-tion transfer schemes (§ 4.2.6 and § 8.3).

For certain selective experiments it may be necessary to generate antiphase mag-netization from a single chemical site without perturbing other resonances in the spectrum. If there is no good estimate of the coupling constant $J_{IS}$, or, more com-monly, if several different couplings are involved, it is very useful to have a proce-dure that is insensitive to the value of $J_{IS}$. Selective radiofrequency pulses have been designed that excite antiphase magnetization directly, without any free precession decay. They are named Janus pulses[29,30] after the Roman god of doors and gates, depicted as facing in two opposite directions at the same time.

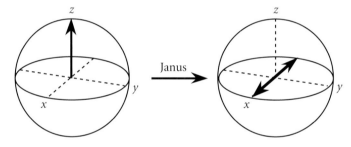

Janus pulses (§ 5.5.2) accept a range of possible values of the coupling constant $J_{IS}$, and are particularly useful in experiments designed to measure long-range het-eronuclear coupling constants.

If we take the modulated echo sequence (§ 2.4.5) and double the free precession delays so that $\tau = 1/(2J_{IS})$ we have the BIRD module.[31] This acts as one of the building blocks of many pulse sequences where it is necessary to discriminate between a direct (large) coupling $^1J_{IS}$ and several long-range (small) couplings $^nJ_{RS}$, $^nJ_{QS}$, etc. The $\alpha$ and $\beta$ vectors representing the large coupling $^1J_{IS}$ are brought to a focus along the $+y$ axis (top), but vectors corresponding to small long-range couplings (bottom) do not diverge significantly in the $\tau$ intervals, and are rotated to the $-y$ axis.

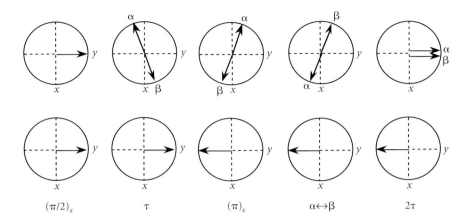

$(\pi/2)_x$     $\tau$     $(\pi)_x$     $\alpha\leftrightarrow\beta$     $2\tau$

This permits a clear discrimination between the two cases, usually exploited by rotating both sets of vectors through $(\pi/2)_x$ to put one along $+z$ and the other along $-z$. In carbon–proton heteronuclear experiments this trick is often used to distinguish between protons in the directly-coupled $^{13}$C–H system, and the much more abundant protons in $^{12}$C–H systems. These BIRD modules are discussed in more detail in § 4.2.8.

## 2.6 Spectroscopy

### 2.6.1 Correlation spectroscopy

The archetype of all two-dimensional experiments is homonuclear correlation spectroscopy (COSY), first proposed by Jeener[32] and analyzed and developed by Aue *et al.*[33] Although it is a particularly simple pulse sequence,

$$(\pi/2)_x - t_1 - (\pi/2)_x \text{ acquire } (t_2) \qquad \textbf{2.20}$$

the actual mechanism of magnetization transfer may not be immediately obvious. The analysis is usually couched in terms of density operator theory or the Cartesian product operator treatment.[3] A more pictorial insight can be obtained by treating it as a polarization transfer experiment, using the vector model and spin populations.

Consider the weakly-coupled homonuclear two-spin ($IS$) case. Since the problem is symmetrical with respect to the $I$ and $S$ spins, we may concentrate on the transfer $I \rightarrow S$ which gives rise to one cross-peak, accepting that there will also be an analogous transfer $S \rightarrow I$ that generates the other cross-peak. We assign equal magnetization vectors $M_0$ to the two $I$-spin transitions, $J_{IS}$ apart. The initial $(\pi/2)_x$ pulse aligns both vectors (labelled $f$ and $s$) along the $+y$ axis and they then precess freely during the evolution period $t_1$, acquiring precession angles:

$$\alpha(f) = (2\pi\delta_I + \pi J_{IS})\, t_1$$
$$\alpha(s) = (2\pi\delta_I - \pi J_{IS})\, t_1 \qquad \textbf{2.21}$$

The second $(\pi/2)_x$ pulse is broken down into a cascade of two selective $(\pi/2)_x$ pulses, the first applied to the $I$ spins and the second to the $S$ spins. If we resolve the $I$-spin vectors along the $+x$ and $+y$ axes of the rotating frame, then the selective pulse applied to the $I$ spins has no effect on the $x$ components but rotates the $y$-components onto the $-z$ axis.

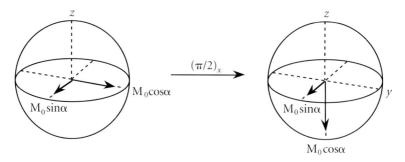

A vector aligned along $-z$ represents a partial population inversion. If $t_1 = 0$, then $\alpha(f) = \alpha(s) = 0$ and both population inversions are complete. When we examine the energy level diagram we see that complete population inversion of both $I$ transitions has no influence on the $S$-spin intensities.

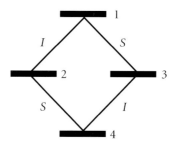

Only if the $I$-spin populations are affected differentially does this change the $S$-spin population differences. If we write the population disturbances of the $I$ spins as $\Delta_{12}$ and $\Delta_{34}$, it is the differential disturbance $\frac{1}{2}|\Delta_{12} - \Delta_{34}|$ that affects the $S$ spins. One $S$-spin transition increases in intensity and the other decreases by the same amount.

$$\tfrac{1}{2}(\Delta_{34} - \Delta_{12}) = \tfrac{1}{2}M_0\left[\cos\alpha(s) - \cos\alpha(f)\right]$$

$$= M_0 \sin\tfrac{1}{2}[\alpha(f) + \alpha(s)]\,\sin\tfrac{1}{2}[\alpha(f) - \alpha(s)] = M_0 \sin(2\pi\delta_I t_1)\,\sin(\pi J_{IS} t_1) \qquad \textbf{2.22}$$

using a standard trigonometrical identity. Fourier transformation as a function of the evolution time $t_1$ gives a profile in the $F_1$ dimension that is an antiphase doublet $(J_{IS})$ at the $I$-spin chemical shift $\delta_I$.

The second selective $(\pi/2)_x$ pulse of the cascade acts as a read pulse, converting the population disturbances into observable $S$-spin signal, which then precesses during the detection period $t_2$ according to

$$M(t_1,t_2) = M_0 \sin(2\pi\delta_I t_1)\sin(\pi J_{IS} t_1)\sin(2\pi\delta_S t_2)\sin(\pi J_{IS} t_2) \qquad \textbf{2.23}$$

In the $F_2$ dimension, this is an antiphase doublet centred at $\delta_S$ with a splitting $J_{IS}$. Thus the cross-peak, which represents polarization transferred from $I$ to $S$, is centred at coordinates $(\delta_I, \delta_S)$ and is split by $J_{IS}$ in both frequency dimensions, giving the characteristic doubly-antiphase square pattern:

Normally the spectrometer phase is adjusted so that these resonances are in the absorption mode. There is also a symmetry-related cross-peak at $(\delta_S, \delta_I)$ attributable to polarization transfer from $S$ to $I$.

So far we have neglected the other two components $M_0\sin\alpha(f)$ and $M_0\sin\alpha(s)$, which lay along the $x$ axis and were unaffected by the second $(\pi/2)_x$ pulse. These terms represent magnetizations that remain at the $I$-spin frequency during both $t_1$ and $t_2$, giving a diagonal response. We might have expected the two $I$-spin resonances to maintain their individual identities throughout $t_1$ and $t_2$, giving just two responses at coordinates $(\delta_I + \frac{1}{2}J_{IS}, \delta_I + \frac{1}{2}J_{IS})$ and $(\delta_I - \frac{1}{2}J_{IS}, \delta_I - \frac{1}{2}J_{IS})$, both on the principal diagonal. This is indeed what happens[34] if the second pulse has a small flip angle $\beta \ll \pi/2$. Had the second pulse, been a $\pi$ pulse, however, its effect on the $S$ spins would have been to interchange the identities of the two $I$-spin resonances, giving two responses at $(\delta_I + \frac{1}{2}J_{IS}, \delta_I - \frac{1}{2}J_{IS})$ and $(\delta_I - \frac{1}{2}J_{IS}, \delta_I + \frac{1}{2}J_{IS})$ displaced from the principal diagonal by $J_{IS}$ Hz.

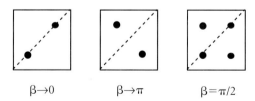

A $\pi/2$ pulse on the $S$ spins has an intermediate effect, spreading the $I$-spin magnetization equally between the four coordinates $(\delta_I \pm \frac{1}{2}J_{IS}, \delta_I \pm \frac{1}{2}J_{IS})$. Summing the magnetization terms $M_0 \sin\alpha(f)$ and $M_0 \sin\alpha(s)$ gives

$$M_0 \sin[(2\pi\delta_I + \pi J_{IS})\, t_1] + M_0 \sin[(2\pi\delta_I - \pi J_{IS})\, t_1] \qquad \textbf{2.24}$$

A standard trigonometrical identity gives

$$2M_0 \sin(2\pi\delta_I t_1)\cos(\pi J_{IS} t_1) \qquad \textbf{2.25}$$

which represents an in-phase doublet ($J_{IS}$) at the chemical shift $\delta_I$ in the $F_1$ dimension. This magnetization continues to precess at $\delta_I \pm \frac{1}{2} J_{IS}$ during the detection period $t_2$, giving an in-phase square array in the COSY spectrum. If the spectrometer phase is adjusted to set cross-peaks in pure absorption, the diagonal peaks are in dispersion in both frequency dimensions.

We see that the vector picture of the COSY experiment gives the same expressions as the product operator treatment (§ 3.3.5), with perhaps a better feel for the mechanism of magnetization transfer. Note that this description avoids the concepts of coherence transfer and mixing pulses because it is based on rearrangement of spin populations on the energy levels. The same arguments can be used to explain heteronuclear correlation spectroscopy (§ 8.3). Correlation spectra of systems with several coupled spins are of course much more complex than this; they are treated in more detail in § 8.2.3.

### 2.6.2 Phase cycles

Most present-day pulse sequences are designed to separate desirable spectral features from unwanted responses, natural or instrumental. The process of extraction may rely on different orders of coherence, spin-echo modulation, the relative magnitudes of coupling constants, or simply a difference in phase (absorption or dispersion). Over the years various phase cycles have been devised to accomplish these ends and we usually rely on the vector picture to explain how they work (§ 6.1.1 and § 6.1.2).

Expressed in the most general terms, a phase cycle involves the generation (in successive scans) of signals with $N$ different phases uniformly distributed around a circle. If $N=2$ this is simply difference spectroscopy (phases 0 and $\pi$). More commonly, $N=4$ and the phases follow the cycle 0, $\pi/2$, $\pi$, $3\pi/2$. Until recently, high resolution spectrometers were not designed to handle receiver phase shifts other than $\pi/2$ or $\pi$, so most phase cycles use $N=4$. But more generally, however, $N$ may be any integer greater than unity, for example three[35] or five.[36] We represent the successive signals by vectors with their respective phases in the $xy$ plane:

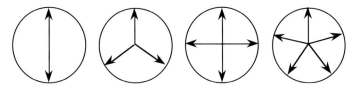

The phase of the spectrometer receiver is programmed to follow the phase of the desired response but not that of the signal to be suppressed, which cancels over the full cycle. Phase cycles may be nested one inside the other but the total number of scans may then become very large and the experimental time rather long. As we shall see in § 6.2, phase cycling is gradually being replaced by pulsed field gradient schemes.[37]

### 2.6.3 Steady-state effects

An insidious phenomenon that can degrade the quality of high resolution spectra, is the establishment of a steady-state regime where the conditions differ significantly from those at equilibrium. This can arise during a phase cycle if the spectroscopist is tempted to repeat the scans without adequate delays for spin–spin and spin–lattice relaxation. The more common of the two steady-state effects occurs when there is partial spin saturation through incomplete spin–lattice relaxation. Take the simple example of difference spectroscopy with just two scans (add and subtract). The first scan has Boltzmann spin populations and therefore the full magnetization $M_0$, but the second scan may start before the spins have fully recovered their equilibrium longitudinal magnetization. The subtraction of undesirable signal components is then imperfect, leaving artifacts in the spectrum. Similar considerations apply to fourfold phase cycles. The remedy is to preface each experimental run with a few dummy scans where pulses are applied as usual but no signals are acquired. This allows a steady-state regime to be established where the perturbations are about the same for each step of the phase cycle. Logical planning of the ordering of the steps in a phase cycle can prove to be an important precaution in such situations.

Steady-state effects with transverse magnetization occur when the pulse interval is comparable with the spin–spin relaxation time $T_2$. They tend to be less serious than longitudinal steady-state effects, but more insidious, since the resulting artifacts vary across the spectral width in a cyclic manner. The vector picture is once again the key to understanding this effect.[38] In a multipulse experiment, if some transverse magnetization remains at the end of the pulse interval $\tau$, then the action of the next pulse is affected. A steady-state regime is established, giving NMR responses that are distorted in phase and relative intensities. Consider the fate of two lines, a and b, at different offsets $\Delta B_a$ and $\Delta B_b$ from the transmitter frequency. In the interval $\tau$ between pulses, their transverse magnetization vectors precess through different angles $(2n\pi+\alpha) = \gamma\Delta B_a\tau$ and $(2n\pi+\beta) = \gamma\Delta B_b\tau$. If we sketch the motion of the two magnetization vectors in projection on the $xy$ plane,

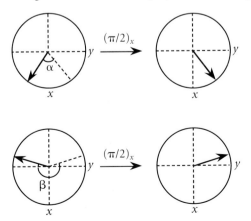

A hard radiofrequency pulse about the $x$ axis cannot change the $x$ component of magnetization, but only the $y$ and $z$ components. Consequently, the steady-state regimes for a and b are quite different. The length of each vector and its orientation in space must adjust itself to be compatible with the motion during the pulse and the free precession during $\tau$. Since the precession angles ($\alpha$ and $\beta$) are different, the two resonance lines (a and b) acquire different phase shifts and different intensity distortions. These perturbations are a cyclic function of the precession angle (which is a function of offset). Transverse steady-state effects can be appreciable for samples with long spin–spin relaxation times in spectrometers with high field/frequency stability.

There is another remarkable consequence of the establishment of a steady-state for transverse magnetization. The dispersal of isochromatic vectors due to $B_0$ spatial inhomogeneity during the free precession interval $\tau$, appears to be refocused by the end of that interval. If the spin–spin relaxation times are long compared with $\tau$, there is a steady-state echo effect, where the signal envelope between pulses exhibits a decay (due to $T_2^*$) followed by an almost symmetrical regrowth.

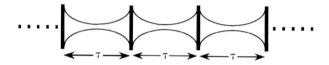

It is as if a conventional spin echo had been divided into two equal halves which were then interchanged. The Fourier transform of such a signal has a vanishing dispersion contribution owing to the (approximate) symmetry. At first sight this seems to present a paradox—how can diverging isochromats converge again to form a focus without the influence of a pulse? The answer is that the converging vectors arise from the effects of several earlier pulses. For example, groups of three successive pulses will excite partial echoes.

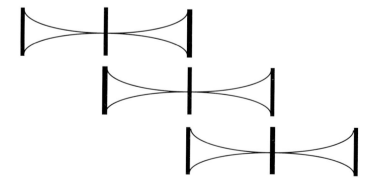

These signals add together to give the steady-state echo envelope. Interactions involving larger numbers of pulses are also involved, since a train of several echoes is observed if the pulse sequence is suddenly interrupted.

Fortunately this is one of those rare occasions where spectrometer imperfections actually improve the situation, since field/frequency instability can interfere with the establishment of the steady-state regime by scrambling the exact values of the resonance offsets from one scan to the next. A similar result can be imposed[38] by introducing an artificial jitter in the timing of the pulse interval $\tau$. Alternatively, pulsed field gradients can be applied to quench the transverse magnetization at the end of the free induction decay.[39] These gradients must be sufficiently intense to allow molecular diffusion to disperse the transverse magnetization, or they should be randomized in duration or intensity to prevent refocusing.

A more positive approach is to accept the steady-state effect on the grounds that it improves the signal-to-noise ratio. The resultant artifacts can be suppressed by performing four experiments with different irradiation frequencies spaced $1/(4\tau)$ apart. This technique[40] has been used to study slowly relaxing insensitive nuclei such as $^{57}$Fe. A similar result can be achieved with a suitable phase cycle.

### 2.6.4 **Implicit Fourier transformation**

The vector model may be used to derive the well-known Fourier transform relationship between a rectangular wave and a sinc function without invoking the Fourier transform concept explicitly. Suppose we have an NMR sample confined in a cylinder aligned in the $z$ direction with sharply defined edges at $-a$ and $+a$. If a linear field gradient $G = dB_0/dz$ is applied, the spin distribution function in the $z$ direction would be described by a rectangular frequency-domain profile.

Field due to applied gradient

If we think in terms of spin isochromats, they would be thin circular discs stacked together in the $z$ dimension and they could be represented by $N$ magnetization vectors each with a slightly different characteristic precession frequency.

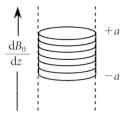

The time evolution of the transient NMR signal can be worked out from first principles. After excitation at time zero, all the vectors are aligned along the $+y$ axis, inducing a strong positive absorption signal (Figure 2.8(a)) which we may take to

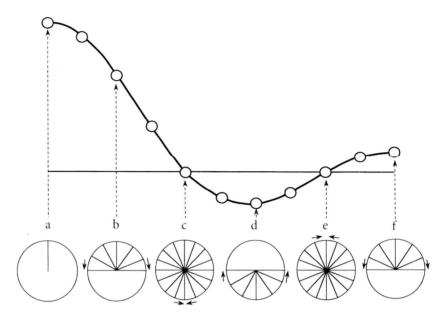

**Figure 2.8** Implicit Fourier transformation. Free precession of nuclei in a sample of restricted length placed in a linear field gradient gives rise to a net signal that evolves as a sinc function, the Fourier transform of a rectangular function.

have unit intensity. The isochromats then diverge in the $xy$ plane through the effects of the applied gradient, and the resultant signal decreases (Figure 2.8(b)). The spread is uniform, reflecting the linearity of the gradient.

Consider first of all a delay $t$ such that the fastest vectors (those from the ends of the sample) precess through $\pm\pi$ radians and just reach the $-y$ axis. Then all the vectors are distributed uniformly around a circle, inducing a zero resultant NMR signal (Figure 2.8(c)). As the fastest vectors overshoot the $-y$ axis, the signal goes negative, reaching an extremum when the maximum precession angle is $\pm 3\pi/2$ radians. Since we may neglect the two-thirds of the vectors that are uniformly distributed around the first $2\pi$ radians, we need only evaluate the resultant of the remaining one-third that are spread around the semicircle where $y$ is negative (Figure 2.8(d)). A standard integral shows this resultant to be $-2/(3\pi)$. The signal then increases with time, passing through zero for a maximum precession of $\pm 2\pi$ radians, where the vectors are again spread uniformly around the circle (twice). The next extremum occurs when $\gamma\Delta Bt = 5\pi/2$ where the resultant signal is $+2/(5\pi)$. If we continue the evolution, plotting out the amplitude of the NMR signal as a function of time we see that it follows a sinc function (Figure 2.8). The sinc function ($\sin x/x$) is well known as the Fourier transform of a rectangular function; it has been derived here as a direct consequence of the vector model, although the underlying mathematics is the Fourier integral.

## 2.7  Limitations of the vector model

### 2.7.1  Multiple-quantum coherence

Experiments involving multiple-quantum coherence cannot readily be couched in terms of the vector model without some rather arbitrary assumptions and unwieldy constructions. The product operator formalism (§ 3.4) is much better suited to the treatment of these problems. Consider the case of a two-spin ($IS$) system that has been prepared by the sequence

$$(\pi/2)_x - \tau - (\pi)_x - \tau - \qquad\qquad \textbf{2.27}$$

with $\tau = 1/(4J_{IS})$ so that the two $S$-spin vectors ($\alpha$ and $\beta$) are aligned along the $\pm x$ axes, the classic preparation for excitation of double-quantum coherence. Although at this time, the antiphase $S$-spin vectors give no NMR signal, any subsequent free precession would induce an amplitude-modulated signal $M_0\sin(\pi J_{IS}t)$, since the $\alpha$ and $\beta$ vectors differ in frequency by $J_{IS}$. This is not a question of an undetectable response; we have simply caught the signal at a null point.

A second $(\pi/2)_x$ pulse is now applied. It may be considered as a cascade of two selective $(\pi/2)_x$ pulses, one applied to the $S$ spins (which has no effect on the $S$-spin vectors since they lie along the $B_1$ field) and the other to the $I$ spins. We know that a $\pi$ pulse on the $I$ spins would have interchanged the $\alpha$ and $\beta$ labels, so we deduce that a $\pi/2$ pulse has an intermediate effect, half of the $S$ spins suffering the inter-change of labels the other half remaining unchanged.

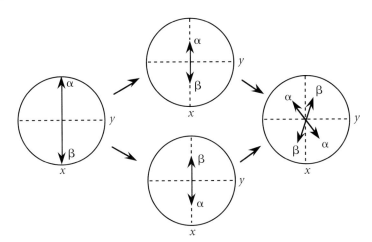

If free precession now occurs, no signal is detected since the vectors are locked into antiphase pairs even though the ($\alpha$–$\alpha$) pair precesses at a different rate from the ($\beta$–$\beta$) pair. This extension of the vector picture is not strictly justifiable, but it does account for the disappearance of the NMR signal after the pulse and the continued precession of some invisible entity. Single-quantum coherence has been destroyed by conversion into double-quantum coherence. In the vocabulary of product operators (§ 3.3.2):

$$2I_zS_x \xrightarrow{\tilde{I}_x} -2I_yS_x$$

$$2I_xS_z \xrightarrow{\tilde{S}_x} -2I_xS_y \qquad \textbf{2.28}$$

With this formalism we revert to algebra and lose much of the feel for a geometrical picture. There seems to be no simple vector model that adequately describes the evolution of multiple-quantum coherence and its subsequent reconversion into a detectable signal, not to mention the effects of radiofrequency phase shifts or applied magnetic field gradients. It has so far proved unproductive to attempt to extend the vector model any further in this direction.

## 2.8  Discussion

Although the NMR phenomenon can be described by some elegant mathematics, the practising chemist is probably more likely to visualize the behaviour of spin systems in a pictorial fashion. This is the strength of the vector model. It is probably true to say that most of the really important advances in spin choreography came about by this route, rather than by solving the Liouville–von Neumann equation. I believe that the vector picture is indispensable to the understanding of high resolution NMR.

## References

1.  F. Bloch, *Phys. Rev.* **102**, 104 (1956).
2.  H. C. Torrey, *Phys. Rev.* **76**, 1059 (1949).
3.  O. W. Sørensen, G. W. Eich, M. H. Levitt, G. Bodenhausen and R. R. Ernst, *Progr. NMR Spectrosc.*, **16**, 163 (1983).
4.  A. Abragam, *The Principles of Nuclear Magnetism,* Clarendon Press, Oxford, 1961.
5.  R. L. Vold, J. S. Waugh, M. P. Klein and D. E. Phelps, *J. Chem. Phys.* **43**, 3831 (1968).
6.  I. Solomon, *Phys. Rev.* **99**, 559 (1955).
7.  S. Forsén and R. A. Hoffman, *J. Chem. Phys.* **39**, 2892 (1963).
8.  N. V. Bloembergen and R. V. Pound, *Phys. Rev.* **95**, 8 (1954).
9.  A Szöke and S. Meiboom, *Phys. Rev.* **113**, 585 (1959).
10. I. Solomon, *C. R. Acad. Sci. Paris* **248**, 92 (1959).
11. J. M. Bernassau and J. M. Nuzillard, *J. Magn. Reson. A.* **104**, 212 (1993).
12. P. Xu, X. L. Wu and R. Freeman, *J. Am. Chem. Soc.* **113**, 3596 (1991).
13. S. R. Hartmann and E. L. Hahn, *Phys. Rev.* **128**, 2042 (1962).
14. E. L. Hahn, *Phys. Rev.* **80**, 580 (1950).
15. H. Y. Carr and E. M. Purcell, *Phys. Rev.* **94**, 530 (1954).
16. E. L. Hahn and D. E. Maxwell, *Phys. Rev.* **84**, 1246 (1951).
17. R. Freeman and H. D. W. Hill, in *Dynamic Nuclear Magnetic Resonance Spectroscopy,* Eds: L. M. Jackman and F. A. Cotton, Academic Press, New York, 1975.
18. G. A. Morris and R. Freeman, *J. Magn. Reson.* **29**, 433 (1978).

19. H. Geen and R. Freeman, *J. Magn. Reson.* **93**, 93 (1991).
20. M. H. Levitt and R. Freeman, *J. Magn. Reson.* **33**, 473 (1979).
21. M. H. Levitt, *Progr. NMR Spectrosc.* **18**, 61 (1986).
22. C. Bauer, R. Freeman, T. Frenkiel, J. Keeler and A. J. Shaka, *J. Magn. Reson.* **58**, 442 (1984).
23. L. Emsley and G. Bodenhausen, *J. Magn. Reson.* **82**, 212 (1989).
24. N. Metropolis, A. W. Rosenbluth, M. N. Rosenbluth, A. H. Teller and E. Teller, *J. Chem. Phys.* **21**, 1087 (1953).
25. N. F. Ramsay, *Phys. Rev.* **100**, 1191 (1955).
26. F. Bloch and A. Siegert, *Phys. Rev.* **57**, 522 (1940).
27. H. Geen, X. L. Wu, P. Xu and R. Freeman, *J. Magn. Reson.* **81**, 646 (1989).
28. G. A. Morris and R. Freeman, *J. Am. Chem. Soc.* **101**, 760 (1979).
29. J. D. Gezelter and R. Freeman, *J. Magn. Reson.* **90**, 397 (1990).
30. Ē. Kupče and R. Freeman, *J. Magn. Reson. A.* **103**, 358 (1993).
31. J. R. Garbow, D. P. Weitekamp and A. Pines, *Chem. Phys. Lett.* **93**, 514 (1982).
32. J. Jeener, Ampere International Summer School, Basko Polje, Yugoslavia, 1971, and reported in *NMR and More; In Honour of Anatole Abragam*, Eds: M. Goldman and M. Porneuf, Les Editions de Physique, Les Ulis, France, 1994.
33. W. P. Aue, E. Bartholdi and R. R. Ernst, *J. Chem. Phys.* **64**, 2229 (1976).
34. A. Bax and R. Freeman, *J. Magn. Reson.* **44**, 452 (1981).
35. G. Bodenhausen, H. Kogler and R. R. Ernst, *J. Magn. Reson.* **58**, 370 (1984).
36. Ē. Kupče and R. Freeman, *J. Magn. Reson. A.* **105**, 310 (1993).
37. R. E. Hurd, *J. Magn. Reson.* **87**, 422 (1990).
38. R. Freeman and H. D. W. Hill, *J. Mag. Reson.* **4**, 366 (1971).
39. R. Kaiser, E. Bartholdi and R. R. Ernst, *J. Chem. Phys.* **60**, 2966 (1974).
40. A. Schwenk, *J. Mag. Reson.* **5**, 374 (1971).

# 3

# Product operator formalism

## 3.1 Introduction

For those who aspire to lead the spins a merry dance by inventing new pulse sequences, the most valuable tool of all is, without doubt, the product operator treatment.[1–5] Whereas the vector model is based firmly on geometry, the product operator treatment relies on algebra, but this has been distilled from the full density operator theory in a manner that avoids the unwieldy nature of density operators applied to multispin systems. Once the basic rules of this algebra have been assimilated, we can adopt several shorthand notations that retain the essential physics without the need to write out all the explicit equations. The beauty of the scheme is that it retains the intuitive geometrical analogies characteristic of the vector model, so that we can focus our attention on the new spin physics while taking the formalism for granted.

The advent of multiple-quantum experiments revealed the Achilles' heel of the vector model. Valiant efforts[6,7] have been made to bend the rules so that multiple-quantum coherence could be represented as a vector picture, but the results have been disappointing. The product operator formalism takes the creation of multiple-quantum coherence (and its reconversion into observable nuclear magnetization) in its stride. What once appeared to be magic, or at least sleight-of-hand, becomes a straightforward consequence of the product operator algebra. These enormous advantages more than compensate for the somewhat clumsy product operator treatment of line-selective radiofrequency excitation.

However, the product operator algebra needs considerable practice before it can be applied with confidence. For this reason three practical examples, homonuclear correlation spectroscopy, spectral editing and pseudocorrelation spectroscopy have been worked through in detail. Beginners may like to use these exercises to gain familiarity with the procedures. Readers more experienced in the product operator formalism will probably prefer to skip rapidly past these sections.

We start by describing the state of the spin system by the density operator $\sigma(t)$, and follow its time evolution in terms of the Liouville–von Neumann equation:

$$\frac{d\sigma(t)}{dt} = -i[\mathcal{H}(t), \sigma(t)] \tag{3.1}$$

where $\mathcal{H}$ is the Hamiltonian. For simplicity we consider only weakly-coupled spin-$\frac{1}{2}$ nuclei, and neglect relaxation. The NMR experiment is broken down into a sequence of consecutive time intervals. Within each interval we may consider a cascade of evolutions due to chemical shift and spin–spin coupling, or a radio-frequency pulse.

The time-dependence of the radiofrequency is removed by transformation into the appropriate rotating frame of reference. A hard radiofrequency pulse may be expressed as a cascade of soft pulses applied to different spins in turn, in any order. An amplitude- or phase-modulated pulse is decomposed into a sequence (a histogram) of many segments, each of constant phase and amplitude. The processes occurring within a given time segment $(k)$ can be expressed as a propagator, $\exp(-i\mathcal{H}_k\tau_k)$, and the time evolution computed as a sequence of such propagators, for example, the density operator at time $t + \tau_1 + \tau_2$ is given by

$$\sigma(t + \tau_1 + \tau_2) = \exp(-i\mathcal{H}_2\tau_2)\exp(-i\mathcal{H}_1\tau_1)\,\sigma(t)\,\exp(+i\mathcal{H}_1\tau_1)\exp(+i\mathcal{H}_2\tau_2). \tag{3.2}$$

The observable absorption-mode signal is then evaluated as the trace:

$$M_y(t) = N\gamma(h/2\pi)\,Tr\left[\sum_k I_{ky}\sigma(t)\right] \tag{3.3}$$

where $N$ is the number of spins per unit volume.

The density operator is expressed as a linear combination of base operators $(B_s)$:

$$\sigma(t) = \sum_s b_s(t)\,B_s \tag{3.4}$$

The key to the whole process is to choose a set of base operators $\{B_s\}$ that retains the clearest physical insight into the spin engineering and which employs relatively simple algebra. Sørensen et al.[1] chose the Cartesian product operators:

$$B_s = 2^{(q-1)}\prod_{k=1}^{n}(I_{kv})^{a_{sk}} \tag{3.5}$$

Here $n$ is the total number of spins in the spin system under consideration, $k$ is an index for the spin, and $v$ represents the $x$, $y$ or $z$ axis. We simply multiply the spin operators together $q$ at a time $(0 \leq q \leq n)$. The exponent $a_{sk}$ is simply a device to define the spins involved in a given product operator; it is unity for the $q$ spins that are involved in the product operator, but zero for the remaining $n-q$ spins. Thus for an $IS$ system where $n = 2$, we consider $q = 0$, 1 and 2. For $q = 0$, the operator is $\frac{1}{2}E$ where $E$ is the unity operator. For $q = 1$ we have the single-spin operators

$$I_x \quad I_y \quad I_z \quad S_x \quad S_y \quad S_z \qquad \textbf{3.6}$$

while for $q = 2$ there are nine two-spin operators:

$$\begin{array}{ccc} 2I_xS_x & 2I_xS_y & 2I_xS_z \\ 2I_yS_x & 2I_yS_y & 2I_yS_z \\ 2I_zS_x & 2I_zS_y & 2I_zS_z \end{array} \qquad \textbf{3.7}$$

We shall see below what these strange quantities actually represent. We may already anticipate from Equation 3.5 that for a three-spin system there will be terms of the form

$$4I_xS_xR_x \qquad \textbf{3.8}$$

and, for a four-spin system, terms like

$$8I_xS_xR_xP_x \qquad \textbf{3.9}$$

## 3.2 Populations, magnetization and coherence

The product operators represent physical quantities relevant to a high resolution NMR spectrum. This can be illustrated in the context of a coupled two-spin system ($IS$). We assume for simplicity that we are dealing exclusively with spin-$\frac{1}{2}$ nuclei and that the spins are weakly coupled.

### 3.2.1 Spin populations

The operators $I_z$ and $S_z$ represent longitudinal magnetization of the $I$ and $S$ spins respectively. They are proportional to the population differences between the relevant energy levels (§ 1.2). At thermal equilibrium, the state of a homonuclear spin system would be represented as

$$I_z + S_z \qquad \textbf{3.10}$$

This is proportional to the total $z$ magnetization of each spin species. If we create a population inversion by applying a nonselective $\pi$ pulse, this expression would be written $-I_z - S_z$. Relaxation effects are normally ignored in product operator treatments.

The Boltzmann distribution of spin populations in a two-spin system can be disturbed in several ways. One of these, characterized by the product operator term $2I_zS_z$, is of particular importance as an intermediate state in many polarization transfer experiments. It is sometimes called longitudinal two-spin order. It has no net polarization and no observable magnetization (but may be converted into detectable signals by a read pulse).

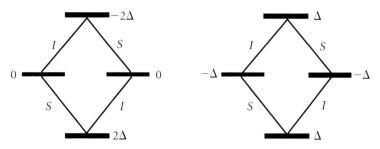

Boltzmann equilibrium          Longitudinal two-spin order

There is a population inversion across one of the $I$-spin transitions and across one of the $S$-spin transitions.

Product operators offer no particular advantage for representing pure population effects, such as spin–lattice relaxation or the nuclear Overhauser effect, but they are useful for describing the magnetization transfer techniques that are so important to modern high resolution NMR spectroscopy.

### 3.2.2 Magnetization

The operators $I_y$ and $S_y$ represent transverse magnetization aligned along the $+y$ axis of the rotating reference frame. In the vector model of a coupled two-spin system, we would represent this situation by two $I$-spin vectors (one for each component of the $J$ doublet) aligned along $+y$, corresponding to an absorption-mode signal. This would be the situation immediately after a $(\pi/2)_x$ excitation pulse. In a similar manner, $I_x$ and $S_x$ indicate transverse magnetization along the $+x$ axis (dispersion mode).

When the transverse magnetization $I_y$ is allowed to precess freely, the magnetization vectors associated with the two lines of the doublet diverge and eventually reach an antiphase configuration. If we ignore chemical shift evolution for the time being (it could be refocused if necessary), we may represent this state by $2I_xS_z$. This is $x$ magnetization of the $I$ spin, antiphase with respect to the $S$ spin. The tiny magnetic fields exerted by the $S$ spin at the site of the $I$ spin have induced a differential precession of the two vectors. A similar term stands for antiphase $I$-spin magnetization along the $\pm y$ axes:

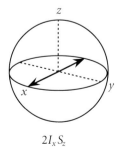

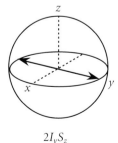

$2I_xS_z$                           $2I_yS_z$

Instead of the familiar in-phase structure $I_y$, these product operator terms generate an up–down pattern of intensities within the multiplet:

This antiphase magnetization is a characteristic feature of many coherence transfer experiments. Note that antiphase magnetization induces no signal in the receiver coil, but it can evolve into an in-phase component that is detectable (see below).

If a third spin $R$ is introduced, there may be antiphase terms and doubly antiphase product operator terms, giving various patterns for the $I$-spin multiplet. We assume that $|J_{IR}| > |J_{IS}|$).

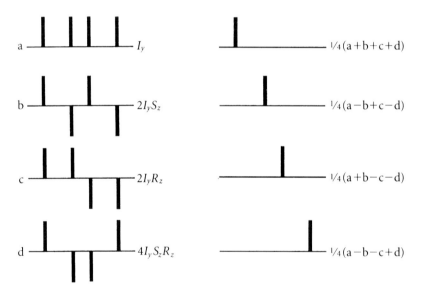

If we wish to represent just one line of the four-line pattern, it is given by the appropriate linear combination of $I_y$, $2I_yS_z$, $2I_yR_z$ and $4I_yS_zR_z$, shown on the right. This formulation is useful for describing experiments where the radiofrequency pulse is line-selective.

### 3.2.3 Coherence

The third type of product operator represents a coherent superposition of states, usually called a coherence for short. (See § 1.3 for comments on the specialized use of this term in NMR.) Although this concept also applies to states separated by $\Delta m = \pm 1$, it is a convenient label for (unobservable) multiple-quantum features involving states separated by $\Delta m \neq \pm 1$. Then we can reserve the term 'magnetization' for single-quantum coherence in order to emphasize that it is detectable as a radiofrequency current in the receiver coil.

The product operator formalism describes how multiple-quantum coherence can be excited, how it is affected by certain spectrometer parameters, and how it may be reconverted into observable magnetization. These product operator terms represent a concerted coherent motion of two or more spins. For the two-spin system there are four such terms: $2I_xS_x$, $2I_xS_y$, $2I_yS_x$ and $2I_yS_y$. In homonuclear systems, linear combinations of these terms represent pure zero- or pure double-quantum coherence.

The most logical approach to multiple-quantum coherence is through the raising and lowering operators, which change the magnetic quantum number by one unit either up or down. For a homonuclear system they are defined by

$$I^+ = I_x + iI_y$$

$$I^- = I_x - iI_y \qquad \textbf{3.11}$$

with similar expressions $S^+$ and $S^-$ for the $S$ spins. Consequently, we may write

$$I_x = \tfrac{1}{2}(I^+ + I^-)$$

$$I_y = \tfrac{1}{2i}(I^+ - I^-) \qquad \textbf{3.12}$$

If we are dealing with a concerted absorption or emission of two quanta in the same sense then the process involves the products $I^+S^+$ and $I^-S^-$. There are two orthogonal forms of double-quantum coherence:

$$I^+S^+ + I^-S^- = 2I_xS_x - 2I_yS_y \qquad \textbf{3.13}$$

$$I^+S^+ - I^-S^- = i\{2I_xS_y + 2I_yS_x\} \qquad \textbf{3.14}$$

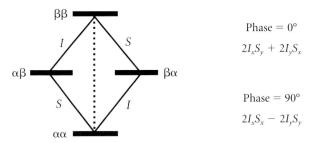

Phase $= 0°$

$2I_xS_y + 2I_yS_x$

Phase $= 90°$

$2I_xS_x - 2I_yS_y$

Double-quantum coherence

If the two spins flip in opposite senses, the relevant products of the raising and lowering operators are $I^+S^-$ and $I^-S^+$:

$$I^+S^- + I^-S^+ = 2I_xS_x + 2I_yS_y \qquad \textbf{3.15}$$

$$I^+S^- - I^-S^+ = i\{2I_yS_x - 2I_xS_y\} \qquad \textbf{3.16}$$

There are thus two orthogonal forms of zero-quantum coherence:

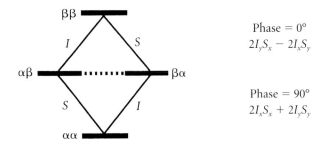

Phase $= 0°$
$2I_yS_x - 2I_xS_y$

Phase $= 90°$
$2I_xS_x + 2I_yS_y$

Zero-quantum coherence

Higher-order multiple-quantum coherences are treated in a later section.

## 3.3 Evolution

Rotations about the $x$ or $y$ axes, caused by radiofrequency pulses, and rotations about the $z$ axis, caused by chemical shifts or spin–spin coupling, are expressed by propagators of the form exp(-i$\mathcal{H}_1\tau_1$). By convention, a sequence of such rotation operators is written in a time-reversed order (the first operation on the right). For simplicity, consider the case of evolution due only to chemical shift ($\Omega$ rad s$^{-1}$), that is to say, a rotation about the $z$ axis. More complex evolutions are simply cascaded.

$$\sigma(t + \tau) = \exp(-i\Omega\tau I_z)\,\sigma(t)\,\exp(+i\Omega\tau I_z) \qquad \textbf{3.17}$$

This is often written in a shorthand form, dropping the imaginary unit i, as

$$\sigma \xrightarrow{\ (\Omega\tau)\,\tilde{I}_z\ } \sigma^+ \qquad \textbf{3.18}$$

To avoid any confusion, we write the rotation angle in parentheses and denote the evolution (super)operator with a tilde ($\tilde{I}_z$) to distinguish it from an operator ($I_z$) that describes a state of the spin system. If we are only considering the effect on transverse magnetization $I_y$, the result is

$$I_y \xrightarrow{\ (\Omega\tau)\ -\tilde{I}_z\ } I_y\cos\Omega\tau + I_x\sin\Omega\tau \qquad \textbf{3.19}$$

(Here the sign of $\tilde{I}_z$ has been changed for reasons that will be explained below.) Equation 3.19 describes free precession of magnetization about the $z$ axis at an angular rate $\Omega$ rad s$^{-1}$.

### 3.3.1 Comparison with the vector model

In the discussion of the vector model of NMR (§ 2.2.3) a particular set of sign conventions was chosen to facilitate the presentation of the trajectories on the unit sphere. This sign convention is entirely consistent with that adopted by Sørensen *et al.*[1] for the rotation of product operators, but some care is necessary when comparing the two treatments.

Consider first chemical shifts. We assume a positive gyromagnetic ratio $\gamma$ (which is the case for most common spin-$\frac{1}{2}$ nuclei, except $^{15}$N and $^{29}$Si). In a frame of reference rotating in synchronism with the transmitter frequency, a nucleus with a Larmor frequency higher than the transmitter frequency would experience a positive offset $\Delta B$. In the vector picture this corresponds to a rotation in the sense

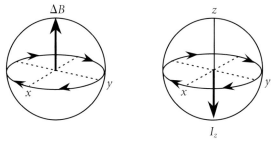

Vector model                   Product operator

In the treatment of Sørensen *et al.*[1] the evolution operator $\tilde{I}_z$ is defined as being antiparallel to $\Delta B$. The sense of rotation (clockwise viewed from the north pole) is the same in both schemes. This is the reason for the negative sign in Equation 3.19.

When considering the effect of radiofrequency pulses we assume initially that they are hard (nonselective). In most experiments the pulse is applied along the $+x$ axis of the rotating frame. In the vector picture this corresponds to a rotation from $+z$ towards $+y$:

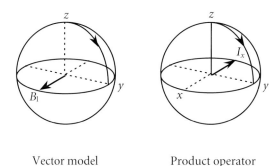

Vector model                   Product operator

In the product operator formalism of Sørensen *et al.*[1] this corresponds to an evolution operator aligned along $-x$, causing the same rotation from $+z$ towards $+y$.

A radiofrequency pulse of flip angle β, acting on longitudinal magnetization, creates transverse magnetization and leaves some residual $z$ magnetization:

$$+I_z \xrightarrow{\quad (\beta)\ -\tilde{I}_x \quad} +I_z \cos\beta + I_y \sin\beta \qquad \textbf{3.20}$$

If $\beta = \pi/2$, we obtain pure $I_y$ magnetization; if $\beta = \pi$, we obtain negative $z$ magnetization (a population inversion).

The third type of evolution is a bilinear rotation caused by the action of the spin–spin coupling $J_{IS}$. This may be written as

$$\sigma(t + \tau) = \exp(-i\pi J_{IS}\tau 2\tilde{I}_z\tilde{S}_z)\, \sigma(t)\, \exp(+i\pi J_{IS}\tau 2\tilde{I}_z\tilde{S}_z) \qquad \textbf{3.21}$$

In the shorthand notation

$$\sigma(t) \xrightarrow{\quad (\pi J_{IS}\tau)\ \ 2\tilde{I}_z\tilde{S}_z \quad} \sigma(t + \tau) \qquad \textbf{3.22}$$

Here we encounter a new type of evolution operator $2\tilde{I}_z\tilde{S}_z$, which also has a specified sense of rotation, equivalent in the vector picture to assigning the relative senses of precession of the α and β vectors. To be compatible with the product operator convention of Sørensen et al.[1] we must employ the evolution operator $-2\tilde{I}_z\tilde{S}_z$ to produce the same result as the vector model. Thus we write:

$$I_y \xrightarrow{\quad (\pi J_{IS}\tau)\ \ -2\tilde{I}_z\tilde{S}_z \quad} I_y\cos(\pi J_{IS}\tau) + 2I_xS_z\sin(\pi J_{IS}\tau) \qquad \textbf{3.23}$$

$$I_x \xrightarrow{\quad (\pi J_{IS}\tau)\ \ -2\tilde{I}_z\tilde{S}_z \quad} I_x\cos(\pi J_{IS}\tau) - 2I_yS_z\sin(\pi J_{IS}\tau) \qquad \textbf{3.24}$$

Consider the case where $\tau = 1/(2J_{IS})$; $y$ magnetization is converted into antiphase $x$ magnetization:

$$I_y \xrightarrow{\quad (\pi/2)\ \ -2\tilde{I}_z\tilde{S}_z \quad} +2I_xS_z \qquad \textbf{3.25}$$

and $x$ magnetization is converted into anti-phase $y$ magnetization:

$$I_x \xrightarrow{\quad (\pi/2)\ \ -2\tilde{I}_z\tilde{S}_z \quad} -2I_yS_z \qquad \textbf{3.26}$$

In the vector picture (neglecting chemical shifts), the β vector has a higher frequency than the reference frame and rotates clockwise towards the $+x$ axis; the α vector rotates in the opposite sense towards the $-x$ axis:

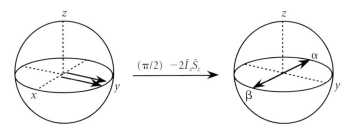

If we allow the evolution to proceed for another equal period $\tau = 1/(2J_{IS})$ then the antiphase vectors along $\pm x$ move into alignment along the $-y$ axis. In the product operator formalism:

$$+2I_xS_z \xrightarrow{(\pi/2) \ -2\tilde{I}_z\tilde{S}_z} -I_y \qquad \textbf{3.27}$$

This is an important step in many coherence transfer sequences because it converts the quantity $2I_xS_z$ into observable magnetization.

The vector picture used in § 2.2, and the product operator treatment therefore have the same sign convention for rotations. However it is important to remember that $\Delta B$ is antiparallel to the evolution operator $\tilde{I}_z$, $B_1$ is antiparallel to the evolution operator $\tilde{I}_x$ (or $\tilde{I}_y$) and that the usual bilinear rotation operator is $-2\tilde{I}_z\tilde{S}_z$ rather than $+2\tilde{I}_z\tilde{S}_z$. Of course, within each model, the sign convention is not critical, provided we are consistent. For most applications of the product operator formalism there is no need to make a direct comparison with the vector model, and the negative signs on the evolution operators can be dropped. Note that this simplification is adopted in the descriptions that follow. Some authors also drop the normalization factors, for example, $2I_zS_z$ becomes $I_zS_z$, and $4I_yS_zR_z$ becomes $I_yS_zR_z$.

### 3.3.2 **Rotations**

The general forms of the product operator transformations are illustrated diagrammatically in Figure 3.1. Particular importance must be attached to signs. The key is

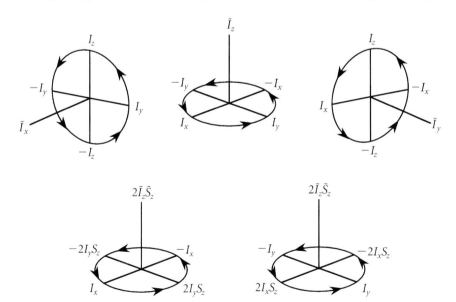

**Figure 3.1** Schematic representation of the rotations caused by the evolution (super)operators $\tilde{I}_x$, $\tilde{I}_z$, $\tilde{I}_y$ and $2\tilde{I}_z\tilde{S}_z$, acting on product operators (shown without a tilde) that describe the state of the spin system. Note that the rotation is always in the same sense when the evolution operators are positive.

to remember that, for a positive evolution operator, when the transformation involves changing indices in their natural order $x$, $y$, $z$ (left-hand sketch) there is no sign change; the reverse order (right) involves a sign change.

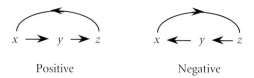

Positive                                    Negative

For example, a $\pi/2$ phase shift of transverse magnetization gives no change in sign (since $x \rightarrow y$):

$$+I_x \xrightarrow{\ (\pi/2)\ \tilde{I}_z\ } +I_y \qquad\qquad \textbf{3.28}$$

A $\pi/2$ excitation pulse about the $x$ axis changes the sign (since $y \leftarrow z$):

$$+I_z \xrightarrow{\ (\pi/2)\ \tilde{I}_x\ } -I_y \qquad\qquad \textbf{3.29}$$

Evolution of an absorption-mode signal into antiphase magnetization along the $\pm x$ axes, through the effect of spin–spin coupling:

$$+I_y \xrightarrow{\ (\pi/2)\ 2\tilde{I}_z\tilde{S}_z\ } -2I_xS_z \qquad\qquad \textbf{3.30}$$

Transfer of antiphase magnetization from the $I$ spins to the $S$ spins:

$$+2I_yS_z \xrightarrow{\ (\pi/2)\ \tilde{I}_x\ } +2I_zS_z \xrightarrow{\ (\pi/2)\ \tilde{S}_x\ } -2I_zS_y \qquad\qquad \textbf{3.31}$$

$$+2I_yS_z \xrightarrow{\ (\pi/2)\ \tilde{S}_x\ } -2I_yS_y \xrightarrow{\ (\pi/2)\ \tilde{I}_x\ } -2I_zS_y \qquad\qquad \textbf{3.32}$$

Note that Equations 3.31 and 3.32 give the same result, reflecting the important rule that the ordering of the propagators is immaterial, since $\tilde{I}_x$ and $\tilde{S}_x$ commute.

Creation of two-spin coherence from antiphase magnetization:

$$+2I_yS_z \xrightarrow{\ (\pi/2)\ \tilde{I}_y\ } +2I_zS_z \xrightarrow{\ (\pi/2)\ \tilde{S}_y\ } +2I_zS_x \qquad\qquad \textbf{3.33}$$

If we compare Equations 3.31 and 3.33 we note the crucial effect of a $\pi/2$ phase shift.

These transformations can also be set out in the form of a conversion chart (Table 3.1). Suppose we are considering the rotation of some product operator $P$ (no tilde) through an angle $\beta$ radians, under the influence of an evolution operator $\tilde{S}$. We look up the corresponding result $R$ in Table 3.1 and write:

$$P \xrightarrow{\ (\beta)\ \tilde{S}\ } P\cos\beta + R\sin\beta \qquad\qquad \textbf{3.34}$$

**Table 3.1** Conversion chart for product operators. An evolution operator (top row, with a tilde) acts on an operator (left margin, without a tilde) that defines the state of the system. The symbol $E$ is the unity operator.

| | $\tilde{I}_x$ | $\tilde{I}_y$ | $\tilde{I}_z$ | $\tilde{S}_x$ | $\tilde{S}_y$ | $\tilde{S}_z$ | $2\tilde{I}_z\tilde{S}_z$ |
|---|---|---|---|---|---|---|---|
| $I_x$ | $\frac{1}{2}E$ | $-I_z$ | $+I_y$ | $\frac{1}{2}E$ | $\frac{1}{2}E$ | $\frac{1}{2}E$ | $+2I_yS_z$ |
| $I_y$ | $+I_z$ | $\frac{1}{2}E$ | $-I_x$ | $\frac{1}{2}E$ | $\frac{1}{2}E$ | $\frac{1}{2}E$ | $-2I_xS_z$ |
| $I_z$ | $-I_y$ | $+I_x$ | $\frac{1}{2}E$ | $\frac{1}{2}E$ | $\frac{1}{2}E$ | $\frac{1}{2}E$ | $\frac{1}{2}E$ |
| $S_x$ | $\frac{1}{2}E$ | $\frac{1}{2}E$ | $\frac{1}{2}E$ | $\frac{1}{2}E$ | $-S_z$ | $+S_y$ | $+2I_zS_y$ |
| $S_y$ | $\frac{1}{2}E$ | $\frac{1}{2}E$ | $\frac{1}{2}E$ | $+S_z$ | $\frac{1}{2}E$ | $-S_x$ | $-2I_zS_x$ |
| $S_z$ | $\frac{1}{2}E$ | $\frac{1}{2}E$ | $\frac{1}{2}E$ | $-S_y$ | $+S_x$ | $\frac{1}{2}E$ | $\frac{1}{2}E$ |
| $2I_zS_z$ | $-2I_yS_z$ | $+2I_xS_z$ | $\frac{1}{2}E$ | $-2I_zS_y$ | $+2I_zS_x$ | $\frac{1}{2}E$ | $\frac{1}{2}E$ |
| $2I_xS_z$ | $\frac{1}{2}E$ | $-2I_zS_z$ | $+2I_yS_z$ | $-2I_xS_y$ | $+2I_xS_x$ | $\frac{1}{2}E$ | $+I_y$ |
| $2I_yS_z$ | $+2I_zS_z$ | $\frac{1}{2}E$ | $-2I_xS_z$ | $-2I_yS_y$ | $+2I_yS_x$ | $\frac{1}{2}E$ | $-I_x$ |
| $2I_zS_x$ | $-2I_yS_x$ | $+2I_xS_x$ | $\frac{1}{2}E$ | $\frac{1}{2}E$ | $-2I_zS_z$ | $+2I_zS_y$ | $+S_y$ |
| $2I_zS_y$ | $-2I_yS_y$ | $+2I_xS_y$ | $\frac{1}{2}E$ | $+2I_zS_z$ | $\frac{1}{2}E$ | $-2I_zS_x$ | $-S_x$ |
| $2I_xS_x$ | $\frac{1}{2}E$ | $-2I_zS_x$ | $+2I_yS_x$ | $\frac{1}{2}E$ | $-2I_xS_z$ | $+2I_xS_y$ | $\frac{1}{2}E$ |
| $2I_xS_y$ | $\frac{1}{2}E$ | $-2I_zS_y$ | $+2I_yS_y$ | $+2I_xS_z$ | $\frac{1}{2}E$ | $-2I_xS_x$ | $\frac{1}{2}E$ |
| $2I_yS_x$ | $+2I_zS_x$ | $\frac{1}{2}E$ | $-2I_xS_x$ | $\frac{1}{2}E$ | $-2I_yS_z$ | $+2I_yS_y$ | $\frac{1}{2}E$ |
| $2I_yS_y$ | $+2I_zS_y$ | $\frac{1}{2}E$ | $-2I_xS_y$ | $-2I_yS_x$ | $+2I_yS_z$ | $\frac{1}{2}E$ | $\frac{1}{2}E$ |

If, as often happens, $\beta = \pi/2$ radians,

$$P \xrightarrow{(\pi/2)\,\tilde{S}} R \qquad\qquad \textbf{3.35}$$

At this point we introduce the commonly used abbreviation of omitting the explicit indication of the flip angle when it is $\pi/2$ radians. This saves time when writing long strings of operations.

Examination of Table 3.1 brings to light some useful general rules:

- An evolution operator for the $I$ spins has no effect on $S$-spin product operators:

$$+S_x \xrightarrow{\tilde{I}_z} +S_x \qquad\qquad \textbf{3.36}$$

- An evolution operator has no effect on an identical product operator

$$+I_x \xrightarrow{\tilde{I}_x} +I_x \qquad\qquad \textbf{3.37}$$

- There is a reciprocity relationship which involves a sign reversal. For example,

$$+I_x \xrightarrow{\tilde{I}_z} +I_y \quad \text{but} \quad +I_z \xrightarrow{\tilde{I}_x} -I_y \qquad\qquad \textbf{3.38}$$

- A rotation through $(2n+1)\pi$ radians (where $n$ is an integer) changes the sign:

$$+I_z \xrightarrow{\ (\pi)\,\tilde{I}_x\ } -I_z \tag{3.39}$$

- The evolution operator $2\tilde{I}_z\tilde{S}_z$ has no effect on two-spin coherence:

$$+2\tilde{I}_x\tilde{S}_x \xrightarrow{\ 2\tilde{I}_z\tilde{S}_z\ } +2I_xS_x \tag{3.40}$$

Once we become familiar with the standard manipulations of product operators, the conversion chart becomes superfluous, except perhaps as a scheme for checking signs.

### 3.3.3 **Refocusing pulses**

The building block

$$[\ -\ \tau\ -\ \pi\ -\ \tau\ -]$$

is much used in modern pulse sequences (§ 4.2). Transverse magnetization is allowed to precess under the influence of chemical shifts and spin–spin coupling, but the former are refocused. Consider, first of all, a homonuclear two-spin ($IS$) system, with chemical shifts $\delta_I$ and $\delta_S$. We write a cascade of propagators; the first three describe coupling and chemical shift evolution (in any order), the next two represent the inversion pulse, and the final three reflect the continued effects of shifts and coupling in the second delay period:

$$\sigma(t) \xrightarrow[\ (2\pi\delta_I\tau)\,\tilde{I}_z\ ]{\ (\pi J_{IS}\tau)\,2\tilde{I}_z\tilde{S}_z\ } \xrightarrow[\ (2\pi\delta_S\tau)\,\tilde{S}_z\ ]{\ (2\pi\delta_I\tau)\,\tilde{I}_z\ } \xrightarrow[\ (\pi J_{IS}\tau)\,2\tilde{I}_z\tilde{S}_z\ ]{\ (2\pi\delta_S\tau)\,\tilde{S}_z\ } \xrightarrow{\ (\pi)\,\tilde{I}_x\ } \xrightarrow{\ (\pi)\,\tilde{S}_x\ } \sigma(t+2\tau) \tag{3.41}$$

The delay is usually set to $\tau = 1/(4J_{IS})$ or to $\tau = 1/(2J_{IS})$ depending on the application.

There are some rather irritating cassette tape recorders that lack a facility for rewinding the tape; if one wants to return to an earlier part of the recording, the cassette has to be removed, reversed, reinserted, driven forward for the required period, then removed again, reversed and reinserted into the recorder. Since it is not possible in practice to reverse the sense of a chemical shift, a similar sequence of manipulations is employed to achieve the effect of a sign inversion. The analog of reversing the cassette is a $\pi$ radian rotation operator, $\exp(i\pi I_x)$, and chemical shift evolution in the true sense is represented by the rotation operator $\exp(+i2\pi\delta_I\tau I_z)$:

$$\exp(-i\pi I_x)\exp(+i2\pi\delta_I\tau I_z)\exp(+i\pi I_x) = \exp(-i2\pi\delta_I\tau I_z) \tag{3.42}$$

This gives the effect of a time-reversed chemical shift. If we now multiply both sides of the equation by the term $\exp(+i\pi I_x)$, and then drop the product $\exp(+i\pi I_x)\exp(-i\pi I_x)$, we have

$$\exp(+i2\pi\delta_I\tau I_z)\exp(+i\pi I_x) = \exp(+i\pi I_x)\exp(-i2\pi\delta_I\tau I_z) \tag{3.43}$$

The reordering of the terms has reversed the chemical shift evolution. There is an equivalent identity for $\delta_S$. If this operation is applied to Equation 3.41, the chemical shift propagators with opposite signs become adjacent and mutually cancel, leaving the simplified evolution

$$\sigma(t)\xrightarrow{(\pi J_{IS}\tau)\,2\tilde{I}_z\tilde{S}_z}\xrightarrow{(\pi)\,\tilde{S}_x}\xrightarrow{(\pi)\,\tilde{I}_x}\xrightarrow{(\pi J_{IS}\tau)\,2\tilde{I}_z\tilde{S}_z}\sigma(t+2\tau)\qquad \textbf{3.44}$$

which is equivalent to:

$$\sigma(t)\xrightarrow{(2\pi J_{IS}\tau)\,2\tilde{I}_z\tilde{S}_z}\xrightarrow{(\pi)\,\tilde{S}_x}\xrightarrow{(\pi)\,\tilde{I}_x}\sigma(t+2\tau)\qquad \textbf{3.45}$$

since, in moving the evolution operator $2\tilde{I}_z\tilde{S}_z$ to the left, we have inverted its sign twice. This leaves the echo modulated by the $J$ coupling (§ 4.2.3). In the vector picture (§ 2.4.5) we would say that both the $I$ and $S$ spins are flipped, leaving the divergence of $J$ vectors unchanged.

The situation for heteronuclear systems is somewhat different. We adopt the usual convention that the $S$ spins are observed. The $\tilde{I}_x$ pulse is now optional. If the $\tilde{I}_x$ and $\tilde{S}_x$ pulses are both included, the formalism outlined above is applicable. If the $\tilde{I}_x$ pulse is retained but the $\tilde{S}_x$ pulse omitted, we have a primitive decoupling technique (sometimes called spin–flip decoupling) where the evolution of the $S$ spins under the $2\tilde{I}_z\tilde{S}_z$ operator is continually refocused by a succession of $I$-spin inversion pulses. We may sketch this as if the evolution operator $(2\tilde{I}_z\tilde{S}_z)$ repeatedly changed its sign:

$$-2\tilde{I}_z\tilde{S}_z\xrightarrow{\qquad}\tilde{I}_x\xleftarrow{\qquad}+2\tilde{I}_z\tilde{S}_z\xrightarrow{\qquad}\tilde{I}_x\xleftarrow{\qquad}-2\tilde{I}_z\tilde{S}_z\xrightarrow{\qquad}\tilde{I}_x$$

$$\text{acq}\xleftarrow{\quad}a\xrightarrow{\quad}|\xleftarrow{\quad}a\xrightarrow{\quad}\text{acq}\xleftarrow{\quad}a\xrightarrow{\quad}|\xleftarrow{\quad}a\xrightarrow{\quad}\text{acq}\xleftarrow{\quad}a\xrightarrow{\quad}|$$

The acquisition of the $S$-spin signal is restricted to the refocusing points, separated by intervals $2a$. With this stroboscopic sampling, the $S$-spin spectrum contains only chemical shifts and no spin–spin splittings. Practical restrictions on the rate of application of hard $\pi$ pulses limit the sampling rate, making the technique unsuitable for decoupling $^{13}\text{C}$ spectra, but elaborations of the method using composite pulses and magic cycles have proved very useful (see § 7.2).

An extension of this idea permits a scaling of the spin–spin splitting (treated according to the vector model in § 4.3.4). The sampling interval of $2a$ remains the same, but the $\tilde{I}_x$ inversion pulse is alternately advanced and delayed with respect to the midpoint by a small interval $b$. Thus for the first sampling interval, where the $\tilde{I}_x$ pulse is advanced, the $S$-spin evolution may be written as

$$\sigma(t)\xrightarrow{(\pi J_{IS}\tau_1)\,2\tilde{I}_z\tilde{S}_z}\xrightarrow{(2\pi\delta_S\tau_1)\,\tilde{S}_z}\xrightarrow{(\pi)\,\tilde{I}_x}$$

$$\xrightarrow{(2\pi\delta_S\tau_2)\,\tilde{S}_z}\xrightarrow{(\pi J_{IS}\tau_2)\,2\tilde{I}_z\tilde{S}_z}\sigma(t+2a)\qquad \textbf{3.46}$$

with $\tau_1=(a-b)$ and $\tau_2=(a+b)$. For the next sampling interval, $\tau_1$ and $\tau_2$ are interchanged, thus delaying the $\tilde{I}_x$ pulse. The $\tilde{I}_x$ operator has no effect on the $S$-spin chemical shift evolution, but it effectively reverses the sign of the $2\tilde{I}_z\tilde{S}_z$ operator, leaving a partially refocused divergence due to $J$ coupling.

$$-2\tilde{I}_z\tilde{S}_z \xrightarrow{\phantom{xxxx}} \tilde{I}_x \xleftarrow{\phantom{xxx}} +2\tilde{I}_z\tilde{S}_z \xrightarrow{\phantom{xx}} \tilde{I}_x \xleftarrow{\phantom{xx}} -2\tilde{I}_z\tilde{S}_z \xrightarrow{\phantom{xx}} \tilde{I}_x$$
$$\text{acq}\leftarrow a-b\rightarrow| \leftarrow a+b \rightarrow \text{acq}\leftarrow a+b\rightarrow| \leftarrow a-b \rightarrow \text{acq}\leftarrow a-b\rightarrow|$$

Note that divergence occurs under the operator $+2\tilde{I}_z\tilde{S}_z$ during the long pulse intervals, but under the operator $-2\tilde{I}_z\tilde{S}_z$ during the short pulse intervals. This results in an apparent divergence at a reduced rate, giving heteronuclear spin–spin splittings in the $S$-spin spectrum that are scaled down by a factor $a/b$.

If the heteronuclear spin echo experiment is performed without the $\tilde{I}_x$ pulse but with the $\tilde{S}_x$ pulse, then the evolution is represented by

$$\sigma(t) \xrightarrow{(+2\pi\delta_S\tau)\,\tilde{S}_z} \xrightarrow{(+\pi J_{IS}\tau)\,2\tilde{I}_z\tilde{S}_z} \xrightarrow{(\pi)\,\tilde{S}_x}$$
$$\xrightarrow{(+\pi J_{IS}\tau)\,2\tilde{I}_z\tilde{S}_z} \xrightarrow{(+2\pi\delta_S\tau)\,\tilde{S}_z} \sigma(t+2\tau) \qquad \textbf{3.47}$$

If we now move the evolution operator $(\pi)\,\tilde{S}_x$ to the left, it changes the sign of both the chemical shift and the spin–spin coupling evolution operators, giving

$$\sigma(t) \xrightarrow{(-2\pi\delta_S\tau)\,\tilde{S}_z} \xrightarrow{(-\pi J_{IS}\tau)\,2\tilde{I}_z\tilde{S}_z} \xrightarrow{(+\pi J_{IS}\tau)\,2\tilde{I}_z\tilde{S}_z} \xrightarrow{(+2\pi\delta_S\tau)\,\tilde{S}_z} \sigma(t+2\tau) \quad \textbf{3.48}$$

and we find that the effects of spin–spin coupling are refocused along with the chemical shift.

### 3.3.4 Family trees

Each step in the evolution of a spin system can multiply the number of product operator terms, and it is therefore useful to represent the result in the form of a branching family tree according to the ideas of van de Ven and Hilbers.[2] These diagrams can rapidly become unmanageable unless a suitable shorthand notation is employed. Consider the action of a pulse of flip angle $\beta$ on equilibrium magnetization $I_z$.

$$I_z \xrightarrow{(\beta)\,\tilde{I}_x} I_z\cos\beta - I_y\sin\beta \qquad \textbf{3.49}$$

In the tree notation this would be written

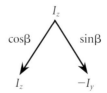

Care is taken to reserve the left-hand branch for the term that remains unchanged (multiplied by the cosine) and the right-hand branch for the term that changes (multiplied by the sine), but to save space, $\cos\beta$ and $\sin\beta$ are usually omitted from the tree diagram. Normally we are interested in only a very small number of the many offspring of the family tree so it is not necessary to carry the trigonometrical terms through every stage of the calculation. The appropriate cosines and sines are simply collected at the end by noting whether left- or right-hand branches are involved. With experience, one can follow just the path that eventually gives the observed signal, and neglect all other side-tracks. But do not venture into this forest unprepared.

## 3.3.5 Correlation spectroscopy (COSY)

These procedures are best illustrated by worked examples. This scheme for correlation spectroscopy may be compared with the analogous treatment by the vector model in § 2.6.1. The COSY pulse sequence[8,9] is well-known but deceptively simple:

$$(\pi/2)_x - t_1 - (\pi/2)_x \text{ acquisition} \qquad \textbf{3.50}$$

We start with $I$-spin polarization and consider coherence transfer to the $S$ spin; the reverse process can be derived by symmetry considerations. Omitting the rotation angles, the cascade of evolution operators may be written

$$\sigma(0) \xrightarrow{\tilde{I}_x} \xrightarrow{\tilde{I}_z} \xrightarrow{\tilde{S}_z} \xrightarrow{2\tilde{I}_z\tilde{S}_z} \xrightarrow{\tilde{I}_x} \xrightarrow{\tilde{S}_x} \xrightarrow{\tilde{I}_z} \xrightarrow{\tilde{S}_z} \xrightarrow{2\tilde{I}_z\tilde{S}_z} \sigma(t) \qquad \textbf{3.51}$$

That is to say, each period of free precession is represented by a cascade of chemical shift ($\tilde{I}_z$ and $\tilde{S}_z$) and spin–spin coupling ($2\tilde{I}_z\tilde{S}_z$) evolution operators (the order is immaterial within a given free precession period). For purely didactic purposes the entire tree is set out in Figure 3.2, since it is useful at this stage to work through some examples, step by painful step. However, we soon realize that large sections of the tree are irrelevant to the problem of determining the final spectrum. For example, the entire left-hand half of the chart gives rise to final product operator terms that correspond to unobservable quantities—either longitudinal magnetization (Long) or multiple-quantum coherence (MQ). In a similar manner, in the right-hand section of the chart we can identify paths that lead only to unobservable antiphase magnetization (Anti).

There are four paths that give rise to observable signals; two of them may be written as

$$+I_z \to -I_y \to +I_x \to +I_x \to +I_x \to +I_x \to +I_x \to +I_x \to +I_x \qquad \textbf{3.52}$$

$$+I_z \to -I_y \to +I_x \to +I_x \to +I_x \to +I_x \to +I_y \to +I_y \to +I_y \qquad \textbf{3.53}$$

They involve no coherence transfer and therefore correspond to the diagonal peaks of the COSY spectrum. When the expressions are written out in full, we have

$$\sigma_1 = +I_x \sin(2\pi\delta_I t_1) \cos(\pi J_{IS} t_1) \cos(2\pi\delta_I t_2) \cos(\pi J_{IS} t_2) \qquad \textbf{3.54}$$

$$\sigma_2 = +I_y \sin(2\pi\delta_I t_1) \cos(\pi J_{IS} t_1) \sin(2\pi\delta_I t_2) \cos(\pi J_{IS} t_2) \qquad \textbf{3.55}$$

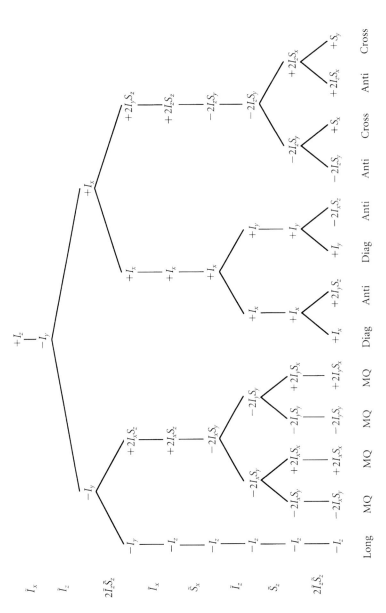

**Figure 3.2** Tree diagram representing the evolution of product operator terms in a homonuclear correlation (COSY) experiment on a two-spin (*IS*) system. The evolution operators are shown in the left-hand column. A left-hand branch implies multiplication by the appropriate cosine term, whereas a right-hand branch implies multiplication by the corresponding sine term. A vertical line implies complete conversion. The final terms are labelled as longitudinal magnetization (Long), multiple-quantum coherence (MQ), diagonal peaks (Diag), antiphase magnetization (Anti) and cross-peaks (Cross).

These may be converted by standard trigonometrical identities into

$$\sigma_1 + \sigma_2 = +\tfrac{1}{4}\{\sin[(2\pi\delta_I + \pi J_{IS})t_1] + \sin[(2\pi\delta_I - \pi J_{IS})t_1]\}$$
$$\times \{I_x \cos[(2\pi\delta_I + \pi J_{IS})t_2] + I_x\cos[(2\pi\delta_I - \pi J_{IS})t_2]$$
$$+ I_y\sin[(2\pi\delta_I + \pi J_{IS})t_2] + I_y\sin[(2\pi\delta_I - \pi J_{IS})t_2]\} \qquad \textbf{3.56}$$

There are four resonances, forming a square pattern with the coordinates

$$(\delta_I - \tfrac{1}{2}J_{IS}, \delta_I + \tfrac{1}{2}J_{IS}) \qquad\qquad (\delta_I + \tfrac{1}{2}J_{IS}, \delta_I + \tfrac{1}{2}J_{IS})$$
$$(\delta_I - \tfrac{1}{2}J_{IS}, \delta_I - \tfrac{1}{2}J_{IS}) \qquad\qquad (\delta_I + \tfrac{1}{2}J_{IS}, \delta_I - \tfrac{1}{2}J_{IS})$$

They fall either exactly on the principal diagonal, or within a distance $|J_{IS}|$. Note that all four terms in Equation 3.56 have the same sign; this indicates that these diagonal peaks are all in the same sense. Normally the spectrometer receiver phase is set so that these four diagonal peaks are in the dispersion mode in both frequency dimensions; the pattern may be sketched as a contour map:

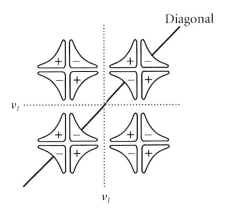

If we trace out the paths leading to observable $S$-spin magnetization, they are

$$+I_z \rightarrow -I_y \rightarrow +I_x \rightarrow +2I_yS_z \rightarrow +2I_zS_z \rightarrow -2I_zS_y \rightarrow -2I_zS_y \rightarrow -2I_zS_y \rightarrow +S_x \qquad \textbf{3.57}$$
$$+I_z \rightarrow -I_y \rightarrow +I_x \rightarrow +2I_yS_z \rightarrow +2I_zS_z \rightarrow -2I_zS_y \rightarrow -2I_zS_y \rightarrow +2I_zS_x \rightarrow +S_y \qquad \textbf{3.58}$$

Note that, because there is an intermediate term $+2I_zS_z$, the transfer may be said to proceed through a population disturbance, as suggested in § 2.6.1. When the trigonometrical terms are reintroduced,

$$\sigma_3 = +S_x \sin(2\pi\delta_I t_1) \sin(\pi J_{IS}t_1) \cos(2\pi\delta_S t_2) \sin(\pi J_{IS}t_2) \qquad \textbf{3.59}$$

$$\sigma_4 = +S_y \sin(2\pi\delta_I t_1) \sin(\pi J_{IS}t_1) \sin(2\pi\delta_S t_2) \sin(\pi J_{IS}t_2) \qquad \textbf{3.60}$$

Standard trigonometrical identities give

$$\sigma_3 + \sigma_4 = +\tfrac{1}{4}\{\cos[(2\pi\delta_I - \pi J_{IS})t_1] - \cos[(2\pi\delta_I + \pi J_{IS})t_1]\}$$
$$\times \{S_x\sin[(2\pi\delta_S + \pi J_{IS})t_2] - S_x\sin[(2\pi\delta_S - \pi J_{IS})t_2]$$
$$- S_y\cos[(2\pi\delta_S + \pi J_{IS})t_2] + S_y\cos[(2\pi\delta_S - \pi J_{IS})t_2]\} \qquad \textbf{3.61}$$

This is again a four-line square pattern with splittings $|J_{IS}|$, but the coordinates are now:

$$(\delta_I - \tfrac{1}{2}J_{IS}, \delta_S + \tfrac{1}{2}J_{IS}) \qquad\qquad (\delta_I + \tfrac{1}{2}J_{IS}, \delta_S + \tfrac{1}{2}J_{IS})$$
$$(\delta_I - \tfrac{1}{2}J_{IS}, \delta_S - \tfrac{1}{2}J_{IS}) \qquad\qquad (\delta_I + \tfrac{1}{2}J_{IS}, \delta_S - \tfrac{1}{2}J_{IS})$$

indicating a transfer of magnetization from $I$ to $S$. Note that, in contrast to the case of the diagonal peaks, there is an alternation in the signs of these terms, which reflects an alternation in the senses of the four lines. With the spectrometer phase set as described above (giving diagonal peaks in dispersion) these cross-peak responses are in the absorption mode in both frequency dimensions. The square four-line pattern can be sketched as a contour map:

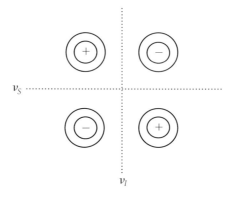

Note that this cross-peak pattern is centred at $(\nu_I, \nu_S)$, away from the principal diagonal of the two-dimensional correlation spectrum.

    We can rewrite Equation 3.61 to represent the symmetrically related $S \rightarrow I$ cross-peak simply by interchanging $I$ and $S$. If we then compare this with the expression for the diagonal peak (Equation 3.56), we see that (apart from the sign alternation) they are related by interchange of sine and cosine throughout. This confirms that if the receiver phase is set so that the cross-peak is in the absorption mode, the diagonal peak will be in the dispersion mode. Finally we may deduce from Equations 3.54, 3.55, 3.59 and 3.60 that the terms responsible for $J$ modulation of the cross-peak, build up with time as $\sin(\pi J_{IS} t_1)$ and $\sin(\pi J_{IS} t_2)$, so that if $J_{IS}$ is small, relatively long intervals $t_1(\max)$ and $t_2(\max)$ must be employed.

    Naturally an analysis by product operators seldom requires this degree of detail. We would normally concentrate on the path that leads directly to $I \rightarrow S$ transfer:

$$+I_z \xrightarrow{\tilde{I}_x} -I_y \xrightarrow{\tilde{I}_z} +I_x \xrightarrow{2\tilde{I}_z\tilde{S}_z} +2I_yS_z \xrightarrow{\tilde{I}_x} +2I_zS_z$$

$$\xrightarrow{\tilde{S}_x} -2I_zS_y \xrightarrow{\tilde{S}_z} +2I_zS_x \xrightarrow{2\tilde{I}_z\tilde{S}_z} +S_y \qquad\qquad \textbf{3.62}$$

With a little experience it is easy to recognize which terms may be safely left out of the evolution and which must be retained.

## 3.4 **Multiple-quantum coherence**

We could probably manage to describe almost all high resolution NMR experiments by the vector model were it not for multiple-quantum effects. Here the product operator formalism comes into its own, not only for illustrating how multiple-quantum coherence is generated and reconverted, but also in explaining the special properties that make multiple-quantum coherence so useful. We shall see in § 6.3.1 how double-quantum filters are used in proton correlation spectroscopy (DQ-COSY)[10] and in natural-abundance $^{13}$C experiments (INADEQUATE).[11,12]

### 3.4.1 **Excitation**

Multiple-quantum coherence between two energy levels $E_1$ and $E_2$ can be excited by several different methods:

- In a slow-passage frequency sweep experiment, with an intense continuous-wave radiofrequency field $B_1$. The required radiofrequency is the appropriate subharmonic, given by $h\nu = (E_1 - E_2)/p$, where $p$ is the order of coherence. Thus for double-quantum coherence ($p=2$) the resonance occurs as we sweep through the mean frequency of two progressively-connected single-quantum transitions.[13]

- By a single soft pulse applied at this same subharmonic frequency, $h\nu = (E_1 - E_2)/p$, so that the radiofrequency source gives up $p$ quanta in a concerted and coherent manner.[14] This is the pulse equivalent of the previous method.

- By a cascade of two or more soft pulses applied successively to a set of connected single-quantum transitions. One implementation of this idea applies a soft $\pi/2$ pulse followed by a sequence of soft $\pi$ pulses to a ladder of connected transitions (the steps are not necessarily progressive).

- By means of a sequence of hard pulses, separated by suitable intervals for free precession (usually containing a refocusing pulse). This is the most common method in present-day spectrometers and is analysed in more detail below. It lends itself naturally to two-dimensional spectroscopy, where the multiple-quantum coherence is monitored indirectly as it precesses during the evolution period.

- Specialized hard pulse sequences that preferentially excite a particular order of coherence[15] or a particular spin coupling topology.[16]

Homonuclear double-quantum coherence is normally excited by the pulse sequence

$$(\pi/2)_x - \tau - (\pi)_x - \tau - (\pi/2)_x \qquad \textbf{3.63}$$

In the language of product operators we would write the evolution as

$$I_z + S_z \xrightarrow{\tilde{I}_x} \xrightarrow{\tilde{S}_x} -I_y - S_y \xrightarrow{2\tilde{I}_z\tilde{S}_z} \xrightarrow{(\pi)\,\tilde{I}_x} \xrightarrow{(\pi)\,\tilde{S}_x} \xrightarrow{2\tilde{I}_z\tilde{S}_z} -2I_xS_z - 2I_zS_x$$

$$\xrightarrow{\tilde{I}_x} \xrightarrow{\tilde{S}_x} +2I_xS_y + 2I_yS_x \qquad \textbf{3.64}$$

Here we have taken a short-cut by recognizing that the $(\pi)_x$ pulses, although they cause sign inversions, are there principally to refocus the chemical shift evolutions, which are therefore omitted from the product operator transformations This leaves only the evolution under the $2\tilde{I}_z\tilde{S}_z$ operator. In the general case, the double-quantum coherence that is generated is given by:

$$\sigma_{DQC} = (2I_xS_y + 2I_yS_x)\sin(2\pi J_{IS}\tau) \qquad \textbf{3.65}$$

When the condition $\tau = 1/(4J_{IS})$ is used, we obtain complete conversion. This is the basis of some of the double-quantum filtration experiments discussed in § 6.3.1.

This is where the product operator formalism provides an insight that is not available through the vector model. The *coup de grâce* that converts antiphase S-spin magnetization into two-spin coherence is administered by the $(\pi/2)_x$ pulse applied to the I spin:

$$S_z \xrightarrow{\tilde{S}_x} -S_y \xrightarrow{(\pi)\,\tilde{S}_x} \xrightarrow{2\tilde{I}_z\tilde{S}_z} -2\tilde{I}_z\tilde{S}_x \xrightarrow{\tilde{I}_x} +2I_yS_x \qquad \textbf{3.66}$$

In a heteronuclear spin system the result is a mixture of zero- and double-quantum coherence because we only excite the term $2I_yS_x$. The reason that pure double-quantum coherence is created in a homonuclear system is that there are two symmetrical paths which together generate $2I_xS_y + 2I_yS_x$. If a selective radiofrequency pulse is used, a homonuclear system can be made to behave like a heteronuclear system in this respect.

### 3.4.2 Three-spin coherence

When a third spin ($R$) is introduced it can do one of two things. It may simply remain passive, creating antiphase versions of the zero- or double-quantum coherence, for example $(4I_xS_xR_z + 4I_yS_yR_z)$ or $(4I_xS_xR_z - 4I_yS_yR_z)$. On the other hand, it may be an active spin, participating in three-spin coherence (§ 1.3.5), either triple-quantum or three-spin single-quantum coherence (combination lines). The pulse sequence for excitation of odd orders of coherence may be written as

$$(\pi/2)_x - \tau - (\pi)_y - \tau - (\pi/2)_y \qquad \textbf{3.67}$$

Suppose we wish to excite three-spin coherence. At the initial stage of evolution the hard pulse is treated as a cascade of soft pulses applied to the $I$, $S$ and $R$ spins in any order:

$$I_z + S_z + R_z \xrightarrow{\tilde{I}_x \quad \tilde{S}_x \quad \tilde{R}_x} -I_y - S_y - R_y \qquad \textbf{3.68}$$

We may then treat the effects on each spin separately, ignoring chemical shift evolution for the reasons given above. There are three evolution operators of the kind $2\tilde{I}_z\tilde{S}_z$, but there is always a third spin ($R$ in this instance) that is not involved. We omit the inactive evolution operator in each case:

$$-I_y \xrightarrow{2\tilde{I}_z\tilde{S}_z} +2I_xS_z \xrightarrow{2\tilde{I}_z\tilde{R}_z} +4I_yS_zR_z \xrightarrow{\tilde{I}_y \quad \tilde{S}_y \quad \tilde{R}_y} +4I_yS_xR_x \qquad \textbf{3.69}$$

$$-S_y \xrightarrow{\quad 2\tilde{I}_z\tilde{S}_z \quad} +2I_zS_x \xrightarrow{\quad 2\tilde{S}_z\tilde{R}_z \quad} +4I_zS_yR_z \xrightarrow{\quad \tilde{I}_y \quad \tilde{S}_y \quad \tilde{R}_y \quad} +4I_xS_yR_x \qquad \textbf{3.70}$$

$$-R_y \xrightarrow{\quad 2\tilde{I}_z\tilde{R}_z \quad} +2I_zR_x \xrightarrow{\quad 2\tilde{S}_z\tilde{R}_z \quad} +4I_zS_zR_y \xrightarrow{\quad \tilde{I}_y \quad \tilde{S}_y \quad \tilde{R}_y \quad} +4I_xS_xR_y \qquad \textbf{3.71}$$

For simplicity we have assumed that the three coupling constants $J_{IS}$, $J_{IR}$ and $J_{RS}$ are equal so that the $\tau$ delays can be set to the optimum value. The sum of the terms 3.69 through 3.71 represents a superposition of triple-quantum coherence and three-spin single-quantum coherence. The two orthogonal forms of triple-quantum coherence may be expressed as

$$I^+S^+R^+ + I^-S^-R^- = \tfrac{1}{2}\{4I_xS_xR_x - 4I_xS_yR_y - 4I_yS_yR_x - 4I_yS_xR_y\} \qquad \textbf{3.72}$$

$$I^+S^+R^+ - I^-S^-R^- = \tfrac{1}{2}\mathrm{i}\{4I_xS_xR_y + 4I_yS_xR_x + 4I_xS_yR_x - 4I_yS_yR_y\} \qquad \textbf{3.73}$$

One of the three combination lines can be represented by the orthogonal components:

$$I^+S^+R^- + I^-S^-R^+ = \tfrac{1}{2}\{4I_xS_xR_x - 4I_yS_yR_x + 4I_yS_xR_y + 4I_xS_yR_y\} \qquad \textbf{3.74}$$

$$I^+S^+R^- - I^-S^-R^+ = \tfrac{1}{2}\mathrm{i}\{4I_yS_xR_x + 4I_xS_yR_x - 4I_xS_xR_y + 4I_yS_yR_y\} \qquad \textbf{3.75}$$

The expressions for other three combination lines are obtained by the appropriate permutations of the raising and lowering operators. By allowing three-spin coherences to precess during the evolution period of a two-dimensional experiment we can derive the corresponding spectra, which have some interesting properties (see § 1.3.5).

Applications of multiple-quantum coherences usually exploit their characteristic properties, particularly the special sensitivity to radiofrequency phase shifts, field gradient pulses and chemical shift evolution. We list some of these features below.

### 3.4.3 Chemical shift evolution

Evolution of $p$-spin coherence involves all $p$ chemical shifts. This is treated as a cascade of chemical shift operators for each spin in turn (in any order). For the example of two-spin coherence,

$$2I_xS_x \xrightarrow{\quad (2\pi\delta_I t)\,\tilde{I}_z \quad (2\pi\delta_S t)\,\tilde{S}_z \quad}$$
$$2\{I_x\cos(2\pi\delta_I t) + I_y\sin(2\pi\delta_I t)\}\{S_x\cos(2\pi\delta_S t) + S_y\sin(2\pi\delta_S t)\} \qquad \textbf{3.76}$$

and

$$2I_yS_y \xrightarrow{\quad (2\pi\delta_I t)\,\tilde{I}_z \quad (2\pi\delta_S t)\,\tilde{S}_z \quad}$$
$$2\{I_y\cos(2\pi\delta_I t) - I_x\sin(2\pi\delta_I t)\}\{S_y\cos(2\pi\delta_S t) - S_x\sin(2\pi\delta_S t)\} \qquad \textbf{3.77}$$

If we subtract Equation 3.77 from Equation 3.76 and use standard trigonometrical identities, we find

$$\sigma_{\mathrm{DQC}} = (2I_xS_x - 2I_yS_y)\cos[2\pi(\delta_I + \delta_S)t] \qquad \textbf{3.78}$$

indicating that the double-quantum coherence has evolved at the sum of the two chemical shifts. The sum of Equation 3.76 and 3.77 gives

$$\sigma_{ZQC} = (2I_xS_x + 2I_yS_y) \cos[2\pi(\delta_I - \delta_S)t] \qquad \textbf{3.79}$$

indicating that the zero-quantum coherence has evolved at the difference of the two chemical shifts. In this context, chemical shift means the frequency offset from the radiofrequency transmitter (the frequency of the rotating reference frame). Consequently the different orders of homonuclear multiple-quantum coherence can be separated by shifting the transmitter frequency by (say) $\Delta$ Hz. Zero-quantum transitions do not move at all, single-quantum transitions are displaced by $\Delta$ Hz, and $p$-quantum transitions move by $p\Delta$ Hz. In this way, by a suitable choice of $\Delta$, it can be arranged that there is no overlap between the subspectra from different orders of coherence.

Normally the transmitter frequency and the receiver frequencies are identical, equal to the frequency of the rotating reference frame. However, if the phase of the receiver is incremented by (say) $\pi/2$ radians at each sampling point, there is an apparent frequency shift of the receiver reference frame (relative to the transmitter frame) of $1/(4\Delta t)$ Hz, where $\Delta t$ is the interval between sampling operations. Since the spectral width is $1/(2\Delta t)$ Hz, this frequency shift amounts to one half of the spectral width in Hz. This permits the separation of the different orders of coherence without shifting the transmitter frequency.

The offset from the transmitter frequency ($\Delta B$) is also affected by a change in the applied magnetic field $B_0$. Consequently, if we impose a magnetic field gradient $dB/dz$, this creates a distribution of $\Delta B$ values or, in terms of Equation 3.78, a distribution of chemical shifts $\delta_I$ and $\delta_S$. This explains why homonuclear double-quantum coherence is twice as sensitive to applied gradients or natural field inhomogeneities as single-quantum coherence and, through Equation 3.79, why homonuclear zero-quantum coherence is immune to such effects. In general the sensitivity to gradients is proportional to $p$, the order of the coherence, not to the number $n$ of spins involved. Combination lines, which represent three-spin single-quantum coherence, behave just like all other single-quantum coherences in this respect. Notice how the product operator treatment formalises concepts that were previously introduced in an *ad hoc* manner (§1.3.2).

### 3.4.4 Radiofrequency phase shifts

As mentioned above, the term 'S-spin chemical shift' means the offset ($\gamma\Delta B/2\pi$) of the S-spin resonance in a reference frame rotating at the transmitter frequency. Consequently, the phase of the S-spin coherence must be measured with respect to the phase of the radiofrequency field $B_1$. If the latter is shifted through an angle $\phi$, then the product operator term for the chemical shift evolution acquires the same phase shift:

$$S_y \xrightarrow{\quad (+\phi)\, \tilde{S}_z \quad} S_y\cos\phi - S_x\sin\phi \qquad \textbf{3.80}$$

A rather less trivial result is obtained by evaluating the four product operator terms for homonuclear double-quantum coherence. The introduction of identities for cos2ϕ and sin2ϕ gives the expression:

$$(2I_xS_x - 2I_yS_y) \xrightarrow{(+\phi)\,\tilde{I}_z} (2I_xS_x - 2I_yS_y)\cos\phi + (2I_yS_x + 2I_xS_y)\sin\phi$$

$$\xrightarrow{(+\phi)\,\tilde{S}_z} (2I_xS_x - 2I_yS_y)\cos2\phi + (2I_xS_y + 2I_yS_x)\sin2\phi \qquad \textbf{3.81}$$

which represent a doubling of the phase shift. For homonuclear zero-quantum coherence, a similar evaluation of four terms leads to the identity:

$$(2I_xS_x + 2I_yS_y) \xrightarrow{(+\phi)\,\tilde{I}_z} (2I_xS_x + 2I_yS_y)\cos\phi + (2I_yS_x - 2I_xS_y)\sin\phi$$

$$\xrightarrow{(+\phi)\,\tilde{S}_z} (2I_xS_x + 2I_yS_y) \qquad \textbf{3.82}$$

that is to say, no phase shift at all. We may infer that a radiofrequency phase shift ϕ translates into a phase shift $p\phi$ for $p$-quantum coherence. This assumes considerable practical importance for the separation of different orders of coherence as, for example, in double-quantum filtration experiments (§ 6.3.1).

### 3.4.5 Evolution under spin–spin coupling

Certain spin–spin couplings are said to be *active* if they connect spins involved in the multiple-quantum coherence. Passive spins are not involved in the creation or reconversion of the multiple-quantum coherence; they are merely spectators. Even the active couplings do not give rise to splittings in the double-quantum spectrum. This is consistent with the invariance of two-spin coherence under the spin–spin coupling evolution operator:

$$2I_xS_x \xrightarrow{2\tilde{I}_z\tilde{S}_z} 2I_xS_x \qquad \textbf{3.83}$$

This property simplifies multiple-quantum spectra, particularly those involving high orders of coherence. For example, the spectrum of three-spin coherences consists of four singlet resonances unless there are passive couplings to other spins outside the active group of three.[17] For more on three-spin coherence spectra, see § 1.3.5.

### 3.4.6 Spectral editing (DEPT)

Assignment is an important aspect of high resolution spectroscopy. Since broadband decoupling of carbon-13 spectra masks the information about the number of directly attached protons, spectral editing techniques have become quite popular. Editing entails the separation of the conventional broadband decoupled $^{13}C$ spectrum into four subspectra corresponding to quaternary, methine, methylene and methyl sites. Although the INEPT experiment[18] has the potential to make this distinction, it is more effectively achieved by a method that relies on heteronuclear multiple-quantum coherence—distortionless enhancement by polarization transfer (DEPT).[19] We use this second worked example as an exercise for those readers who may be applying the product operator formalism for the first time.

The DEPT scheme excites heteronuclear two-spin coherence (zero- and double-quantum coherence) but in reconverting this into observable $S$-spin magnetization, makes a distinction between $IS$, $I_2S$ and $I_3S$ spin systems. The analysis of the pulse sequence may not be obvious at first sight, and rumour has it that it was discovered experimentally, by accident.

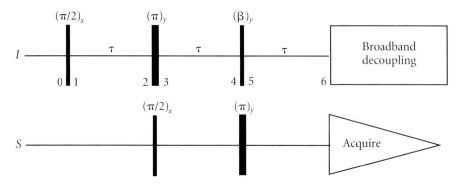

In manipulations where the evolution is complex, it is often useful to indicate the key points in the sequence by numbers. Normally these points would be immediately before and immediately after each radiofrequency pulse. Thus we can write the evolution up to point 3 as

$$+I_z \xrightarrow{\tilde{I}_x} -I_y \xrightarrow{2\tilde{I}_z\tilde{S}_z} +2I_xS_z \xrightarrow{(\pi)\,\tilde{I}_y} -2I_xS_z \xrightarrow{\tilde{S}_x} +2I_xS_y \qquad \textbf{3.84}$$

This analysis has ignored $I$-spin chemical shift evolution. If we were to take this into account we would have a second contribution:

$$+I_z \xrightarrow{\tilde{I}_x} -I_y \xrightarrow{\tilde{I}_z} +I_x \xrightarrow{2\tilde{I}_z\tilde{S}_z} +2I_yS_z \xrightarrow{(\pi)\,\tilde{I}_y} +2I_yS_z \xrightarrow{\tilde{S}_x} -2I_yS_y \qquad \textbf{3.85}$$

This would result in heteronuclear two-spin coherence that is some linear combination of $+2I_xS_y$ and $-2I_yS_y$. However, we can safely neglect the chemical shifts because we know that they will be refocused. The $I$-spin chemical shifts evolve only during the period 1 to 4, while $S$-spin shifts evolve during the period 3 to 6. Consequently both evolutions are refocused by the $\pi$ pulses at the midpoints of these two periods. Because of the rapid proliferation in product operator terms (the branching tree), it is very important to identify the paths that are important and to neglect the rest. This is often the most tricky aspect of the method.

We concentrate on three different spin systems: $IS$, $I_2S$, $I_3S$. The quaternary sites $S$ can be neglected for the present since they are essentially unaffected by the $I$-spin manipulations (only very weak long-range couplings are involved). Between points 3 and 4, heteronuclear two-spin coherence evolves unchanged in the $IS$ system. Although the $I_2S$ and $I_3S$ systems normally contain equivalent protons, we now label them $I$, $I'$ and $I''$ so that we can consider the operations in an appropriate pulse cascade.

$IS$ system:
$$+2I_xS_y \xrightarrow{\;2\tilde{I}_z\tilde{S}_z\;} +2I_xS_y \qquad\qquad \textbf{3.86}$$

$I_2S$ system:
$$+2I_xS_y \xrightarrow{\;2\tilde{I}'_z\tilde{S}_z\;} -4I_xI'_zS_x \qquad\qquad \textbf{3.87}$$

$$+2I'_xS_y \xrightarrow{\;2\tilde{I}_z\tilde{S}_z\;} -4I_zI'_xS_x \qquad\qquad \textbf{3.88}$$

$I_3S$ system:
$$+2I_xS_y \xrightarrow{\;2\tilde{I}'_z\tilde{S}_z\;} -4I_xI'_zS_x \xrightarrow{\;2\tilde{I}''_z\tilde{S}_z\;} -8I_xI'_zI''_zS_y \qquad\qquad \textbf{3.89}$$

$$+2I'_xS_y \xrightarrow{\;2\tilde{I}_z\tilde{S}_z\;} -4I_zI'_xS_x \xrightarrow{\;2\tilde{I}''_z\tilde{S}_z\;} -8I_zI'_xI''_zS_y \qquad\qquad \textbf{3.90}$$

$$+2I''_xS_y \xrightarrow{\;2\tilde{I}_z\tilde{S}_z\;} -4I_zI''_xS_x \xrightarrow{\;2\tilde{I}'_z\tilde{S}_z\;} -8I_zI'_zI''_xS_y \qquad\qquad \textbf{3.91}$$

We see that at point 4, all the terms consist of heteronuclear two-spin coherences (which may be antiphase or doubly antiphase). At this stage the experiment resembles the multiple-quantum coherence method first introduced by Müller.[20] We now consider the effect of the $(\pi)_y$ pulse applied to the $S$ spins to refocus the $S$-spin chemical shift. Its only other effect is on the $I_2S$ system (the only one with $S_x$ operators), changing the signs:

$$-4I_xI'_zS_x \xrightarrow{\;(\pi)\,\tilde{S}_y\;} +4I_xI'_zS_x \qquad\qquad \textbf{3.92}$$

$$-4I_zI'_xS_x \xrightarrow{\;(\pi)\,\tilde{S}_y\;} +4I_zI'_xS_x \qquad\qquad \textbf{3.93}$$

The key step in the editing process occurs through the operation of the $(\beta)_y$ pulse between points 4 and 5. Its action depends in a crucial way on the number of equivalent protons affected by this pulse: one, two or three. We use the general expression for the effect of a radiofrequency pulse of arbitrary flip angle $\beta$, and drop any new terms that correspond to two-spin coherence, because we are interested only in antiphase $S$-spin magnetization at point 5. For the $IS$ system the result is quite straightforward:

$$+2I_xS_y \xrightarrow{\;(\beta)\,\tilde{I}_y\;} -2I_zS_y \sin\beta \qquad\qquad \textbf{3.94}$$

For the $I_2S$ system, where there are two equivalent protons, we consider the $\beta$ pulse as a cascade:

$$+4I_xI'_zS_x \xrightarrow{\;(\beta)\,\tilde{I}_y\;} -4I_zI'_zS_x \sin\beta \xrightarrow{\;(\beta)\,\tilde{I}'_y\;} -4I_zI'_zS_x \sin\beta \cos\beta \qquad\qquad \textbf{3.95}$$

$$+4I_zI'_xS_x \xrightarrow{\ (\beta)\ \tilde{I}_y\ } +4I_zI'_xS_x\ \cos\beta \xrightarrow{\ (\beta)\ \tilde{I}'_y\ } -4I_zI'_zS_x\ \sin\beta\ \cos\beta \qquad \textbf{3.96}$$

These two terms can be summed to give $-4I_zI'_zS_x\ \sin2\beta$. Note that several other terms, for example $4I_xI'_zS_x\ \sin^2\beta$, have been dropped at this stage.

The $I_3S$ system employs a cascade of three $\beta$ pulses:

$$-8I_xI'_zI''_zS_y \xrightarrow{\ (\beta)\ \tilde{I}_y\ } +8I_zI'_zI''_zS_y\ \sin\beta \xrightarrow{\ (\beta)\ \tilde{I}'_y\ }$$

$$+8I_zI'_zI''_zS_y\ \sin\beta\ \cos\beta \xrightarrow{\ (\beta)\ \tilde{I}''_y\ } +8I_zI'_zI''_zS_y\ \sin\beta\ \cos^2\beta \qquad \textbf{3.97}$$

Similar manipulations of $-8I_zI'_xI''_zS_y$ and $-8I_zI'_zI''_xS_y$ give the same result. These three terms may now be summed to give $+8I_zI'_zI''_zS_y\ 3\sin\beta\ \cos^2\beta$. Once again, many of the products have been dropped since they involve multiple-quantum coherences. A standard identity gives an alternative form:

$$8I_zI'_zI''_zS_y\ 3\sin\beta\ \cos^2\beta = 8I_zI'_zI''_zS_y\ \{\tfrac{3}{4}\ (\sin\beta + \sin3\beta)\} \qquad \textbf{3.98}$$

Evolution for a further period $\tau$ (between points 5 and 6) under the spin coupling operators, refocuses these anti-phase signals, converting all three into observable magnetization $+S_x$. This is to permit the use of decoupling during acquisition.

$$IS:\quad -2I_zS_y\ \sin\beta \xrightarrow{\ 2\tilde{I}_z\tilde{S}_z\ } +S_x\ \sin\beta \qquad\qquad \textbf{3.99}$$

$$I_2S:\quad -4I_zI'_zS_x\ 2\sin\beta\ \cos\beta \xrightarrow{\ 2\tilde{I}_z\tilde{S}_z\ } -2I'_zS_y\ 2\sin\beta\ \cos\beta \xrightarrow{\ 2\tilde{I}'_z\tilde{S}_z\ } +S_x\ 2\sin\beta\ \cos\beta \quad \textbf{3.100}$$

$$I_3S:\quad +8I_zI'_zI''_zS_y\ 3\sin\beta\ \cos^2\beta \xrightarrow{\ 2\tilde{I}_z\tilde{S}_z\ } -4I'_zI''_zS_x\ 3\sin\beta\ \cos^2\beta$$

$$\xrightarrow{\ 2\tilde{I}'_z\tilde{S}_z\ } -2I''_zS_y\ 3\sin\beta\ \cos^2\beta \xrightarrow{\ 2\tilde{I}''_z\tilde{S}_z\ } +S_x\ 3\sin\beta\ \cos^2\beta \qquad \textbf{3.101}$$

The three spin systems therefore show distinctly difference dependences on the flip angle $\beta$, permitting an eventual separation of the subspectra:

$$M(IS) = +S_x\ \sin\beta \qquad\qquad \textbf{3.102}$$

$$M(I_2S) = +2S_x\ \sin\beta\ \cos\beta \qquad\qquad \textbf{3.103}$$

$$M(I_3S) = +3S_x\ \sin\beta\ \cos^2\beta \qquad\qquad \textbf{3.104}$$

Now, if we set $\beta = \pi/2$, the pure methine ($IS$) subspectrum is obtained. With $\beta = \pi/4$, all three spin systems give positive signals. With $\beta = 3\pi/4$ the methine ($IS$) and methyl ($I_3S$) resonances are upright while the methylene ($I_2S$) resonances are inverted. We see that suitable linear combinations of these results give the pure subspectra, although allowances must be made for a variation in $J_{IS}$ values. Deviations from the timing condition $\tau = 1/(2J_{IS})$, which inevitably arise because the $J_{IS}$ values span an appreciable range, can cause phase anomalies for coupled acquisition. It is now common to employ refinements of the DEPT sequence that are more tolerant of the natural variation of $J_{IS}$ values, for example the SEMUT sequence.[21]

## 3.4.7 Composite rotations

To make the best use of limited space, the classic juke box stored its records vertically (let us say, normal to the $+y$ axis) but needed to rotate them (through $\pi/2$ radians about, say, the $+x$ axis) in order to play them on a horizontal turntable (rotating about the $+z$ axis). It would then return them to the vertical stack.

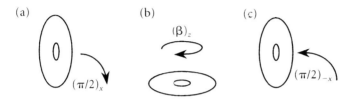

We might argue that this is equivalent to playing the discs while they were still stacked vertically (that is, rotating about the $+y$ axis). In the vocabulary of rotation operators (which are written in time-reversed order),

$$\exp(-i\tfrac{1}{2}\pi I_x)\,\exp(+i\beta I_z)\,\exp(+i\tfrac{1}{2}\pi I_x) = \exp(i\beta I_y) \qquad \textbf{3.105}$$

Translated into operations on the density matrix, we have

$$\sigma \xrightarrow{\;+\tilde{I}_x\;} \xrightarrow{\;(\beta)\tilde{I}_z\;} \xrightarrow{\;-\tilde{I}_x\;} \sigma^+ \qquad \textbf{3.106}$$

which is equivalent to

$$\sigma \xrightarrow{\;(\beta)\tilde{I}_y\;} \sigma^+ \qquad \textbf{3.107}$$

In this manner, we have achieved a rotation about some desired axis by a cascade of three successive rotations involving the other two orthogonal axes.

This can be put to practical use to implement an arbitrary phase shift of a particular coherence, which can be useful if the spectrometer only provides phase shifts in multiples of $\pi/2$. For example, the various orders of coherence in a three-spin system can be separated by advancing the phase in steps of $\pi/3$ radians at the beginning of the evolution period in a two-dimensional experiment.[22] For each increment in $t_1$, there is an incremental rotation about the $+z$ axis of $\pi/3$ radians. Signals derived from zero-quantum coherence are unaffected, but single-quantum signals are shifted by one-third of the spectral width, double-quantum signals by two-thirds, and triple-quantum signals by the entire spectral width. Spectra from the different orders can be completely separated if the spectral width is suitably chosen.

Levitt[23] has used the concept of a composite $z$ rotation to explain the DEPT flip angle effect by another approach. The $(\beta)_y$ pulse is replaced by the sandwich

$$(\beta)_y = (\pi/2)_x\,(\beta)_z\,(\pi/2)_{-x} \qquad \textbf{3.108}$$

In this scenario, the first pulse creates homonuclear multiple-quantum coherence—double-quantum coherence within the $I_2S$ groups, and a mixture of single- and triple-quantum coherence in the $I_3S$ groups. The second pulse then acts simply as a radiofrequency phase shift ($\beta$ radians) changing the phase of single-quantum coherence by $\beta$, double-quantum coherence by $2\beta$, and triple-quantum coherence by $3\beta$. This gives the separation of signals from the sites of different multiplicity. The third pulse recreates antiphase (and doubly antiphase) $S$-spin magnetization which then evolves under the spin coupling superoperator in the same manner as before. This picture accords well with the alternative formulations of Equations 3.103 and 3.104:

$$M(I_2S) = S_x \sin 2\beta \qquad\qquad \textbf{3.109}$$

$$M(I_3S) = \tfrac{3}{4}S_x \left(\sin\beta + \sin 3\beta\right) \qquad\qquad \textbf{3.110}$$

which emphasize that the various coherences are affected in proportion to $p\beta$ where $p$ is the coherence order. In some ways this is a more satisfying description than the direct flip angle effect.

## 3.5  Soft radiofrequency pulses

### 3.5.1  Pseudocorrelation spectroscopy

As our third worked example we examine a technique which produces COSY-like spectra but without using any evolution period and without two-dimensional Fourier transformation. The point of this $\psi$-COSY experiment[24] is to demonstrate the close relationship between correlation spectroscopy and double resonance (§ 8.2.1). The evolution dimension $F_1$ is explored directly by stepping the frequency of a soft pulse in very small increments through the frequency range of interest. The practical advantage is that any particularly interesting region of the spectrum may be examined under very fine digital resolution without the need to acquire data from the rest of the chemical shift range. Essentially we are implementing a zoom feature to take a closer look at the fine structure. The disadvantage is that the method has a much lower intrinsic sensitivity than conventional correlation spectroscopy since it does not benefit from the multiplex advantage in the $F_1$ dimension.

The first requirement is a soft radiofrequency pulse that is so highly selective that only one transition is excited at any one time. For line widths of the order of 1 Hz, the pulse duration would be of the order of 1 s. This makes for a very slow experiment if the frequency increments are small. A suitable pulse shape is the first half of a Gaussian curve[25] since this avoids the introduction of frequency-dependent phase shifts into the excitation spectrum.

There are two possible approaches to the problem of line-selective excitation—either we modify the evolution superoperator or we modify the operator describing the state of the system. We choose the latter. The equilibrium longitudinal magnetization of the high-frequency line of the $I$-spin doublet would then be represented by

$$\sigma_1 = \tfrac{1}{2}(I_z + 2I_zS_z) \qquad\qquad \textbf{3.111}$$

The $z$ magnetization of the low-frequency line would be

$$\sigma_2 = \tfrac{1}{2}(I_z - 2I_zS_z) \qquad\qquad \textbf{3.112}$$

A soft $(\pi/2)_x$ radiofrequency pulse about the $+x$ axis at exact resonance for the high-frequency line gives:

$$\tfrac{1}{2}(I_z + 2I_zS_z) \xrightarrow{\ \tilde{I}_x\ } -\tfrac{1}{2}[I_y + 2I_yS_z] \qquad\qquad \textbf{3.113}$$

This is followed immediately by a hard $(\pi/2)_x$ pulse that affects both transitions of both spins, although we still focus on the high-frequency $I$-spin line:

$$-\tfrac{1}{2}(I_y + 2I_yS_z) \xrightarrow{\ \tilde{I}_x\ } -\tfrac{1}{2}(I_z + 2I_zS_z) \xrightarrow{\ \tilde{S}_x\ } -\tfrac{1}{2}(I_z - 2I_zS_y) \qquad \textbf{3.114}$$

The first term represents unobservable (inverted) $I$-spin populations while the last term $(2I_zS_y)$ corresponds to antiphase $S$-spin magnetization in the absorption mode. This is exactly the same as the corresponding $F_2$ trace through a cross peak of the conventional correlation spectrum (§ 8.2.3).

The equivalents of diagonal peaks arise when the soft excitation pulse is slightly off-resonance for one of the $I$-spin transitions. Most product operator calculations employ hard pulses where off-resonance effects can be safely neglected, but soft pulse excitation can involve a severe tilt of the effective field in the rotating frame. The treatment is essentially the same as for the vector model. If we define the tilt angle as

$$\tan\theta = \Delta B/B_1 \qquad\qquad \textbf{3.115}$$

then it varies from zero at exact resonance to $\pi/2$ radians at very large offsets. The required transformation is a composite rotation, equivalent to a rotation of the reference frame through $\theta$ about the $+y$ axis towards $+z$, application of the usual $I_x$ pulse, and a return of the reference frame to its original position:

$$\sigma \xrightarrow{(+\theta)\,\tilde{I}_y} \xrightarrow{\ \tilde{I}_x\ } \xrightarrow{(-\theta)\,\tilde{I}_y} \sigma^+ \qquad\qquad \textbf{3.116}$$

This can be used to calculate the off-resonance excitation of a single $I$-spin transition:

$$\tfrac{1}{2}(I_z + 2I_zS_z) \xrightarrow{(+\theta)\,\tilde{I}_y} \tfrac{1}{2}(I_z + 2I_zS_z)\cos\theta + \tfrac{1}{2}(I_x + 2I_xS_z)\sin\theta \xrightarrow{\ \tilde{I}_x\ }$$

$$-\tfrac{1}{2}(I_y + 2I_yS_z)\cos\theta + \tfrac{1}{2}(I_x + 2I_xS_z)\sin\theta \xrightarrow{(-\theta)\,\tilde{I}_y}$$

$$-\tfrac{1}{2}(I_y + 2I_yS_z)\cos\theta + \tfrac{1}{2}(I_x + 2I_xS_z)\sin\theta\cos\theta + \tfrac{1}{2}(I_z + 2I_zS_z)\sin^2\theta \qquad \textbf{3.117}$$

We see that the first term will eventually give rise to the off-resonance absorption-mode signal, whereas the last term will create only longitudinal magnetization, so we can focus on the second term and examine the effect of the hard $(\pi/2)_x$ pulse:

$$+ \tfrac{1}{2}\{[I_x + 2I_xS_z] \sin\theta \cos\theta \xrightarrow{\;\bar{I}_x\;} \xrightarrow{\;\bar{S}_x\;} + \tfrac{1}{2}[I_x - 2I_xS_y] \sin\theta \cos\theta \qquad \textbf{3.118}$$

This leaves only a dispersion-mode signal $(+\tfrac{1}{4}I_x \sin2\theta)$ while the rest has been converted into unobservable multiple-quantum coherence. Note that this dispersion signal reaches a maximum when $\theta = \pi/4$ radians and it changes sign with the sense of the offset (the sign of $\theta$). The overall result is that the observed signals from ψ-COSY are essentially the same as the diagonal and cross-peaks of conventional COSY.

## 3.6  Discussion

Any manipulation of a weakly-coupled multispin system involving multiple-quantum coherence is best handled by the product operator formalism. One has only to attempt to explain of the DEPT experiment[13] by means of a vector picture but without the aid of product operators to see how indispensable the latter have become. Many other coherence transfer schemes can be profitably treated by this powerful algebra, and very often only the briefest shorthand notation is sufficient to indicate the direction of the spin evolution. Probably more important still is the use of product operators as an aid to finding new twists to existing experiments, or as a tool for the discovery of entirely new procedures, since a very small change (for example, the phase of one pulse) can initiate a completely different behaviour. Where multiple-quantum effects are absent, the product operator formalism exactly parallels the vector picture and for tricky new situations it can be helpful to consider both descriptions side by side.

Shorthand notations are not without their inherent dangers, but the product operator formalism is surprisingly convenient for describing modern 'spin gymnastics'. A little effort expended on learning how to manipulate this algebra (and to keep track of the signs) is soon rewarded. Sometimes a scientific paper written in an obscure foreign language turns out to be comprehensible only through the mathematical equations. The same can be true of some complex NMR pulse sequences—only the product operator formulation allows us to follow the essentials of the experiment.

## References

1. O. W. Sørensen, G. W. Eich, M. H. Levitt, G. Bodenhausen and R. R. Ernst, *Progr. NMR Spectrosc.*, **16**, 163 (1983).
2. F. J. M. van de Ven and C. W. Hilbers, *J. Magn. Reson.* **54**, 512 (1983).
3. P. K. Wang and C. P. Slichter, *Bull. Magn. Reson.* **8**, 3 (1986).
4. R. R. Ernst, G. Bodenhausen and A. Wokaun, *Principles of Nuclear Magnetic Resonance in One and Two Dimensions,* Clarendon Press, Oxford, 1987.

5. C. P. Slichter, *Principles of Magnetic Resonance*, Springer Verlag, Berlin, 1990.

6. G. L. Hoatson and K. J. Packer, *Mol. Phys.* **40**, 1153 (1980).

7. G. Bodenhausen, *Progr. NMR Spectrosc.*, **14**, 137 (1981).

8. J. Jeener, Ampere International Summer School, Basko Polje, Yugoslavia, 1971, reported in *NMR and More. In Honour of Anatole Abragam* Eds: M. Goldman and M. Porneuf, Les Editions de Physique, Les Ulis, France, 1994.

9. W. P. Aue, E. Bartholdi and R. R. Ernst, *J. Chem. Phys.* **64**, 2229 (1976).

10. U. Piantini, O. W. Sørensen and R. R. Ernst, *J. Am. Chem. Soc.* **104**, 6800 (1982).

11. A. Bax, R. Freeman and T. A. Frenkiel, *J. Am. Chem. Soc.* **103**, 2102 (1981).

12. A. Bax, R. Freeman, T. A. Frenkiel and M. H. Levitt, *J. Magn. Reson.* **43**, 478 (1981).

13. W. A. Anderson, R. Freeman and C. A. Reilly, *J. Chem. Phys.* **39**, 1518 (1963).

14. S. Vega and A. Pines, *J. Chem. Phys.* **66**, 5624 (1977).

15. W. S. Warren, S. Sinton, D. P. Weitekamp and A. Pines, *Phys. Rev. Lett.* **43**, 1791 (1979).

16. M. H. Levitt and R. R. Ernst, *Chem. Phys. Lett.* **100**, 119 (1983).

17. X. L. Wu, P. Xu and R. Freeman, *J. Magn. Reson.* **88**, 417 (1990).

18. G. A. Morris and R. Freeman, *J. Am. Chem. Soc.* **101**, 760 (1979).

19. D. M. Doddrell, D. T. Pegg and M. R. Bendall, *J. Magn. Reson.* **48**, 323 (1982).

20. L. Müller, *J. Am. Chem. Soc.* **101**, 4418 (1979).

21. O. W. Sørensen, D. Dønstrup, H. Bildsøe and H. J. Jakobsen, *J. Magn. Reson.* **55**, 347 (1983).

22. R. Freeman, T. A. Frenkiel and M. H. Levitt, *J. Magn. Reson.* **44**, 409 (1981).

23. M. H. Levitt, *Progr. NMR Spectrosc.*, **18**, 61 (1986).

24. S. Davies, J. Friedrich and R. Freeman, *J. Magn. Reson.* **75**, 540 (1987).

25. J. Friedrich, S. Davies and R. Freeman, *J. Magn. Reson.* **75**, 390 (1987).

# 4

# Spin echoes

## 4.1 Echo phenomena

One of the most versatile tools available to the NMR spectroscopist is the spin echo experiment.[1] Not only does it pervade the fields of solid- and liquid-state NMR, ESR, and magnetic resonance imaging, but it has acted as a catalyst for the invention of a host of new physical techniques introduced over the last four decades. Today we take it for granted that nuclear spins can be manipulated in much the same way that a conjuror performs tricks with a pack of cards, but if we examine the origin of these ideas we see that it was the spin echo experiment that inspired these fascinating studies in spin choreography. This chapter will restrict itself to the theme of spin echoes in high resolution liquid-phase NMR where the impact of this idea has been extremely influential.

Many everyday analogies have been advanced to help describe the basic spin echo phenomenon. Some authors favour a race-track picture in which runners of different abilities are strung out around a circular track and then recalled by a second signal from the starter's gun so that they all reach the starting line together. Others visualize a colony of ants running round the edge of a pancake which then gets a 180° flip. We may even adapt the playing card analogy hinted at above. It is notoriously difficult to shuffle an ordered pack of cards to get a reasonably random distribution. In fact we may argue that with a finite number of shuffling operations, the result is never really random because in principle a reversal of the manipulations would restore the original ordered state. This idea of time reversal is implicit in the idea of the spin echo refocusing. Indeed it has been suggested by Waugh[2] that the echo phenomenon appears to challenge the validity of the second law of thermodynamics since it is possible to prepare a nuclear spin system in such a way that, although it appears to be in an equilibrium state, at some later time its macroscopic magnetization (the degree of order) grows spontaneously from a previously negligible level. To the untutored observer this would appear just as miraculous as a gas mixture separating of its own accord into its pure components. Spontaneous mixing of two gases is a perfectly normal process, but it cannot be reversed because it is not possible to reverse all the molecular velocities in a gas mixture to make the particles retrace their previous paths. Of course, the NMR spectroscopist would realize that this magic sample possesses a long spin–spin relaxation time, and has been prepared by means of a two-pulse sequence; the spontaneous ordering is simply a spin echo. If we could examine a very small volume element of the sample, we would find that throughout the whole period the spins are precessing in phase, giving a local macroscopic magnetization. This is masked in the bulk sample by interference effects between different volume elements. The sample is not really in an equilibrium state at all, and the second law remains inviolate.

This leads us to a very useful concept, coined by Abragam,[3] of a spin isochromat. The NMR sample is considered to be made up of a mosaic of tiny volume elements, each so small that any spatial variation of the applied magnetic field across that element can be safely neglected, yet still big enough to contain a large number of nuclear spins. The spatial inhomogeneity endows each volume element with a slightly different field intensity and therefore a slightly different nuclear precession frequency. The nuclear magnetization of each element is represented by a small vector $m_i$ (an isochromat) and the total magnetization of the sample is given by summing over all such vectors:

$$M = \sum_i m_i \qquad\qquad \textbf{4.1}$$

Immediately after a $(\pi/2)_x$ excitation pulse all the isochromatic vectors are aligned along the $+y$ axis and an NMR signal corresponding to the full magnetization is detected. Free precession for a period $\tau$ allows each isochromat to go its own way at its own pace, so there is a fanning out of vectors in the $xy$ plane due to the spatial inhomogeneity of the applied magnetic field (or as a result of any field gradients applied deliberately). The detected signal is the resultant of all these vectors; it decays with time, a process usually represented by an instrumental damping term, usually written $\exp(-t/T_2^*)$, although of course it is unlikely to be a pure exponential decay. After several time constants the signal reaches a negligible level.

## 4.1.1 **Hahn spin echoes**

An effect similar to time reversal can now be achieved in several different ways. Hahn used a second $(\pi/2)$ pulse:

$$(\pi/2)_x — \tau — (\pi/2)_x — \tau — \text{echo} \qquad \textbf{4.2}$$

Consider a representative vector $m_i$ which precesses through an angle $\theta_i$ in the first $\tau$ interval. This vector may be resolved into $m_i \cos \theta_i$ along the $y$ axis and $m_i \sin\theta_i$ along the $x$ axis. The former is rotated to $-z$ by the second $(\pi/2)_x$ pulse and suffers no further precession (we shall return to a consideration of this component later). The other component is unaffected by the $(\pi/2)_x$ pulse and precesses through a further angle $\theta_i$, generating a component $m_i \sin^2\theta_i$ along the $-y$ axis at time $2\tau$, (the $x$ components cancel if the field distribution is symmetrical). This is the Hahn echo. It has an amplitude

$$\sum_i m_i \sin^2\theta_i$$

and since the mean value of $\sin^2\theta$ over a uniform distribution around a circle is 0.5, it represents only 50% of the available nuclear magnetization. As a consequence of the $\sin^2\theta_i$ dependence, we may take a short cut when visualizing the vector picture of the Hahn echo, concentrating only on bundles of isochromats that precess in the first $\tau$ interval through an angle $\theta_i$ close to $\pm(n + \frac{1}{2})\pi$ radians, where $n$ is an integer. These are the ones that contribute most strongly to the echo. To a fair approximation the others can be neglected, permitting a simple pictorial representation (Figure 4.1). The bundles moving clockwise and counterclockwise reach the $\pm x$ axes at time $\tau$ and are virtually unaffected by the second $(\pi/2)_x$ pulse. After further precession through $\pm(n + \frac{1}{2})\pi$ radians they converge on the $-y$ axis at time $2\tau$ to form the Hahn echo. It is interesting to note that the echo arises only because 50% of the magnetization was 'withdrawn from play' by the second $(\pi/2)_x$ pulse; otherwise there would have been cancellation, and no echo.

Figure 4.2 shows a more comprehensive analysis of the Hahn echo in three

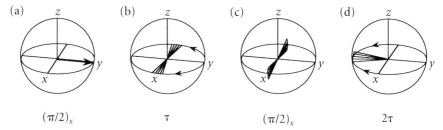

| (a) | (b) | (c) | (d) |
|---|---|---|---|
| $(\pi/2)_x$ | $\tau$ | $(\pi/2)_x$ | $2\tau$ |

**Figure 4.1** Schematic illustration of the formation of a Hahn spin echo. Since the relative contribution of isochromats to the echo amplitude follows a $\sin^2\theta$ dependence, the dominant effect arises from magnetization vectors that precess through angles $\theta$ near $\pm(n + \frac{1}{2}) \pi$ radians in the first $\tau$ interval. These bundles of vectors are scarcely affected by the second $(\pi/2)_x$ pulse and in the second $\tau$ interval precess through a further $\pm(n + \frac{1}{2}) \pi$ radians to converge on the $-y$ axis to form the echo.

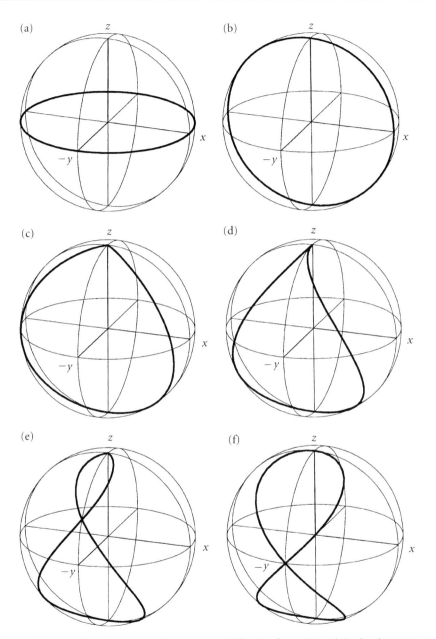

**Figure 4.2** Formation of the Hahn 8-ball echo. (a) After the first $\tau$ interval the isochromats are distributed around a great circle in the $xy$ plane. The locus of the tips of these vectors is shown as a heavy line. (b) After the second $(\pi/2)_x$ pulse, the locus is rotated into the $xz$ plane. The time evolution of this locus is shown (c) after a further interval $\tau/4$, (d) after $\tau/2$, (e) after $3\tau/4$, and (f) after $\tau$, when it forms a figure-of-eight. Vectors exactly at the poles are stationary. The resultant of all the vectors terminating on the figure-of-eight lies along the $-y$ axis.

dimensions by following the evolution of the locus of the tips of the isochromatic vectors on the surface of the unit sphere. We assume a uniform distribution of isochromats around a circle in the $xy$ plane, that is to say, a virtually complete decay of the free induction signal before the second $(\pi/2)_x$ pulse. Immediately after this pulse the locus becomes a great circle in the $xz$ plane. The various parts of this locus then precess at different rates (but with fixed points at the north and south poles). Note that vectors near the north pole are fast-moving, whereas those near the south pole move rather slowly. For the moment we neglect vectors that precess more than $\pm\pi$ radians in the period $\tau$, and simply concentrate on those dispersed around a circle (Figure 4.2(b)). We may then follow the evolution of the locus at times $\tau/4$, $\tau/2$, $3\tau/4$ and $\tau$ after the second pulse, showing the twisting of the circle into a characteristic figure-of-eight pattern on the unit sphere (the 8-ball echo). The resultant of all these vectors is directed along the $-y$ axis. Any vector that initially precessed through $2n\pi + \phi$ radians in the first $\tau$ interval (where $n$ is an integer) terminates at the same point on the figure-of-eight as a vector that only precessed through $\phi$ radians.

Further evolution of the 8-ball isochromats simply leads to a progressive divergence and consequent loss of detected signal. However, there is one extraordinary situation where an entire train of multiple Hahn echoes is observed. It only occurs where the nuclear spins are in such high concentration that they make an appreciable contribution to the bulk magnetic susceptibility of the sample. This strange phenomenon is treated in § 11.4.2.

### 4.1.2  Stimulated echoes

The analysis of the Hahn spin echo phenomenon outlined above indicates that 50% of the available magnetization is somehow lost or at least misplaced. This was explained by resolving the vectors along the $x$ and $y$ axes and by neglecting the latter. It turns out that these neglected vectors give rise to a new kind of echo (a stimulated echo[1]) if a third $(\pi/2)_x$ pulse is applied at some later time $(\tau + T)$. As before, we assume that $\tau$ is sufficiently long that there is an almost uniform dispersion of isochromats around the $xy$ plane. We then resolve these vectors into their $x$ and $y$ components and concentrate only on the latter (Figure 4.3(b)) because these will be mainly responsible for the stimulated echo. They have precessed through angles close to $n\pi$ radians during the $\tau$ interval. There are two opposed bundles of echoes—those with $n$ even, aligned roughly along $+y$, and those with $n$ odd, aligned roughly along $-y$. The second $(\pi/2)_x$ pulse rotates these vectors into the $\pm z$ axis, and in the two-pulse experiment they would remain there and never be observed. The residual transverse components decay through $B_0$ inhomogeneity, leaving behind pure longitudinal magnetization. At first sight this hardly seems a recipe for refocusing. The interesting point is that this longitudinal magnetization is stored in such a manner that the individual isochromats retain a coded memory of their respective spatial locations, such that at a later time they form this new type of echo. During the storage period $T$, transverse magnetization is dispersed and the

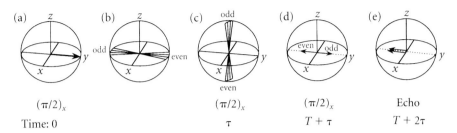

**Figure 4.3** Formation of a stimulated echo. Since the contributions to the echo amplitude follow a $\cos^2\theta$ dependence, we concentrate on bundles of isochromats (b) that precess through angles $\theta$ near to $\pm n\pi$ radians in the first $\tau$ interval, where $n$ is an integer. One bundle has $n$ even and the other $n$ odd. After rotation into the $\pm z$ directions (c) the transverse components decay, leaving only longitudinal magnetization which is then rotated towards the $\pm y$ axes (d). During the last $\tau$ interval, the even vector precesses though a whole number of complete revolutions, whereas the odd vector precesses from $+y$ to $-y$, giving a stimulated echo at $T+2\tau$.

two opposed bundles become simple vectors representing pure $z$ magnetization. The third $(\pi/2)_x$ pulse rotates the resultant 'even' vector to the $-y$ axis and the 'odd' vector to the $+y$ axis (Figure 4.3(d)). The former now precesses through a whole number of complete revolutions while the latter precesses through an odd number of half revolutions; they come into coincidence along $-y$. If a typical vector precesses through $\theta_i$ radians in a given interval, its contribution to the stimulated echo is proportional to $\cos^2\theta_i$. This justifies confining the description to the bundles of vectors illustrated in Figure 4.3, since they make the dominant contribution. The convergence of the strongest magnetization vectors onto the $-y$ axis at time $T+2\tau$ constitutes the stimulated echo (Figure 4.3(e)).

We see that the spatial information that eventually leads to echo formation is stored in the $T$ interval by a type of spin sorting process (§ 1.2.3)—the odd and even bundles of isochromats are stored along $+z$, and $-z$ respectively. During the $T$ interval they exist in nonequilibrium states and are affected by spin–lattice relaxation. The even magnetization vectors are attenuated, since they have suffered a population inversion, whereas the odd vectors are scarcely affected, being near equilibrium. This allows the determination of the spin-lattice relaxation time $T_1$ from the dependence of the stimulated echo amplitude on the interval $T$.[1]

## 4.1.3 Carr–Purcell spin echoes

The Hahn echo and the stimulated echo are not exact refocusing schemes. Carr and Purcell[4] simplified this picture considerably by using a $\pi$ pulse for refocusing and this has now become the standard procedure. This brings all of the available magnetization to a focus. For simplicity all the isochromats may be assumed to have equal magnetizations represented by unit vectors, initially aligned along the $+y$ axis of the rotating frame by the action of the $(\pi/2)_x$ pulse. The subsequent precessional motion is confined to the $xy$ plane. Unlike the case of the Hahn echo, the refocusing

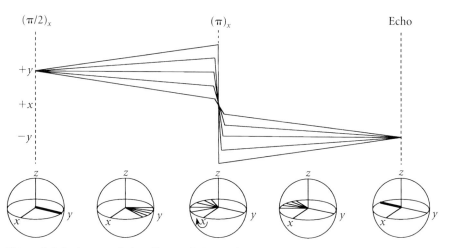

**Figure 4.4** A phase evolution diagram following five representative isochromats in the Carr–Purcell spin echo experiment. The $(\pi)_x$ pulse rotates the vectors into mirror image positions with respect to the $xz$ plane. They reach a focus along the $-y$ axis.

effect is not dependent on waiting for a nearly uniform distribution of isochromats in the $xy$ plane; refocussing can be effected at any time. At time $\tau$ a $(\pi)_x$ pulse is applied, rotating each vector into its mirror-image configuration with respect to the $xz$ plane. Each isochromat continues to precess at its own characteristic rate and in the same sense. If it had accumulated a phase divergence of $+\theta_i$ radians at time $\tau$ it now starts from an angle $-\theta_i$ with respect to the $-y$ axis and reaches that axis at time $2\tau$. The motion is reminiscent of the opening and closing of a fan. All isochromats are brought to an exact focus along the $-y$ axis. There can be no stimulated echo if the $(\pi)_x$ pulse is ideal since all the magnetization vectors are brought to an exact focus in the Carr–Purcell experiment. Figure 4.4 shows a phase evolution diagram of five representative vectors, emphasizing the analogy with the refocusing of divergent rays of light by a lens. Since the refocusing process is simpler and more efficient than for the Hahn echo, most spin echo experiments are now carried out by the Carr–Purcell scheme.

### 4.1.4 Gradient echoes

In magnetic resonance imaging the spin echo experiment is performed in an applied field gradient, and refocusing is achieved by reversing the sense of this gradient without any further radiofrequency pulse. Normally the applied gradients vastly exceed any natural field inhomogeneity in this application. This serves as the closest analogy to time reversal since each isochromat does indeed retrace its path in the second $\tau$ interval. Pulsed field gradients are now widely used as a device for purging undesirable magnetization components, or for separation of different orders of coherence (§ 6.2.2).

## 4.1.5 Molecular diffusion

In each of these cases, the key factor is that each isochromat moves in the same sense and at exactly the same rate before and after the refocusing event. If there is any appreciable macroscopic motion in the direction of the gradient, this timekeeping is perturbed and the isochromats start to lose their identities, blurring the focus. This is used as a method of measuring diffusion in liquids. Carr and Purcell introduced a scheme to defeat the effects of diffusion by refocusing with rapidly repeated $\pi$ pulses, giving a train of spin echoes. If the repetition rate is sufficiently high, refocusing occurs before diffusion can cause a significant shift in the precession frequencies of the isochromats, and the resulting echo is not appreciably attenuated.

## 4.1.6 Meiboom–Gill modification

There is a danger inherent in all multiple-refocusing schemes—any imperfections in the $\pi$ pulses may well have a cumulative effect, leading to a progressive attenuation or even a modulation of the echo amplitudes. A solution was proposed by Meiboom and Gill[5] who showed that if the refocusing pulses are applied about the $y$ axis rather than the $x$ axis (a $\pi/2$ phase shift with respect to the initial excitation pulse) then pulse length errors are compensated on even-numbered echoes and are not therefore cumulative. (A similar effect can be achieved if the $\pi$ refocusing pulses are applied along the $x$ axis but alternated in phase.) Figure 4.5 shows this diagrammatically,

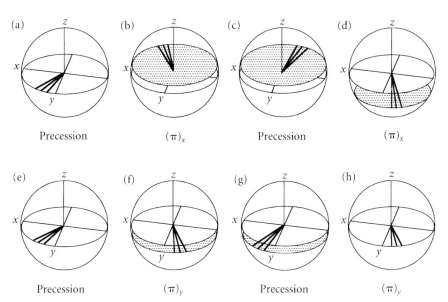

**Figure 4.5** Effect of a poorly calibrated (long) $\pi$ pulse in a Carr–Purcell spin echo experiment. In the upper sequence the magnetization vectors are rotated into a plane (shown shaded) slightly above the equator (b) and a second $(\pi)_x$ pulse carries them into a plane roughly twice as far below the equator (d) giving a cumulative error. In the lower sequence, which incorporates the Meiboom–Gill phase shift, the initial displacement (f) is compensated after the second $(\pi)_y$ pulse (h).

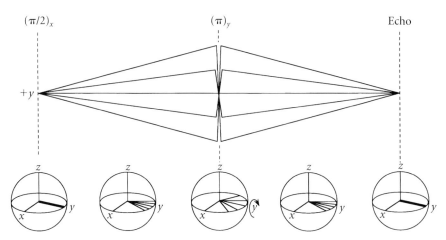

**Figure 4.6** A phase evolution diagram following five representative isochromats in the Carr–Purcell–Meiboom–Gill spin echo experiment. In contrast to Figure 4.4, the $(\pi)_y$ pulse rotates the vectors into mirror image positions with respect to the $yz$ plane. They reach a focus along the $+y$ axis.

supposing that the $\pi$ pulses have been set too long. When $(\pi)_x$ pulses are used, the free precession takes place in a plane slightly above the equatorial plane in the second $\tau$ period. A second $\pi$ pulse aggravates this effect and the next free precession takes place in a plane that is almost twice as far below the equatorial plane (Figure 4.5(d)). Pulse length errors are therefore cumulative. In contrast, when $(\pi)_y$ pulses are used, the error is compensated for even-numbered echoes (Figure 4.5(h)). This simple innovation, which has been widely adopted, is known as the Carr–Purcell–Meiboom–Gill (CPMG) sequence.

Whereas the Carr–Purcell method brings isochromats to a focus along the $\pm y$ axes on alternate echoes, the Meiboom–Gill modification refocuses exclusively along the $+y$ axis. This can be appreciated in Figure 4.6 which shows the phase evolution diagram for the CPMG sequence. The fact that a $\pi/2$ phase shift inverts the sense of the first echo was an important ingredient in the design of one of the first phase cycles (§ 6.1.2) for elimination of certain artifacts in two-dimensional spectroscopy. Phase cycling has since developed into a widely-used refinement in many multiple-pulse experiments.

The Meiboom–Gill conjuring trick served to reinforce a growing suspicion that spin manipulation could become a complex science, anticipating the explosion of new pulse sequences introduced in the last two decades. We now realize that a subtle change in pulse flip angle, an additional pulse, or the appropriate phase shift may profoundly affect the outcome of an experiment in spin physics.

### 4.1.7 Spin–spin relaxation

A factor that has so far been neglected is the loss of magnetization through spin–spin relaxation, reducing the detected signal at the peak of the echo from $M$ to

$M\exp(-2\tau/T_2)$. It is important to realize that this process is irreversible; the time-keeping of the individual spins is determined by the local distribution and the spin states ($\alpha$ or $\beta$) of all the neighbouring spins, and these are essentially random. The spin echo experiment thus distinguishes between the reversible signal loss through dephasing of isochromats ($T_2^*$) and the irreversible loss through dephasing of individual spins ($T_2$). A Carr–Purcell spin echo train is the classic experiment for the determination of the spin–spin relaxation time $T_2$.

### 4.1.8 Rotary echoes

Rumour has it that Hahn discovered the spin echo by accident when his radiofrequency pulse unit malfunctioned. If so, it is to his great credit that he subsequently worked out the intricacies of the focusing process and presented the rest of us with the key to opening up a whole new area of spin physics. Solomon[6] came across rotary echoes by a similar accident. For reasons that no longer seem important, he was experimenting with the application of dc pulses to the radiofrequency master crystal of an early NMR spectrometer while monitoring the transient nutation of the signal in the presence of the radiofrequency field $B_1$. For certain conditions the signal persisted for a much longer time than usual and also acquired a distinct modulation. It would have been easy to put this strange new effect down to some instrumental quirk, but Solomon realized that he had discovered the analogue of the spin echo in the rotating frame, the rotary echo. It is now the spatial inhomogeneity of the radiofrequency field which causes dephasing. Since $B_1$ is large compared with the inhomogeneities of the static field $B_0$, the latter may be neglected in this situation. We may postulate a new type of isochromat which differs from its neighbours by a shift in the $B_1$ intensity rather than the $B_0$ intensity. The corresponding vectors diverge in the $yz$ plane (rather than the $xy$ plane) as they rotate about the $B_1$ field which is applied along the $x$ axis (Figure 4.7). The analogue of time reversal is phase inversion of $B_1$, thus reversing the direction of rotation, and rewinding the divergent vectors (Figures 4.7(f–j)). It so happened in Solomon's experiments that the dc pulses had caused a phase shift of the transmitter frequency and this had been adjusted intuitively to $\pi$ radians. The exact focus point is of course the $+z$ axis but maximum NMR signals are detected as the bundle of vectors passes through the $\pm y$ axes. Continued rotation leads to renewed divergence, and a train of rotary echoes can be engineered by repeating the phase inversion of the $B_1$ field.

### 4.1.9 Steady-state echoes

As Solomon's experience demonstrates, spin echoes are no fleeting phenomenon, requiring careful adjustment of operating conditions. Whenever more than one radiofrequency pulse is applied to a spin system within a time interval comparable with the spin–spin relaxation time, some kind of echo effect is likely to occur. When a repeated train of radiofrequency pulses is used, steady-state echoes may be generated.[7,8] It turns out that these effects are extremely widespread (but seldom recognized) in high resolution NMR spectroscopy of liquids since the spin–spin

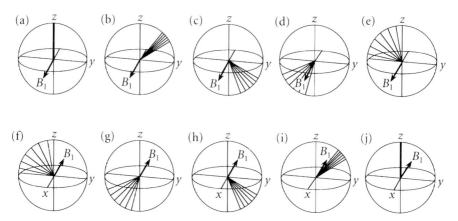

**Figure 4.7** Formation of a rotary echo: (a)–(e) show how the spatial inhomogeneity of the radiofrequency field $B_1$ causes a divergence of isochromats in the $yz$ plane. If the sense of the $B_1$ field is suddenly inverted (f) then the divergence is reversed, and after an equal interval, all isochromats come to a focus (j). The echo is detected as the narrow fan of isochromats passes through the $+y$ and $-y$ axes.

relaxation times often exceed the commonly used pulse intervals (§ 2.6.3). Steady-state effects show up as phase and intensity distortions that differ for different lines in the spectrum, but they may be masked by residual instrumental instabilities (see below).

We may picture how steady-state echoes come about by returning to the case of the Hahn echo excited by a pair of $\pi/2$ pulses. If a third $\pi/2$ pulse is applied at $2\tau$ (the centre of the Hahn echo) we would then expect to see a stimulated echo at $3\tau$ (since $T = \tau$) and another Hahn echo at $4\tau$. Clearly the possibilities for refocusing multiply rapidly with the number of pulses. If a regular train of pulses is applied, a steady-state is established[9,10] where each pulse is accompanied by an echo arising from the effects of earlier pulses. When the radiofrequency pulse sequence is extinguished, echoes persist for a time of the order of the spin–spin relaxation time. In the steady-state regime the signal decays through instrumental ($T_2^*$) effects after each pulse but then grows back again prior to the next pulse in what amounts to a time-reversed free induction decay. The approximate mirror symmetry of the signal observed between the two pulses ensures that its Fourier transform spectrum features only pure absorption peaks. Dispersion signals vanish, being associated with the negligible antisymmetric component of the time-domain signal. Similar results are obtained by Fourier transformation of the whole echo[11] (both the rising and falling parts) from a Carr–Purcell experiment.

In high resolution spectroscopy of carbon-13, steady-state echo effects can be a decided nuisance in some situations.[10] Each chemically shifted site experiences different steady-state conditions, determined principally by the pulse interval and the particular offset of that group from the transmitter frequency. This perturbs the intensity and phase of each resonance in a different manner and it is no longer possible to adjust for pure absorption-mode lines throughout the spectrum. Fortunately it is relatively easy to upset the steady-state by small instabilities in the field/frequency regulation or by deliberately introducing small random errors in the interpulse delay. This is one of the rare instances where spectrometer shortcomings are actually beneficial.

### 4.1.10 Rotational echoes

The vast majority of high resolution studies are performed with a spinning sample. In liquid phase NMR this is to defeat part of the spatial inhomogeneity of the applied field, while in the solid state it is to eliminate anisotropic interactions such as the dipolar coupling. Basically the spins are tricked into believing that they see only the mean field measured over one complete rotation. If we think of the experiment in the time domain there is a rotational echo effect. Isochromats that are all in phase immediately after the excitation pulse, lose phase coherence because they are in different applied fields. After one complete rotation of the spinner, all the isochromats at a given radius from the spinner axis have passed through the same combination of applied fields and are therefore once again in phase. This is a rotational echo and it repeats at regular multiples of the spinner period. The envelope of the free induction decay is modulated at the spinner frequency and after Fourier transformation gives rise to the well-known spinning sidebands.

It is hardly surprising that rotational echoes may interfere with conventional echoes in a rather messy way as the spinner modulation and the radiofrequency pulse repetition rate move in and out of synchronism. Often it is better to perform spin-echo experiments with a stationary sample (this is less of a penalty in modern spectrometers than in the earlier machines). Alternatively the pulses may be deliberately synchronized with the sample rotation using a trigger signal from the spinner tachometer. Two complete rotations are allowed to occur between refocusing pulses so that each volume element moves through the same sequence of applied fields during the defocusing and refocusing intervals.

In high resolution solid state experiments with spinning at the magic angle (54.7°), the spinner rates are much higher (several kHz) and the spinning is essential to the method. Both the chemical shift anisotropy and the dipole–dipole interaction are affected by the spinning, in some cases to different extents, leading to separation of the two types of interaction. Both interactions can give rotational echoes which appear as sidebands in the spectrum. Stroboscopic sampling during the detection period can be used to circumvent the rotational echo effect and remove the sidebands.

### 4.1.11 Pseudoechoes

There are quite a few instances in high resolution NMR where we would like to eliminate dispersion-mode responses and where the usual phasing routines are not applicable. For example, certain two-dimensional spectra derived from phase-modulated time-domain signals have a bizarre lineshape made up of a two-dimensional absorption signal superimposed on a two-dimensional dispersion signal. This phase-twist structure (§ 8.7.1) has some unfortunate properties, including long dispersion tails and a vanishing projection in certain directions.

The disappearance of all dispersion contributions from the Fourier transform of a whole echo stems from the symmetry about the echo centre.[11] Now it is possible to bestow a related symmetry on a free induction decay by shaping it to resemble a spin echo. Ideally we would first eliminate the $T_2^*$ decay by multiplying by a suitably rising function, perhaps an exponential. Then we would reweight the signal with an envelope that had symmetrically rising and falling parts, for example a time-shifted Gaussian or a sine bell. This pseudoecho behaves much like a true echo except for the fact that all the component frequencies are in phase at the beginning rather than at the centre, introducing a frequency-dependent phase shift. If this is corrected, a pure absorption-mode signal is obtained.[12] Note that since the dispersion ($u$-mode) is suppressed, the absolute-value mode display $(u^2 + v^2)^{\frac{1}{2}}$ gives a pure absorption lineshape.

## 4.2  Echo modulation

So far we have considered only an isolated spin, and the Bloch equations are quite adequate for describing its motion of nutation and free precession. An important new phenomenon occurs when there are two spin species $I$ and $S$ coupled by a scalar interaction $J_{IS}$. Then the echo amplitude, studied as a function of time, is found to be modulated at the frequency $\frac{1}{2}J_{IS}$ Hz. The behavior may be described either by the vector picture (§ 2.4.5) or the product operator treatment (§ 3.4.5), a shorthand notation for density operator calculations. It is instructive to work through both of these alternative treatments.

### 4.2.1  Bloch vector picture

This treatment relies principally on geometrical arguments. We consider a homonuclear $IS$ system and focus attention on the $I$ spin (Figure 4.8). The spectrum is a $J$ doublet and may be represented by two equal vectors $\alpha$ and $\beta$, corresponding to the two spin states of the $S$ spin. It is important to recall that a $\pi$ pulse on the $S$ spins interchanges these labels; this interchanges the speeds at which these vectors precess in the rotating frame.

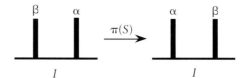

Since all pulses are of very short duration (microseconds), we may safely neglect spin coupling effects during the pulse; during free precession it is a justifiable extension

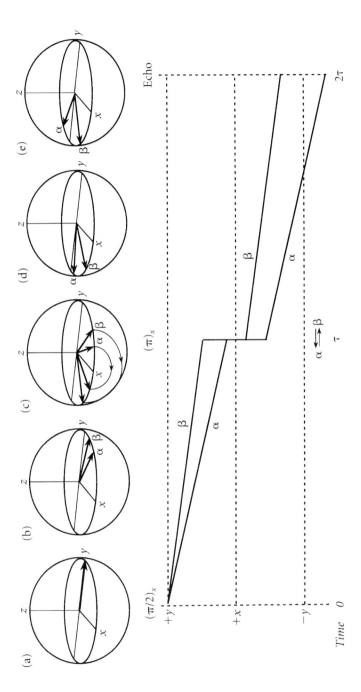

**Figure 4.8** Vector representation of spin echo modulation. The two *I*-spin vectors α and β correspond to the two states of the *S* spin. Their divergence (b) would normally be refocused by the (π)ₓ pulse, but the simultaneous spin inversion pulse on the *S* spins interchanges the α and β labels so that the phase divergence persists thoughout the period 2τ.

of the Bloch picture to treat the $\alpha$ and $\beta$ vectors independently as if they were fictitious spin-$\frac{1}{2}$ species.[3]

The initial $\pi/2$ pulse aligns these two vectors along the $+y$ axis. Free precession for a period $\tau$ involves the $I$-spin resonance offset $\delta$ and the spin–spin coupling $J$. Under these two influences, the two vectors rotate in the $xy$ plane (Figure 4.8(b)) at angular frequencies $2\pi(\delta + \frac{1}{2}J)$ and $2\pi(\delta - \frac{1}{2}J)$. Without prejudice to the argument we may label the faster vector $\alpha$ and the slower one $\beta$. The hard $\pi$ pulse is considered to be a cascade of soft $\pi$ pulses applied to the $I$ and $S$ spins separately, and in either order. The $(\pi)_x$ pulse on the $I$ spins rotates the $\alpha$ and $\beta$ vectors into mirror image locations with respect to the $xz$ plane (Figure 4.8(c)). The $\pi$ pulse on the $S$ spins inverts the spin states of $S$, thus interchanging the $\alpha$ and $\beta$ labels of the $I$ spins. There follows a second period $\tau$ of free precession during which the $\alpha$ and $\beta$ vectors precess through the same angles as before and in the same sense. Had it not been for the label interchange brought about by the $\pi$ pulse in the $S$ spins, the faster $\alpha$ vector would have caught up with the slower $\beta$ vector just as they both reached a focus along the $-y$ axis at time $2\tau$. However the interchange of labels places the faster $\alpha$ vector ahead of the slower $\beta$ and they continue to diverge due to $J$ coupling, although the chemical shift effect is brought to a focus along $+y$. (Expressing this more mathematically, we would say that the spin inversion operator leaves the bilinear term $J_{IS}\, I_z S_z$ unchanged.) The configuration at time $2\tau$ is such (Figure 4.8(e)) that the resultant $I$-spin signal in the detector is $M\cos(2\pi J\tau)$, the chemical shift term having been cancelled. If the evolution is followed as a function of time, the echo is modulated at a frequency $\frac{1}{2}J_{IS}$ Hz.

### 4.2.2 Product operator treatment

The advantage of this method (§ 3.4.5) is that a set of algebraic rules can be followed without the need to keep track of the underlying physics. In practice it is usually much less tedious not to work blindly though the algebra, but to introduce simplifications based on an understanding of the spin physics. A product operator analysis can be likened to a branching tree, and the number of terms increases rapidly with the number of successive operations and the number of coupled spins. Purely as an exercise, we first take the pedestrian route and then note how it may be simplified.

We consider a weakly coupled two-spin system $(IS)$, and for simplicity neglect the fate of $S$ magnetization, since the problem is symmetrical with respect to $I$ and $S$. We retain the convention, discussed in § 3.3.1, where all evolution operators carry a negative sign, so we anticipate exactly the same result as the vector treatment above. The initial $(\pi/2)_x$ pulse generates transverse magnetization:

$$I_z \xrightarrow{\quad (\pi/2)\ -\tilde{I}_x \quad} + I_y \qquad \textbf{4.3}$$

The first period $\tau$ of free precession is described by an operator $(-\tilde{I}_z)$ which causes a rotation through $2\pi\delta\tau$ radians, and an operator $(-2\tilde{I}_z\tilde{S}_z)$ representing the divergence through $\pm\pi J\tau$ radians. The relevant terms just prior to the $\pi$ pulse are

$$+I_y\cos(2\pi\delta\tau)\cos(\pi J\tau)$$
$$+2I_xS_z\cos(2\pi\delta\tau)\sin(\pi J\tau)$$
$$+I_x\sin(2\pi\delta\tau)\cos(\pi J\tau)$$
$$-2I_yS_z\sin(2\pi\delta\tau)\sin(\pi J\tau)\qquad\textbf{4.4}$$

The hard $\pi$ pulse is considered as a cascade of two soft $\pi$ pulses, one applied to the $I$ spins, giving

$$I_y \xrightarrow{\;(\pi)\;-\tilde{I}_x\;} -I_y$$
$$2I_yS_z \xrightarrow{\;(\pi)\;-\tilde{I}_x\;} -2I_yS_z \qquad\textbf{4.5}$$

and one applied to the $S$ spins, giving

$$2I_xS_z \xrightarrow{\;(\pi)\;-\tilde{S}_x\;} -2I_xS_z$$
$$2I_yS_z \xrightarrow{\;(\pi)\;-\tilde{S}_x\;} -2I_yS_z \qquad\textbf{4.6}$$

All other terms remain unaffected. Free precession for a second period $\tau$ under the evolution operators $-\tilde{I}_z$ and $-2\tilde{I}_z\tilde{S}_z$ results an expression consisting of 16 terms:

$$-I_y\cos^2(2\pi\delta\tau)\cos^2(\pi J\tau)$$
$$-I_x\cos(2\pi\delta\tau)\sin(2\pi\delta\tau)\cos^2(\pi J\tau)$$
$$-2I_xS_z\cos^2(2\pi\delta\tau)\sin(\pi J\tau)\cos(\pi J\tau)$$
$$+2I_yS_z\cos(2\pi\delta\tau)\sin(2\pi\delta\tau)\sin(\pi J\tau)\cos(\pi J\tau)$$
$$+I_x\sin(2\pi\delta\tau)\cos(2\pi\delta\tau)\cos^2(\pi J\tau)$$
$$-I_y\sin^2(2\pi\delta\tau)\cos^2(\pi J\tau)$$
$$-2I_yS_z\sin(2\pi\delta\tau)\cos(2\pi\delta\tau)\sin(\pi J\tau)\cos(\pi J\tau)$$
$$-2I_xS_z\sin^2(2\pi\delta\tau)\sin(\pi J\tau)\cos(\pi J\tau)$$

$$-2I_xS_z\cos^2(2\pi\delta\tau)\sin(\pi J\tau)\cos(\pi J\tau)$$
$$+2I_yS_z\cos(2\pi\delta\tau)\sin(2\pi\delta\tau)\cos(\pi J\tau)\sin(\pi J\tau)$$
$$+I_y\cos^2(2\pi\delta\tau)\sin^2(\pi J\tau)$$
$$+I_x\cos(2\pi\delta\tau)\sin(2\pi\delta\tau)\sin^2(\pi J\tau)$$
$$-2I_yS_z\sin(2\pi\delta\tau)\cos(2\pi\delta\tau)\cos(\pi J\tau)\sin(\pi J\tau)$$
$$-2I_xS_z\sin^2(2\pi\delta\tau)\cos(\pi J\tau)\sin(\pi J\tau)$$
$$-I_x\sin(2\pi\delta\tau)\cos(2\pi\delta\tau)\sin^2(\pi J\tau)$$
$$+I_y\sin^2(2\pi\delta\tau)\sin^2(\pi J\tau)\qquad\textbf{4.7}$$

Fortunately these reduce (through standard identities) to the simple expression

$$-I_y\cos(2\pi J\tau)\;-\;2I_xS_z\sin(2\pi J\tau)\qquad\textbf{4.8}$$

which represents echo modulation[13,14] at a frequency equal to $\frac{1}{2}J_{IS}$ Hz.

This tedious operation is greatly simplified if we take account of the fact that this is a refocusing experiment and that certain terms unwind during the first $\tau$ period and rewind during the second, just as if there had been an actual time-reversal. If we write the Hamiltonian

$$\mathcal{H} = 2\pi\delta_I I_z + 2\pi\delta_S S_z + \pi J\,2I_zS_z \qquad\textbf{4.9}$$

and consider the effect of the inversion operators $(\pi)\,\tilde{S}_x$ and $(\pi)\,\tilde{I}_x$, it is clear that the first two terms of $\mathcal{H}$ are inverted, but that the third term is unchanged. This allows us to neglect the changes caused by the chemical shifts, since they merely change sign at time $\tau$ (the equivalent of a time reversal) and are refocused at time $2\tau$. This is a considerable simplification, for now we may write symbolically

$$+I_z \xrightarrow{\;(\pi/2)\,-\tilde{I}_x\;} +I_y \xrightarrow{\;(\pi)\,-\tilde{S}_x\;} +I_y \xrightarrow{\;(\pi)\,-\tilde{I}_x\;} -I_y$$

$$\xrightarrow{\;(\phi)\,-2\tilde{I}_z\tilde{S}_z\;} -I_y\cos\phi \;-2I_xS_z\sin\phi \qquad\qquad \textbf{4.10}$$

where $\phi = 2\pi J\tau$. This gives (as before) the two terms

$$-I_y\cos(2\pi J\tau) - 2I_xS_z\sin(2\pi J\tau) \qquad\qquad \textbf{4.11}$$

It is clear that the vector and product operator descriptions of echo modulation are exactly equivalent.

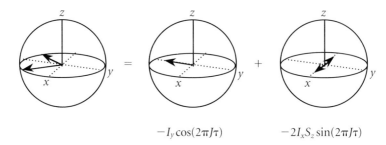

$$-I_y\cos(2\pi J\tau) \qquad\qquad -2I_xS_z\sin(2\pi J\tau)$$

### 4.2.3 Conditions for echo modulation

Both the vector and product operator treatments emphasize the importance of the inversion of the $S$ spin in causing echo modulation. In heteronuclear systems, or in weakly-coupled homonuclear systems where a selective $\pi$ pulse is applied, the $S$ spin remains unaffected and no echo modulation is observed. In homonuclear systems, or in heteronuclear systems where a $\pi$ pulse is deliberately applied to the $S$ spin at the correct time, the echoes are modulated. This provides the facility to switch the modulation on or off by inserting soft or hard $\pi$ pulses at the appropriate frequencies. We shall see below that echo modulation is the key feature in a large family of spin manipulation techniques used for high resolution NMR.

When a repeated sequence of spin echoes is generated, the $J$ modulation persists throughout the echo train. However, there is an exception for homonuclear systems if the $\pi$ pulse repetition rate is high compared with the chemical shift difference between the $I$ and $S$ spins. As the interval between $\pi$ pulses decreases, the effective chemical shift difference is reduced, but the coupling remains constant, so the spin system becomes strongly coupled. The echo modulation becomes more complex and eventually disappears (compare the sequence of spectra $AX \rightarrow AB \rightarrow A_2$ in § 1.5.1).

### 4.2.4 Spin echo difference spectroscopy

Echo modulation offers a powerful method for separating the spectra of coupled spin systems from those of systems that are not coupled.[15] The 1% abundant isotope $^{13}C$ is a case in point; we might wish to study the weak proton satellite spectrum, but the proton spectrum from $^{12}C$ molecules is 180 times more intense and may well obscure the features of interest. Let the protons be represented as the $I$ spins and $^{13}C$ as the $S$ spins. The spin echo sequence may be written as

$I$ spins: $\qquad\qquad\qquad (\pi/2)_x - \tau - (\pi)_x - \tau - \text{echo}$
$S$ spins (even scans): $\qquad\qquad (\pi)$ $\qquad\qquad\qquad\qquad$ **4.12**

This is a situation where echo modulation is deliberately introduced (on even-numbered scans only) by the application of the $\pi$ pulse on the $S$ spins. We shall call the two $^{13}C$ satellites $\alpha$ and $\beta$, reflecting the two $S$-spin states. The filtration effect may be described either by the product operator formalism or the vector picture. We adopt the latter approach.

Figure 4.9 first follows the evolution of vectors in the rotating frame in the absence of the $\pi$ pulse on the $S$ spins. During this odd-numbered scan, the $\alpha$ and $\beta$ vectors diverge by $\pm\pi/2$ radians in the period $\tau = 1/(2J_{IS})$. A mirror image arrangement is achieved by the $(\pi)_x$ pulse on the $I$ spins, setting up the vectors so that they precess to the $-y$ axis in the second period $\tau$. On the even-numbered scans, the

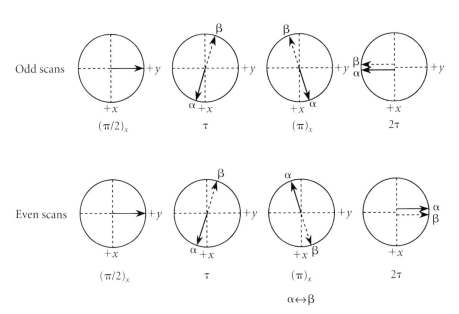

**Figure 4.9** Detection of weak satellite lines by spin-echo difference spectroscopy. On odd-numbered scans, the $\alpha$ and $\beta$ vectors diverge through $\pm\pi/2$ radians in the first interval $\tau$. They are then rotated through $\pi$ radians about the $x$ axis into mirror-image locations, and after an equal period $\tau$ reach the $-y$ axis. On even-numbered scans a $\pi$ pulse on the $S$ spins interchanges $\alpha$ and $\beta$, causing convergence onto the $+y$ axis at time $2\tau$. The vectors of the strong parent signals, with zero (or very small) coupling to the $S$ spins, execute essentially the same evolution on both odd and even scans, always terminating along $-y$. These signals are eliminated by difference spectroscopy.

same sequence is repeated with the crucial addition of the $\pi$ pulse on the $S$ spins, which has the effect of interchanging $\alpha$ and $\beta$ vectors so that they now come to a focus along the $+y$ axis at time $2\tau$. Protons that are not coupled to $^{13}$C (or are only coupled by a small long-range interaction) undergo the same evolution on both odd and even scans, and are simply rotated to the $-y$ axis at time $2\tau$. Difference spectroscopy cancels these parent signals, leaving only the satellites from directly-bonded protons.

The modulated-echo pulse sequence shown above can be used, with slight modifications, to play all kinds of useful tricks in high resolution NMR spectroscopy as can be seen in the next three sections.

## 4.2.5 Multiplicity of $^{13}$C resonances

Normally $^{13}$C spectra are recorded with proton broadband decoupling in order to simplify the spectra and improve the sensitivity. Unfortunately this discards the information about the number of protons directly attached to each carbon site. For this reason several techniques have been developed to determine multiplicity, that is to say, to make an assignment of methyl, methylene, methine and quaternary sites.

One simple illustrative example of multiplicity determination exploits spin echo modulation.[16] The roles of the spins in the pulse sequence shown in Equation 4.12 are now reversed:

$S$ spins ($^{13}$C): $\qquad (\pi/2)_x — \tau — (\pi)_x — \tau —$ acquire
$I$ spins ($^{1}$H): $\qquad\qquad\qquad\quad (\pi) \qquad\qquad$ decouple $\qquad$ **4.13**

The form of the echo modulation depends on the number of attached protons. If it is a quaternary site, the modulation arises only from small long-range couplings and may be neglected on the short timescale of this experiment. The $^{13}$C spin echoes from the CH, CH$_2$ and CH$_3$ sites are modulated by the $^{13}$C–H coupling. Neglecting the $^{13}$C chemical shift, we can describe this modulation by considering precessing vectors with relative intensities given by the binomial coefficients (1:1, 1:2:1 and 1:3:3:1).

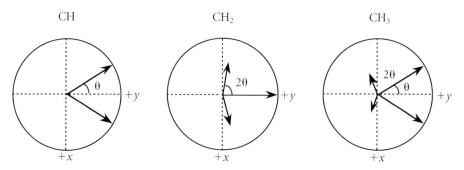

where $\theta = 2\pi J_{IS}\tau$. When the decoupler is switched on, we observe only the resultant vector along the $+y$ axis (normalized to $M_0$ in each case).

$$M(CH) = M_0\cos\theta$$
$$M(CH_2) = \tfrac{1}{2} M_0 (1 + \cos2\theta) = M_0\cos^2\theta$$
$$M(CH_3) = \tfrac{1}{4} M_0 (3\cos\theta + \cos3\theta) = M_0\cos^3\theta \qquad \textbf{4.14}$$

If the time at which decoupling and acquisition start is set to the condition $2\tau = 1/(J_{IS})$, then $\theta = \pi$ and $\cos\theta = -1$. Then the CH and $CH_3$ sites have inverted signals, whereas the $CH_2$ signals remain positive (as do the quaternary sites). The spectrum is consequently labelled according to parity, that is to say, whether the carbon sites carry an odd or even number of attached protons. Figure 4.10 shows a 100 MHz $^{13}C$ spectrum of estrone methyl ether with this type of multiplicity determination.[16] All the methylene (2) and quaternary sites (0) show positive signals, whereas the methyl (3) and methine (1) sites have inverted resonances. This single experiment often suffices to assign the $^{13}C$ sites even though it does not distinguish $CH_2$ from quaternary, or CH from $CH_3$. Elaborations of this method permits separation of all four multiplicities.

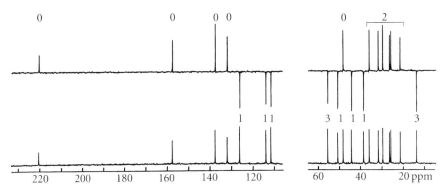

**Figure 4.10** Determination of the number of directly attached protons at the $^{13}C$ sites in estrone methyl ether. Above: the spectrum derived from the modulated $^{13}C$ spin echoes with the condition $\tau = 1/(2J_{IS})$. Sites with an even number of attached protons (0 or 2) give upright signals, whereas those with an odd number (1 or 3) generate inverted signals. Below: the conventional decoupled $^{13}C$ spectrum.

## 4.2.6 Insensitive nuclei enhanced by polarization transfer

A slight extension of the echo-modulation scheme gives the INEPT[17] sequence:

$$I \text{ spins:} \qquad (\pi/2)_x - \tau - (\pi)_x - \tau - (\pi/2)_y$$
$$S \text{ spins:} \qquad (\pi)_x - \tau - (\pi/2)_x \text{ acquire} \qquad \textbf{4.15}$$

where $I$ represents protons and $S$ represents (for example) the low-abundance $^{13}C$ or $^{15}N$ nuclei. The idea is to bestow on the insensitive nucleus the advantage of the more favourable polarization and relaxation properties of the protons. As in the example examined above, we focus attention on the proton satellites, represented by the vectors $\alpha$ and $\beta$. With the timing condition set at $\tau = 1/(4J_{IS})$, these vectors are prepared so that they subtend and angle of $\pi/2$ radians at time $\tau$ (Figure 4.11(b)).

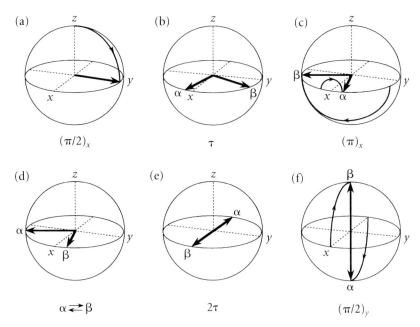

**Figure 4.11** Vector diagram of the INEPT experiment. (a) The *I*-spin vectors $\alpha$ and $\beta$ correspond to the two quantum states of the $S$ spin. (b) They are allowed to precess for a time $\tau$ until they subtend an angle of $\pi/2$ radians. (c) They are rotated into mirror-image locations with respect to the $xz$ plane. (d) A $\pi$ pulse on the $S$ spins interchanges the $\alpha$ and $\beta$ labels. (e) Further free precession carries the $\alpha$ and $\beta$ vectors into an antiphase alignment. (f) A $(\pi/2)_y$ pulse rotates these vectors along the $\pm z$ axes.

The refocusing $I$ pulse and the spin-inversion $S$ pulse set up the configuration illustrated in Figure 4.11(d), and free precession for a further time $\tau$ brings these vectors into an antiphase alignment along the $\pm x$ axes (Figure 4.11(e)). Then the $(\pi/2)_y$ pulse rotates the vectors into the $\pm z$ directions (Figure 4.11(f)). Note that the phase shift from $x$ to $y$ is essential to the success of this method. This entails a special kind of population disturbance known as longitudinal two-spin order (§ 3.2.1) and represented by $2I_zS_z$ in the product operator vocabulary.

If we fix attention on the case where the $S$ spins are $^{13}$C, the fact that they share common energy levels with the protons means that any disturbance of proton populations inevitably affects them with a multiplicative factor $\gamma_I/\gamma_S = 4$. This is illustrated in Figure 4.12, where the proton transition H($\alpha$) suffers a population inversion, disturbing the population differences across the $^{13}$C transitions. The final $(\pi/2)_x$ pulse on the $^{13}$C spins merely acts as a read pulse, converting the population disturbance into an antiphase signal in the detector. The resulting change in the $^{13}$C signal is stronger than the natural signal by a factor $\gamma_I/\gamma_S$. This is a sensitivity enhancement of four, exceeding the maximum nuclear Overhauser enhancement ($+3$). For the case that the $S$ spins represent $^{15}$N the enhancement is tenfold, far

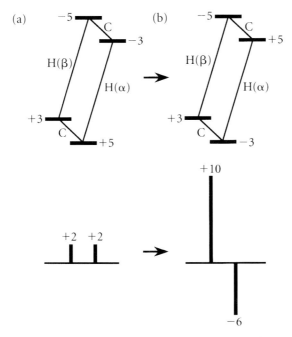

**Figure 4.12** Energy-level diagram for the INEPT experiment with $I$ = $^1$H and $S$ = $^{13}$C. (a) The population numbers have been calculated so that the $^1$H intensities are four times stronger than the $^{13}$C intensities. (b) The condition represented by Figure 4.11(f), where one $^1$H transition has a population inversion while the other is at Boltzmann equilibrium. Now the $^{13}$C intensities have changed by +8 and –8, an enhancement of sensitivity equal to $(\gamma_I/\gamma_S)$.

greater than the maximum nuclear Overhauser effect (−4). Note that the observed signal has a characteristic up–down intensity profile.

That is not quite the end of the story. Since protons have appreciably faster spin–lattice relaxation rates than $^{13}$C or $^{15}$N, the experiments can be repeated at a rapid cadence, giving a further improvement in signal-to-noise by more efficient time averaging. The same principle produces an even greater effect if deuterium is substituted for protons. At first sight one would imagine that the much lower gyro-magnetic ratio of deuterium $(\gamma_H/\gamma_D = 6.5)$ would render it ineffective for $^2$D → $^{13}$C or $^2$D → $^{15}$N polarization transfer because of the much lower Boltzmann popula-tions. However, the spin–lattice relaxation rate of the quadrupolar deuterium nucleus is so much faster than that of $^1$H, that there is still a net enhancement of the low-abundance nuclei if the timing delays are optimized.[18]

A useful extension[19] of the basic INEPT sequence permits refocusing of the $S$ spin vectors before detection, by the addition of a further group of pulses:

$$I \text{ spins:} \quad (\pi/2)_x - \tau_1 - (\pi)_x - \tau_1 - (\pi/2)_y - \tau_2 - (\pi) - \tau_2 - \text{decouple}$$
$$S \text{ spins:} \qquad\qquad (\pi) - \tau_1 - (\pi/2)_x - \tau_2 - (\pi)_x - \tau_2 - \text{acquire} \qquad \textbf{4.16}$$

If broadband decoupling is attempted in the standard INEPT experiment, the
$S$-spin signal disappears through mutual cancellation of antiphase components of
the multiplets. The added refocusing stage brings the multiplet components into the
same phase and thus permits the use of broadband $I$-spin decoupling. When there
are two or more equivalent $I$ spins attached to a given $S$ spin, the optimum setting
for $\tau_2$ is different from that for $\tau_1$. This is important in the common situation of
$H \rightarrow {}^{13}C$ transfer, when there are CH, $CH_2$ and $CH_3$ groups involved.

The refocused INEPT experiment may also be implemented without decoupling,
in order to record the $S$-spin multiplet structure. However the multiplets exhibit
phase and intensity anomalies, that is to say, the components cannot be phased for
pure absorption and the relative intensities are not equal to the expected binomial
coefficients. Normality can be restored by adding a $(\pi/2)_y$ purging pulse on the $I$
spins just before signal acquisition.[20] In this mode (called INEPT+), the $S$-spin
spectrum can be properly phased and the multiplets have intensities given by the
appropriate binomial coefficients.

Polarization transfer methods such as INEPT are now widely employed for the
study of low-abundance $S$ nuclei. The experiment can be reversed simply by reassigning the $I$ and $S$ labels in sequence [4.15], thus transferring polarization from
the insensitive nucleus to protons, which have a much higher inherent sensitivity.
This suggests the idea of combining two such transfers, first from protons to the $S$
spins (followed by evolution period to probe the $S$-spin frequencies) and then back
again to protons to take advantage of its more favourable detection properties. This
two-dimensional scheme[21] is sometimes called the Overbodenhausen experiment
or heteronuclear single-quantum correlation (HSQC). We shall see in § 8.3.3 that
this is predicted to give an improvement (over direct detection of the $S$ spins) of the
order of $(\gamma_I)^{5/2}$, which is approximately 32 when $S = {}^{13}C$, and 300 when $S = {}^{15}N$.
The pulse sequence may be written as

$I$ spins: $(\pi/2)_x - \tau - (\pi)_x - \tau - (\pi/2)_y - t_1 - (\pi/2)_x - \tau' - (\pi)_x - \tau' - \text{acquire}$
$S$ spins: $\qquad\qquad (\pi) - \tau - (\pi/2)_x - t_1 - (\pi/2)_x - \tau' - (\pi) - \tau' - \text{decouple}$

**4.17**

If required, an $I$-spin $\pi$ pulse in the centre of the evolution period ($t_1$) serves to
decouple the heteronuclear splitting of the $S$ spins in the $F_1$ dimension. The original
application was for indirect detection of natural abundance ${}^{15}N$ resonances but the
technique is now also widely employed for ${}^{13}C$ spectroscopy.

### 4.2.7  Multiple-quantum coherence

The INEPT experiment manipulates magnetization vectors in a manner that is
easily followed by the vector picture, and the HSQC technique extends this to
round trip polarization transfer ($I \rightarrow S \rightarrow I$) which improves sensitivity by exploiting the high initial polarization and high detection efficiency of the $I$ spins (usually
protons). These experiments reflect a particular geometrical frame of mind. It is
interesting to note that at about the same time (1979) Müller[22] used the alternative

algebraic approach to devise a competitive technique that cannot be explained by the vector model at all. It achieves the same degree of sensitivity enhancement by a heteronuclear multiple-quantum experiment. This scheme starts with the preparation of antiphase $I$-spin magnetization, just as in the INEPT technique:

$$+I_z \xrightarrow{\tilde{I}_x} -I_y \xrightarrow{(\pi/4)\,2\tilde{I}_z\tilde{S}_z} \xrightarrow{(\pi)\,\tilde{I}_x} \xrightarrow{(\pi)\,\tilde{S}_x} \xrightarrow{(\pi/4)\,2\tilde{I}_z\tilde{S}_z} -2I_xS_z \qquad \textbf{4.18}$$

but it continues in a quite different direction, creating a mixture of heteronuclear zero- and double-quantum coherence:

$$-2I_xS_z \xrightarrow{(\pi/2)\,\tilde{S}_x} +2I_xS_y \qquad \textbf{4.19}$$

This is allowed to evolve for a variable period $t_1$ of a two-dimensional experiment. Normally the multiple-quantum frequencies would appear in the $F_1$ frequency dimension, but by introducing a proton $\pi$ pulse at the midpoint of the evolution period, the proton contribution is removed, leaving only the chemical shift of the low-sensitivity nucleus ($S$). One relatively simple implementation may be expressed as:

$I$ spins: $(\pi/2) - \tau - \quad -\tfrac{1}{2}t_1 - \pi - \tfrac{1}{2}t_1 - \quad -\tau' -$ acquire
$S$ spins: $\quad\quad (\pi/2) \quad - \quad t_1 \quad - \quad (\pi/2)_x - \tau' -$ decouple $\qquad \textbf{4.20}$

Antiphase magnetization is created in the first $\tau$ period and the $S$-spin pulse converts this into multiple-quantum coherence which evolves during $t_1$. The final $\tau$ period serves to refocus antiphase signals so that broadband decoupling can be used. Elaborations by Bax $et\ al.$,[23] usually called heteronuclear multiple-quantum coherence (HMQC) experiments, are now very widely employed (§ 8.3.4).

Related experiments may be performed in homonuclear spin systems. The product operator formalism gives

$$2I_zS_x + 2I_xS_z \xrightarrow{(\pi/2)\,\tilde{I}_x} \xrightarrow{(\pi/2)\,\tilde{S}_x} -2I_yS_x - 2I_xS_y \qquad \textbf{4.21}$$

which represents pure double-quantum coherence. One of the most valuable applications of this form of invisible coherence exploits the fact that it requires the presence of two coupled spins; a single isolated spin cannot possibly show the effect. Now there is considerable chemical interest in the scalar coupling between two $^{13}C$ spins, but its detection is fraught with practical problems. In natural abundance samples, only one molecule in about 8100 has a pair of $^{13}C$ nuclei at the chosen sites. Molecules with only a single $^{13}C$ spin are 90 times more abundant, and their signals tends to mask those of the coupled spin system. However, if we momentarily create double-quantum coherence, the single-spin species are left out in the cold. Now double-quantum coherence is twice as sensitive to a radiofrequency phase shift as single-quantum coherence (§ 3.4.4), so by suitably cycling the phase of the hard read pulse we can filter the signal of interest from the undesirable contribution (§ 6.1.4). This is the basis of the (one-dimensional) INADEQUATE experiment[24] which provides four-line spectra from coupled $^{13}$C pairs, untrammelled by the conventional single-spin spectrum. The sequence may be written as

$$(\pi/2)_x - \tau - (\pi)_y - \tau - (\pi/2)_x - \Delta - (\pi/2)_\phi, \text{acquire } (\psi) \qquad \textbf{4.22}$$

As before, $\tau = 1/(4J_{IS})$. Double-quantum coherence is excited at the start of the short fixed delay $\Delta$ (which is included simply to allow radiofrequency phase switching) and is then reconverted into antiphase magnetization by a $\pi/2$ pulse. The phase of acquisition $(\psi)$ is rotated in $-\pi/2$ steps as $\phi$ is incremented in $+\pi/2$ steps, thus selecting signals resulting from double-quantum coherence and rejecting the rest. This phase cycle is analysed in detail in § 6.1.4.

By replacing the fixed $\Delta$ delay with an incremented evolution period $t_1$, this experiment is converted into a two-dimensional version[25] where the spectra from adjacent pairs of $^{13}$C spins are spread out in the $F_1$ dimension according to their double-quantum frequencies. This has proved to be a powerful method of establishing the connectivity of the carbon framework of a molecule (§ 8.4.1). It is only a pity that the sensitivity is inherently poor owing to the low abundance of doubly-labelled species in nature. A small amount of random artificial enrichment in $^{13}$C goes a long way, since it affects both coupled sites.

## 4.2.8 Bilinear rotation decoupling (BIRD)

Pines[26] and co-workers introduced the novel idea of creating a module that could be slipped into any existing pulse sequence to discriminate between protons attached to the $^{13}$C nuclei and the much more abundant protons attached to $^{12}$C. This BIRD operator is also an adaptation of the basic spin echo sequence and it comes in two distinct forms. We shall call the first of these BIRD-CP since it is derived from the Carr–Purcell sequence.

$I$ spins:     $\qquad\qquad (\pi/2)_x - \tau - (\pi\ )_x - \tau - (\pi/2)_x$
$S$ spins:     $\qquad\qquad\qquad\quad (\pi)$ $\qquad\qquad\qquad\qquad\qquad$ **4.23**

With the usual timing condition $\tau = 1/(2J_{IS})$, the proton $\alpha$ and $\beta$ vectors undergo the following sequence of rotations:

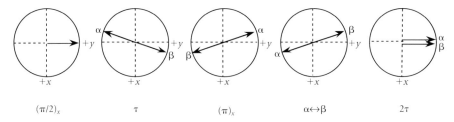

$\qquad (\pi/2)_x \qquad\qquad\qquad \tau \qquad\qquad\qquad (\pi)_x \qquad\qquad\quad \alpha\leftrightarrow\beta \qquad\qquad\quad 2\tau$

They are then brought to the $-z$ axis by the final $(\pi/2)_x$ pulse, a population inversion. In contrast, protons attached to $^{12}$C show no divergence during the two $\tau$ intervals. They have their chemical shift refocused and suffer only a $2\pi$ rotation as a result of the BIRD module. This is thus a powerful method for discriminating between protons attached to $^{13}C$ (which are interesting) and those attached to $^{12}C$ (which are a nuisance).

We call the second form of this module BIRD-MG since it resembles the Meiboom–Gill modification of the basic spin echo sequence. Once again a small change (a $\pi/2$ phase shift of one $\pi$ pulse) makes a fundamental difference.

I spins: $\qquad\qquad\qquad (\pi/2)_x - \tau - (\pi)_y - \tau - (\pi/2)_x$

S spins: $\qquad\qquad\qquad\qquad\qquad (\pi)$ **4.24**

With the same timing condition as above, $\tau = 1/(2J_{IS})$ the vectors follow the sequence

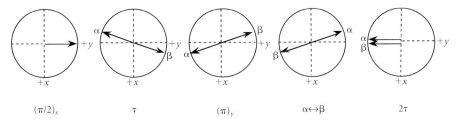

$\qquad (\pi/2)_x \qquad\qquad\qquad \tau \qquad\qquad\qquad (\pi)_y \qquad\qquad\qquad \alpha\leftrightarrow\beta \qquad\qquad\qquad 2\tau$

By rotating the $\alpha$ and $\beta$ vectors about the $+y$ axis rather than the $+x$ axis, this reverses the mode of discrimination; protons attached to $^{13}C$ are now rotated into the $+z$ axis, whereas the protons bonded to $^{12}C$ are inverted. This is often very useful as the initial step of the HSQC or HMQC scheme, since the main sequence can be started after a delay $t = T_1 \ln 2$, just as the abundant $^{12}C$ protons are passing through the null condition after population inversion. The BIRD module is also used as a refocusing pulse that makes a distinction between direct and long range $^{13}C$–H couplings.

There is an extension of bilinear rotation decoupling, called TANGO, that acts as a $\pi/2$ excitation pulse instead of a population inversion pulse.[27] Suppose we wish to excite the protons directly bonded to $^{13}C$ while leaving the distant protons and the $^{12}C$ protons unaffected. The TANGO sequence is

I spins: $\qquad\qquad (3\pi/4)_x - \tau - (\pi)_x - \tau - (\pi/4)_x$ acquire

S spins: $\qquad\qquad\qquad\qquad\qquad \pi$ **4.25**

We may label the two proton vectors $\alpha$ and $\beta$. Unlike most other spin echo sequences, TANGO operates with the magnetization vectors confined to the surface of a cone making an angle of $45°$ with respect to the $z$ axis (Figure 4.13). The interval $\tau$ is set equal to $1/(2J_{CH})$ which allows the transverse components of the $\alpha$ and $\beta$ vectors to become diametrically opposed (Figure 4.13(b)). The $(\pi)_x$ pulse carries them into the upper half of the cone (Figure 4.13(c)) and the $^{13}C$ inversion pulse interchanges the $\alpha$ and $\beta$ labels (Figure 4.13(c). Further precession for an equal period $\tau$ refocuses the proton chemical shift and brings the $\alpha$ and $\beta$ vectors into coincidence again (Figure 4.13(e)), and the final $(\pi/4)_x$ pulse aligns them along the $+y$ axis (Figure 4.13(f)). Vectors that have no appreciable spin–spin coupling to $^{13}C$

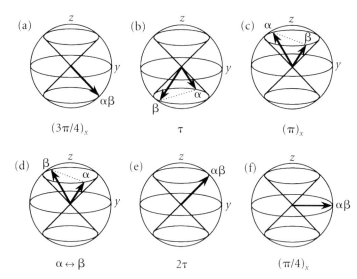

**Figure 4.13** TANGO sequence for excitation of protons directly bonded to $^{13}$C without affecting distant protons. Directly coupled protons are represented by the vectors $\alpha$ and $\beta$. They are first rotated through $3\pi/4$ radians about the $+x$ axis (perpendicular to the plane of the paper). They precess on the surface of a 45° cone where chemical shift and the direct spin–spin coupling carry them into positions (b) where their transverse components are diametrically opposed. They are then rotated through $\pi$ radians about the $+x$ axis (c) and their spin state labels are exchanged (d). Further free precession (e) brings them into coincidence and a final $\pi/4$ pulse about the $+x$ axis places them along $+y$. All other proton vectors are rotated through a total of $2\pi$ radians about the $+x$ axis and therefore return to the $+z$ axis.

have their chemical shifts refocused but do not diverge appreciably during the period $2\tau$. They execute a $2\pi$ rotation about the $x$ axis and are left aligned along $+z$ at the end of the sequence.

### 4.2.9  J spectroscopy

Echo modulation provides the key to one of the most difficult operations in high resolution NMR—the separation of chemical shifts and spin–spin couplings. For a homonuclear $IS$ system the chemical shift can be refocused by a $\pi$ pulse, while the spin echo remains modulated by spin–spin coupling. A train of spin echoes exhibits the modulation discussed in § 4.2, represented by two counter-rotating vectors, or by the product operator expression given in § 4.2.2:

$$-I_y\cos(2\pi J\tau) - 2I_xS_z\sin(2\pi J\tau) \qquad \textbf{4.26}$$

Fourier transformation of this modulation gives a new kind of spectrum that has no frequency displacement due to chemical shielding, merely a doublet of splitting $J_{IS}$ centred at zero frequency. In anything more complex than a two-spin system, the result is best studied by two-dimensional spectroscopy (§ 8.5.2), using the pulse sequence

$$(\pi/2)_x - \tfrac{1}{2}t_1 - (\pi)_y - \tfrac{1}{2}t_1 - \text{acquire } (t_2) \qquad\qquad \textbf{4.27}$$

Signal acquisition starts at the peak of the spin echo. Fourier transformation as a function of both $t_1$ and $t_2$ generates a two-dimensional $J$ spectrum consisting of a long thin strip only a few Hz wide in the $F_1$ dimension but several ppm long in the $F_2$ dimension. The spin multiplets from individual chemical sites are disposed along 45° diagonals, rather like a half-open Venetian blind.

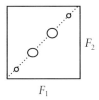

At first sight this would appear the ideal vehicle for separating shifts and coupling constants[28] since a projection in the 45° direction onto the $F_2$ axis should give a singlet response at each chemical shift frequency, while the appropriate 45° sections should give the individual multiplets free from any overlap between neighbours (§ 8.5.2). In practice there is a very serious drawback, which arises from the peculiar two-dimensional lineshape characteristic of this experiment. This "phase twist" line shape is discussed in detail in § 8.7.1. The property that concerns us here is the fact that the 45° projection vanishes through self-cancellation.[29] Several different solutions have been suggested for solving the lineshape problem.[30–34] One promising approach records the two-dimensional $J$ spectrum in a form where the multiplets lie on both the 45° and 135° diagonals, forming a St Andrew's cross.[34]

This particular symmetry property is then used to find each two-dimensional multiplet and separate it from the rest. This gives the desired proton chemical shifts. A decoupled proton spectrum of 4-androsten-3,17-dione is shown in Figure 4.14. It has resonances at the proton chemical shifts but no spin–spin splittings (§ 8.5.3). An alternative scheme employs a lineshape transformation to remove the dispersion-mode contributions to the phase-twist response, so that the 45° projection no longer vanishes.[35]

In heteronuclear systems, true broadband decoupling is available and the separation of shifts from heteronuclear couplings is much more readily implemented:

$I$ spins: $\qquad\qquad\qquad\qquad\qquad (\pi) - \tfrac{1}{2}t_1 - \text{decouple}$
$S$ spins: $\qquad (\pi/2)_x - \tfrac{1}{2}t_1 - (\pi)_y - \tfrac{1}{2}t_1 - \text{acquire } (t_2) \qquad\qquad \textbf{4.28}$

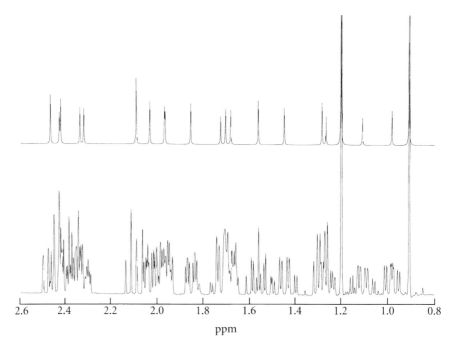

**Figure 4.14** (Above) The decoupled 400 MHz proton spectrum of 4-androsten-3,17-dione obtained by processing the two-dimensional $J$ spectrum as described in the text. The projection onto the $F_2$ axis (bottom) shows the coupled spectrum. The two tall (truncated) peaks are from methyl groups.

Note the extra $\pi$ pulse to flip the $I$-spin states. Applied to $^{13}$C spectroscopy, complete separation can be achieved with C–H splittings in the $F_1$ dimension and $^{13}$C shifts in the $F_2$ dimension (§ 8.5.1).

## 4.2.10  Coherence transfer echoes

When coherence precesses at one frequency during the evolution period $t_1$ but, through coherence transfer, is converted to another frequency during the detection period $t_2$, a new echo phenomenon occurs.[36] If there is a change in coherence order ($p$), or if the transfer is to another nuclear species with a different gyromagnetic ratio, field inhomogeneity effects cause the isochromats to diverge at one rate during $t_1$ but converge at a different rate during $t_2$. The timing of the resulting coherence transfer echo may then appear unusual, either advanced or delayed with respect to the point where it might normally have been expected. (Compare this with the use of pulsed field gradients, described in § 4.3.1 and § 6.2.2.) For example, if a proton coherence evolves for a time $\tau$, followed by transfer to $^{13}$C, the coherence transfer echo occurs after an overall delay of $5\tau$, because $\gamma_I/\gamma_S = 4$. In a homonuclear experiment, if multiple-quantum coherence of order $p$ is converted into single-quantum coherence, the timing of the coherence transfer echo is delayed, because the isochromats converge $p$ times more slowly than they originally diverged.

This phenomenon influences the lineshapes observed in certain two-dimensional spectra when field inhomogeneity is the dominating factor. The $T_2^*$ broadening acts along the principal diagonal of slope $\arctan(\gamma_S/\gamma_I)$, while field inhomogeneity is exactly refocused in the perpendicular direction, and the natural linewidth is observed (provided that molecular diffusion can be neglected).

### 4.2.11 Echoes and antiechoes

In a two-dimensional coherence transfer experiment, the signal observed during the detection period $t_2$ is amplitude-modulated as a function of the evolution time $t_1$, and it may seem a little surprising that a spin echo effect occurs under such circumstances, because we associate echoes with divergence and convergence of the phases. To understand this, we decompose the amplitude modulation during the evolution period into a sum of two equal counter-rotating phase modulations. One of these induces a signal component that precesses in opposite senses during $t_1$ and $t_2$, thus generating the coherence transfer echo. The other precesses in the same sense in both periods, continuing to be influenced by the field inhomogeneity in the normal manner and hence no echo is formed. It is useful to label these two responses the echo and the antiecho (sometimes called the $n$ and $p$ type responses). The echo signal can be favoured over the antiecho by its different dependence on the flip angle of the conversion pulse.[37] This can be useful for determining the sense of precession of multiple-quantum signals. By using a flip angle of $3\pi/4$ to select the echo component, the sign of the double-quantum frequency can be determined in the INADEQUATE experiment[25] allowing the transmitter to be set in the centre of the spectrum to save data storage.

## 4.3 Refocusing experiments

If the concept of refocusing is considered in a more general sense there are several other techniques that are related to the spin echo principle.

### 4.3.1 Pulsed field gradients

For many years, artifact suppression and the separation of desirable from unwanted signals was accomplished by phase cycles, sometimes of considerable complexity (§ 6.1). While this is acceptable when a protracted period of multiscan averaging is required for sensitivity enhancement, it is very wasteful of spectrometer time in situations where the signal-to-noise is already adequate. Furthermore, difference spectroscopy and phase cycling rely implicitly on good spectrometer stability, otherwise imperfect subtraction leaves behind appreciable artifacts. For these reasons, pulsed field gradients[38–41] are now increasingly employed as an alternative to phase cycling (§ 6.2). The reproducibility of gradients is now so high that even the most demanding applications are better handled by pulsed gradient technology. One example is water suppression (§ 11.2.1), where not only is the suppression ratio very high but the artifacts known as $t_1$ noise are also eliminated.[41]

Basically the method employs two carefully matched gradients, the first to defocus all signal components and the second to refocus the signal of interest while the undesirable components continue to dephase. For example, homonuclear double-quantum coherences may be selected by applying a gradient for fixed time $\Delta$ during evolution and an equal intensity gradient for a time $2\Delta$ just prior to detection, since double-quantum coherence is twice as sensitive to an applied gradient (§ 3.4.3). The gradient pulse should have high intensity and a short duration so that molecular diffusion can be neglected. Much work has gone into the practical problem of generating intense gradients that are actively shielded to avoid the effects of eddy currents in neighbouring conductors.

### 4.3.2 Composite pulses

As discussed in § 2.5.3, a radiofrequency pulse of finite intensity does not behave ideally for spins that are some distance from resonance. Rotation takes place about a tilted effective field in the rotating frame and the magnetization trajectories diverge away from the $zy$ plane, generating increasing amounts of dispersion-mode signal. Levitt[42] was the first to realize that this effect could be compensated by designing a composite pulse, a sandwich of three pulses where the imperfections of the first is corrected by the similar imperfections of the last. The first of the family was a composite $\pi$ pulse:

$$R = (\pi/2)_x \, (\pi)_y \, (\pi/2)_x \qquad \textbf{4.29}$$

We can appreciate that this exploits the spin echo concept if the approximation is made that the imperfect (tilted) $\pi/2$ pulses can be broken down into perfect $\pi/2$ pulses ($^*$) and short periods ($\Delta$) of free precession:

$$R = (\pi/2)_x^* - \Delta - (\pi)_y - \Delta - (\pi/2)_x^* \qquad \textbf{4.30}$$

If the $\pi$ pulse is also perfect, the phase errors that accumulate during the first $\Delta$ interval are exactly refocused in the second $\Delta$ interval. In practice the $\pi$ pulse is also imperfect, and some second-order errors remain, but the method is quite effective for compensating resonance offset effects. It also corrects for pulse length errors (from miscalibration or from spatial inhomogeneity), but it cannot correct for both types of imperfection simultaneously (§ 2.5.4).

This sandwich pulse was the prototype for an entire family of composite pulses for excitation, spin inversion, refocusing etc. An inversion pulse that is particularly effective in compensating resonance offset effects was discovered by Shaka:[43]

$$R = (\pi/2)_x \, (\pi)_{-x} \, (3\pi/2)_x \qquad \textbf{4.31}$$

this has been used for broadband decoupling (see below, § 4.3.3).

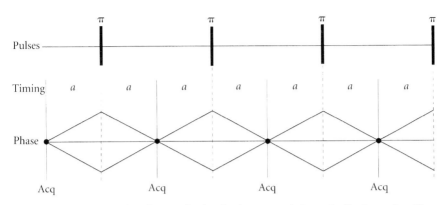

**Figure 4.15** Phase evolution diagram for the $S$-spin vectors during spin–flip decoupling. The $\pi$ pulses applied to the $I$ spins refocus the phase divergence due to $J_{IS}$. Because the $S$-spin free induction signal is sampled stroboscopically at the refocusing points, no splitting is detected.

### 4.3.3 **Broadband decoupling**

Sometimes a rather naive idea can be developed into a useful methodology. This was the case with spin–flip decoupling[44] which turns out to be quite impractical for the usual high resolution operating conditions, but which nevertheless triggered the discovery of a powerful method of broadband decoupling. The method employs a regular sequence of $I$-spin (proton) $\pi$ pulses to refocus the $\alpha$ and $\beta$ vectors of the $S$ spin ($^{13}$C) in a repetitive fashion. If the $S$-spin sampling is confined to the focus points, then the $J_{IS}$ splitting is no longer in evidence and the spectrum appears to have been decoupled (§ 3.3.3). Figure 4.15 shows the phase evolution diagram for this experiment. Unfortunately it is not feasible to generate sufficiently intense $\pi$ pulses at rates that are high enough to make this a practical proposition for $^{13}$C spectroscopy where the sampling rate must be of the order of 40 kHz. Furthermore, pulse errors would be cumulative.

Levitt's discovery of composite $\pi$ pulses[42] changed this picture radically, since they made it possible to cover a wide range of proton shifts with a much weaker pulse intensity. The composite pulses were repeated so rapidly that there were no windows for free precession between $\pi$ pulses. Unfortunately even very small pulse imperfections become important when a very long repetitive sequence is used. To counter this difficulty, Levitt devised magic cycles and supercycles that compensated pulse imperfections to high order.[45,46] Waugh[47] developed the theory to explain these effects, and this led to improved sequences, such as WALTZ-16 decoupling[43,48] and more recent schemes designed to cover even wider bands.[49–54] Broadband decoupling is treated in detail in § 7.4.

### 4.3.4 **J scaling**

Instead of completely collapsing the $J_{CH}$ splittings in a $^{13}$C spectrum, it can be advantageous to reduce them by a known factor so that the multiplicity can be

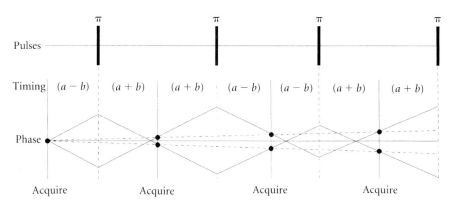

**Figure 4.16** Phase evolution diagram for the $S$-spin vectors during $J$ scaling. The slight displacement ($b < a$) of the $\pi$ pulses causes a slow divergence of the $S$-spin vectors (dashed lines) when monitored stroboscopically. This corresponds to a scaled-down splitting of the $S$-spin signal, equal to $(b/a)\,J_{IS}$.

observed without too much danger of overlap between adjacent multiplets. This is known as $J$ scaling. If the original scheme for spin–flip decoupling is slightly modified (§ 4.3.3), the focusing can be made only partial.[44] The $I$ spins (protons) are subject to a sequence of $\pi$ pulses where the intervals alternate between $2(a-b)$ and $2(a+b)$ with $a > b$. The $S$-spin ($^{13}$C) acquisition is performed at regular intervals $2a$. Figure 4.16 shows the timing scheme and the phase evolution diagram for the two $S$-spin vectors. The $\pi$ pulses brings these vectors repeatedly to a focus but this is slightly offset from the $S$-spin sampling points, leading to an apparent slow divergence of these vectors (dashed lines). As far as the sampled signal is concerned it appears to be coupled to the $I$ spins with a reduced coupling constant $(b/a)J_{IS}$ Hz. The scaling factor $(b/a)$ is chosen large enough to resolve the multiplet structure, but still small enough that adjacent $S$-spin responses do not overlap.

This rather inconvenient mode of data acquisition can be improved by adapting the WALTZ-8 broadband decoupling sequence in a time-shared scheme[55] where the decoupler is switched off for a short period after every complete WALTZ-8 cycle. The scaling factor is given by $t_{off}/(t_{off} + t_{on})$ and would typically be set at about 0.1 so that all long-range C–H splittings disappear within the linewidth, overlap between adjacent multiplets is minimized, but the multiplicity due to direct C–H coupling is still resolved.

### 4.3.5  Pure-phase pulses

Excitation pulses that are frequency-selective have found many applications in high-resolution spectroscopy.[56,57] It is a considerable advantage if these soft pulses are designed to refocus magnetization trajectories so that the natural divergence is corrected and they are brought to the $+y$ axis over an appreciable range of resonance offsets. The excited region of the spectrum then appears in the pure absorption mode without the need for phase correction routines. The refocusing effect can be achieved for a modest range of offsets by using a $3\pi/2$ Gaussian pulse[58] for this has obvious analogies with a spin echo sequence made up of a soft $\pi/2$ pulse followed immediately by a soft $\pi$ pulse. Much better performance is obtained by amplitude modulated soft pulses designed specifically for this purpose. The BURP family of pulses, analysed in detail in § 5.2.6, achieves pure phase over an appreciable operating bandwidth.[59]

Magnetization trajectories provide considerable insight into the operation of self-focusing pulses. If relaxation can be neglected, they move on the surface of the unit sphere, and when they are in the hemisphere where $y$ is negative, resonance offset causes divergence in one sense, but this is compensated by the opposite divergence when in the $+y$ hemisphere. We shall see in § 5.2.6 how a simulation of magnetization trajectories can be used to illustrate the refocusing effect of BURP pulses.

## 4.4 **Discussion**

This chapter has set out to show that the spin echo concept pervades the whole of magnetic resonance methodology. A whole new science has grown up, based on the manipulation of nuclear spins, but the inspiration clearly comes from the early work of Hahn,[1] Carr and Purcell,[4] and Meiboom and Gill.[5] What has followed is essentially a set of ingenious extensions of their ideas, aided an abetted by the availability of computer-controlled spectrometers that make spin choreography[60] a relatively straightforward reprogramming operation, rather than a risky adventure with a hot soldering iron.

# References

1. E. L. Hahn, *Phys. Rev.* **80**, 580 (1950).
2. J. S. Waugh, *Pulsed Magnetic Resonance: NMR, ESR and Optics* Ed: D. M. S. Bagguley, Clarendon Press, Oxford, 1992, Chapter 6.
3. A. Abragam, *The Principles of Nuclear Magnetism*, Clarendon Press, Oxford, 1961.
4. H. Y. Carr and E. M. Purcell, *Phys. Rev.* **94**, 630 (1954).
5. S. Meiboom and D. Gill, *Rev. Sci. Instr.* **29**, 688 (1958).
6. I. Solomon, *Phys. Rev. Lett.* **2**, 301 (1959).
7. R. Bradford, C. Clay and E. Strick, *Phys. Rev.* **84**, 157 (1951).
8. H. Y. Carr, *Phys. Rev.* **112**, 1693 (1958).
9. R. R. Ernst and W. A. Anderson, *Rev. Sci. Instr.* **37**, 93 (1966).
10. R. Freeman and H. D. W. Hill, *J. Magn. Reson.* **4**, 366 (1971).
11. A. Bax, A. F. Mehlkopf and J. Smidt, *J. Magn. Reson.* **35**, 373 (1979).
12. A. Bax, R. Freeman and G. A. Morris, *J. Magn. Reson.* **43**, 333 (1981).
13. E. L. Hahn and D. E. Maxwell, *Phys. Rev.* **88**, 1070 (1952).
14. R. Freeman and H. D. W, Hill, in *Dynamic Nuclear Magnetic Resonance Spectroscopy*, Eds: L. M. Jackman and F. A. Cotton, Academic Press, New York, 1975, Chapter 5.
15. R. Freeman, T. H. Mareci and G. A. Morris, *J. Magn. Reson.* **42**, 341 (1981).
16. F. K. Pei and R. Freeman, *J. Magn. Reson.* **48**, 318 (1982).
17. G. A. Morris and R. Freeman, *J. Am. Chem. Soc.* **101**, 760 (1979).

18.  P. L. Rinaldi and N. J. Baldwin, *J. Magn. Reson.* **104**, 5791 (1982).

19.  D. P. Burum and R. R. Ernst, *J. Magn. Reson.* **39**, 163 (1980).

20.  O. W. Sørensen and R. R. Ernst, *J. Magn. Reson.* **51**, 477 (1983).

21.  G. Bodenhausen and D. J. Ruben, *Chem. Phys. Lett.* **69**, 185 (1980).

22.  L. Müller, *J. Am. Chem. Soc.* **101**, 4481 (1979).

23.  A. Bax, R. H. Griffey and B. L. Hawkins, *J. Magn. Reson.* **55**, 301 (1983).

24.  A. Bax, R. Freeman and S. P. Kempsell, *J. Am. Chem. Soc.* **102**, 4849 (1980).

25.  A. Bax, R. Freeman and T. A. Frenkiel, *J. Am. Chem. Soc.* **103**, 2102 (1981).

26.  J. R. Garbow, D. P. Weitekamp and A. Pines, *Chem. Phys. Lett.* **93**, 514 (1982).

27.  S. Wimperis and R. Freeman, *J. Magn. Reson.* **58**, 348 (1984).

28.  W. P. Aue, J. Karhan and R. R. Ernst, *J. Chem. Phys.* **64**, 4226 (1976).

29.  K. Nagayama, P. Bachmann, K. Wüthrich and R. R. Ernst, *J. Magn. Reson.* **31**, 133 (1977).

30.  B. Blümich and D. Ziessow, *J. Magn. Reson.* **49**, 151 (1982).

31.  A. J. Shaka, J. Keeler and R. Freeman, *J. Magn. Reson.* **56**, 294 (1984).

32.  P. Xu, X. L. Wu and R. Freeman, *J. Am. Chem. Soc.* **113**, 3596 (1991).

33.  P. Xu, X. L. Wu and R. Freeman, *J. Magn. Reson.* **95**, 132 (1991).

34.  M. H. Woodley and R. Freeman, *J. Magn. Reson. A.* **109**, 103 (1994).

35.  M. H. Woodley and R. Freeman, *J. Magn. Reson. A.* **111**, 225 (1994).

36.  A. A. Maudsley, A. Wokaun and R. R. Ernst, *Chem. Phys. Lett.* **55**, 9 (1978).

37.  T. H. Mareci and R. Freeman, *J. Magn. Reson.* **48**, 158 (1982).

38.  W. P. Aue, E. Bartholdi and R. R. Ernst, *J. Chem. Phys.* **64**, 2229 (1976).

39.  A. Bax, P. G. de Jong, A. F. Mehlkopf and J. Smidt, *Chem. Phys. Lett.* **69**, 567 (1980).

40.  P. Barker and R. Freeman, *J. Magn. Reson.* **64**, 334 (1985).

41.  R. E. Hurd, *J. Magn. Reson.* **87**, 3442 (1990).

42.  M. H. Levitt and R. Freeman, *J. Magn. Reson.* **33**, 473 (1979).

43.  A. J. Shaka, J. Keeler and R. Freeman, *J. Magn. Reson.* **53**, 313 (1983).

44.  R. Freeman, S. P. Kempsell and M. H. Levitt, *J. Magn. Reson.* **35**, 447 (1979).

45.  M. H. Levitt and R. Freeman, *J. Magn. Reson.* **43**, 502 (1981).

46.  M. H. Levitt, R. Freeman and T. Frenkiel, *J. Magn. Reson.* **50**, 157 (1982).

47.  J. S. Waugh, *J. Magn. Reson.* **50**, 30 (1982).

48.  A. J. Shaka, J. Keeler, T. Frenkiel and R. Freeman, *J. Magn. Reson.* **52**, 335 (1983).

49.  A. J. Shaka, P. B. Barker and R. Freeman, *J. Magn. Reson.* **64** , 547 (1985).

50.  N. Sunitha Bai, N. Hari and R. Ramachandran, *J. Magn. Reson. A.* **106**, 241 (1994).

51.  T. Fujiwara, T. Anai, N. Kurihara and K. Nagayama, *J. Magn. Reson. A.* **104**, 103 (1993).

52.  M. R. Bendall, *J. Magn. Reson. A.* **112**, 126 (1995).

53.  E. Kupče and R. Freeman, *J. Magn. Reson. A.* **115**, 273 (1995).

54.  E. Kupče and R. Freeman, *J. Magn. Reson. A.* **117**, 246 (1995).

55.  G. A. Morris. G. L. Nayler, A. J. Shaka, J. Keeler and R. Freeman, *J. Magn. Reson.* **58**, 155 (1984).

56.  H. Kessler, S. Mronga and G. Gemmecker, *Magn. Reson. Chem.* **29**, 527 (1991).

57.  R. Freeman, *Chem. Rev.*, **91**, 1397 (1991).

58.  L. Emsley and G. Bodenhausen, *J. Magn. Reson.* **82**, 221 (1989).

59.  H. Geen and R. Freeman, *J. Magn. Reson.* **93**, 93 (1991).

60.  R. Freeman, Spin Choreography in *Pulsed Magnetic Resonance: NMR, ESR, and Optics. A recognition of E. L. Hahn.* Ed. D. M. S. Bagguley, Clarendon Press, Oxford, 1992.

# 5

# Soft radiofrequency pulses

## 5.1 Introduction

It is sometimes fruitful to violate accepted tenets of NMR spectroscopy, for example the essentially universal use of short intense radiofrequency pulses for excitation. We may define such hard pulses as those where $\gamma B_1/2\pi >> \Delta F$, where $\Delta F$ is the total range of chemical shifts of the nucleus in question. In contrast, when there is any degree of frequency-selectivity, the pulses are said to be soft. We shall use this as a general term that comprises line-selective, multiplet-selective and band-selective pulses, with their different pulse widths, usually in the range between 1 and $10^{-4}$ s.

Selective NMR experiments are not new. The old continuous-wave spectrometers readily accommodated experiments where a particular line or group of lines was singled out while the remainder were rejected. Most double-resonance or double-quantum studies fell into this category. Alexander[1] introduced a soft pulse experiment as early as 1961, and selective spin–spin and spin–lattice relaxation measurements were performed some time before the Fourier transform revolution got under way. What really inhibited soft pulse investigations was the realization in the mid-1970s that two-dimensional spectroscopy provides the same information for all chemical sites in a single experiment (§ 8.2). These 2D techniques use the evolution period to label the NMR frequencies in the $F_1$ dimension, and hence behave as an entire family of selective excitation experiments. So successful were these new methods that practising spectroscopists did not seriously search for alternatives until about a decade later. The incentive was a growing realization that multidimensional spectroscopy could be quite wasteful of spectrometer time if only a few pieces of information were required to solve the problem in hand. Unfortunately the 2D sledgehammer was always handy and the nut suffered its usual fate.

### 5.1.1 Magnetization trajectories

The action of a soft pulse is most easily visualized in terms of the vector model of nuclear magnetization (§ 2.5.2). At Boltzmann equilibrium the magnetization is represented by a vector $M_0$ aligned along the $+z$ axis of the rotating frame. For most purposes relaxation during the pulse can be neglected.

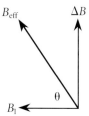

This vector rotates about an effective field $B_{eff}$, the resultant of the radiofrequency field intensity $B_1$ and the resonance offset $\Delta B$, about an axis which is tilted at an angle $\theta$ with respect to the $+x$ axis:

$$\theta = \arctan(\Delta B/B_1) \qquad \textbf{5.2}$$

$$B_{eff} = (B_1{}^2 + \Delta B^2)^{\frac{1}{2}} \qquad \textbf{5.2}$$

$$\alpha = \alpha_0 B_{eff}/B_1 \qquad \textbf{5.3}$$

where $\alpha_0$ is the nominal flip angle on resonance. At exact resonance ($\Delta B = 0$) with the pulse duration $\tau$ adjusted to the condition $\gamma B_1 \tau = \pi/2$ radians, $M$ is simply rotated about the $+x$ axis from $+z$ to $+y$, giving a pure absorption-mode signal.

A typical family of magnetization trajectories is shown in Figure 5.1. For moderate offsets where $\Delta B < B_1$, the component in the $xy$ plane remains fairly constant, but the phase error $\phi = \arctan (M_x/M_y)$ is appreciable. In this situation $\phi$ is an approximately linear function of $\Delta B$, and it may compensated after detection by standard software, but in more complex experiments the phase gradient may not be acceptable. At larger offsets ($\Delta B > B_1$) the trajectories begin to loop back towards

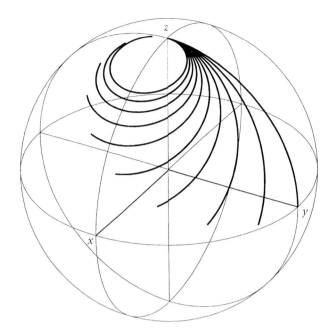

**Figure 5.1** Magnetization trajectories calculated for a 1 ms soft rectangular pulse with $\gamma B_1/2\pi$ = 250 Hz. As the resonance offset increases from 0 to 900 Hz in 100 Hz steps, the effective field becomes progressively tilted in the $xz$ plane. Whereas the first trajectory (zero offset) terminates on the $+y$ axis (pure absorption) there is an increasing contribution of dispersion mode as the offset increases. For the final trajectory the detected signal is almost zero.

the $+z$ axis and the detected $xy$ component of the signal after the pulse falls off quite rapidly. At still larger offsets ($\Delta B > 4B_1$) the magnetization vector begins to execute cyclic trajectories and these get closer and closer to the $+z$ axis as $\Delta B$ increases. This generates oscillatory sidelobes on the detected response, and although they get weaker at large offsets, they still have appreciable amplitudes some considerable distance from resonance (Figure 5.2). Sidelobes are a disaster for many selective excitation experiments, since they can extend well into regions where we would prefer to have no excitation at all. A monotonically decreasing function is much more desirable in this context.

## 5.2  Pulse shaping

### 5.2.1  Gaussian pulses

A soft pulse is derived from a hard pulse by reducing the $B_1$ intensity and by increasing the pulse duration accordingly. In the approximation that the spin response is linear, the overall frequency-domain excitation pattern can be described by a sinc function, the Fourier transform of the rectangular envelope. This suggests

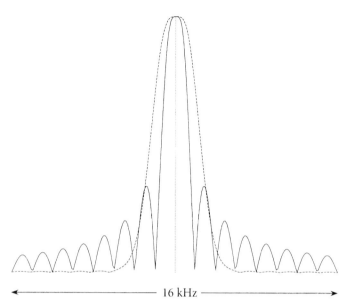

16 kHz

**Figure 5.2** Frequency-domain excitation profiles (absolute-value mode) of a rectangular pulse (full line) and a Gaussian pulse (dotted line) of the same duration (1 ms). The rectangular pulse creates an extensive series of sidelobe responses at offsets where the Gaussian excitation is quite negligible. The slight undulation in the tails of the Gaussian excitation arises from truncation of the pulse envelope at the 2% level.

that it is the discontinuities in the pulse envelope that cause the trouble. It was found that sidelobes can be eliminated by rounding off the sharp leading and trailing edges of the pulse, that is to say, by shaping the envelope in some suitable manner. One of the most effective shaping functions for this purpose is the Gaussian envelope,[2] given by the expression

$$S(t) = \exp[-a(t - t_0)^2]$$                 **5.4**

where $t_0$ is the centre of the pulse envelope. In the linear approximation (see § 5.2.4), the frequency-domain excitation pattern (absolute-value mode) is also a Gaussian and this has the useful property that the excitation falls off very fast in the tails. This is a consequence of a general theorem of Fourier transforms[3]—if a function $S(t)$ can be differentiated $k$ times before the derivative exhibits a discontinuity, its Fourier transform $S(f)$ falls off inversely as the $k$th power of frequency in the tails. The off-resonance behaviour is therefore far better than that of a rectangular pulse of the same duration. This can be appreciated from the calculated excitation profiles (Figure 5.2). Note that at all offsets beyond the first sidelobe of the sinc-like function, the response from the equivalent Gaussian pulse is negligible. This does not mean that the Gaussian pulse has no effect on spins at these offsets. It actually carries the magnetization vectors away from the $+z$ axis but then brings them back again, completing a closed loop that resembles a tear drop. Only at extremely large offsets can it be said that the soft Gaussian pulse has no effect at all.

The time-domain envelope of a Gaussian has to be truncated somewhere in the tails, usually in the region of 2% to 5% of the maximum intensity. If the truncation is too severe, it introduces some sinc function character into the excitation profile. The slight undulations visible in Figure 5.2 (dashed curve) arise from truncation of the tails of the time-domain envelope at 2% of maximum amplitude. Typical magnetization trajectories for a Gaussian pulse are illustrated in Figure 5.3. In common with most soft pulses, the Gaussian pulse excites a spectrum with an appreciable phase gradient, because magnetization that is off-resonance rotates about a tilted effective field. This is sometimes represented in terms of a time-shift, as if the divergence of vectors started at a point within the pulse envelope near its centre.

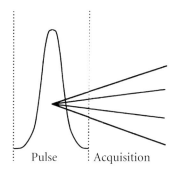

Pulse      Acquisition

Normally the resulting phase gradient can be compensated by the usual linear phase correction software. The excitation profile is then quite suitable for many types of selective excitation experiment (Figure 5.2).

A useful extension of simple Gaussian shaping is to multiply by an even-order polynomial, giving (for example) the Hermite function,[4]

$$S(t) = [1 - 0.8(t - t_0)^2] \exp[-a(t - t_0)^2] \qquad \textbf{5.5}$$

This generates a bull-nose excitation pattern with a flatter top than that obtained with a pure Gaussian.

## 5.2.2 Half-Gaussian pulses

If the purpose of soft pulse excitation is to initiate coherence transfer, it is usually the absorption mode response that is important and the form of the dispersion-mode profile is less relevant. In such cases it may be worth considering the use of a soft pulse shaped as the rising half of a Gaussian curve,[5] given by

$$S(t) = \exp[-a(t - t_0)^2] \qquad \text{for } t \leq t_0$$
$$S(t) = 0 \qquad\qquad\qquad \text{for } t > t_0 \qquad \textbf{5.6}$$

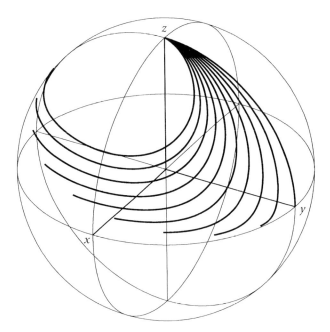

**Figure 5.3** Magnetization trajectories calculated for a 1 ms soft Gaussian pulse with $\gamma B_1/2\pi$ = 250 Hz. The resonance offset increases from 0 to 900 Hz in 100 Hz steps. This diagram may be compared with that for a rectangular pulse in Figure 5.1. The first trajectory gives pure positive absorption, while the final one corresponds to the first negative absorption lobe.

The absorption-mode excitation pattern approximates a Gaussian curve in the frequency domain with its characteristic rapid monotonic fall-off in the tails and no sidelobe responses. This may seem surprising in view of the large discontinuity at $t = t_0$. One rationalization is to picture the half Gaussian as the superposition of a full Gaussian (symmetrical about $t = t_0$) and an antisymmetrical version:

$$S(t) = +\exp[-a(t - t_0)^2] \quad \text{for } t \leq t_0$$
$$S(t) = -\exp[-a(t - t_0)^2] \quad \text{for } t > t_0 \qquad \textbf{5.7}$$

If we can assume that the absorption mode excitation profile originates with the symmetrical part of the pulse while the dispersion mode arises from the antisymmetrical part, this suggests an explanation. The discontinuity at $t = t_0$ simply creates a very broad dispersion-mode excitation profile. If necessary, these undesirable dispersion components may be purged by recording two scans; in the first we append a hard $+(\pi/2)_y$ pulse, and in the second we use a hard $-(\pi/2)_y$ pulse.[6] This returns the $x$ components of magnetization to the $\pm z$ axis.

## 5.2.3 **Pulse shape design**

Magnetic resonance imaging and *in vivo* spectroscopy have made the greatest use of soft pulses that are shaped to give a specific excitation profile, and a major part of the literature on pulse design has emerged from this field. For example, the key initial step in many imaging applications is slice selection—the excitation of a thin slice normal to the applied field gradient, leaving everything else essentially unperturbed. To obtain a signal from the selected slice that has the minimum contamination by signals from adjacent regions, the excitation profile should have sharp edges. If, in addition, the excitation within the slice is required to be uniform, the ideal profile would be the so-called top-hat function. Much use has been made of sinc-function pulse envelopes in imaging applications because the corresponding excitation profile is roughly rectangular. Usually the sinc function is truncated symmetrically in each tail at a suitable zero-crossing point.

The purpose of soft pulse design is to achieve some particular target profile for the excitation. This may include a region of very uniform excitation flanked by other regions of negligible excitation, the top-hat profile. Or it may prove necessary to suppress all dispersion-mode components, leaving only pure absorption signals. Alternatively it may require the inversion of signals in one frequency range, with normal signals in another range, the antisymmetric Janus excitation profile (§ 5.5.2). Rather surprisingly, certain soft pulse schemes can be more effective for wideband excitation than a single hard radiofrequency pulse, and it is relatively easy to insert a rejection notch for solvent suppression purposes (§ 5.5.1). In practice the target profile may be very complex indeed, as in template excitation described in § 5.5.3.

### 5.2.4 Linearity of the NMR response

Linearity requires that the excited signal be directly proportional to the pulse flip angle $\alpha$. Under these conditions, the frequency-domain excitation profile is given by the Fourier transform of the pulse shape. This approximation is only justified for flip angles small enough that $\sin\alpha \approx \alpha$, so it already breaks down for a $\pi/2$ pulse, and is quite invalid for $\pi$ pulses. Nevertheless, there is a temptation to invoke Fourier transform arguments to give a rough first approximation to the excitation profile. By exploiting the inverse Fourier transform, one could then derive the approximate pulse envelope required to generate a given form of excitation spectrum. As long ago as 1973, Tomlinson and Hill[7] adopted this principle in their paper on tailored excitation. They proposed that the desired frequency-domain pattern (for example the top-hat profile mentioned above) could be expressed as a histogram, Fourier transformed and then used to phase-modulate a sequence of hard radiofrequency pulses.[7,8]

We must exercise some care in deciding to what extent a particular soft pulse has a linear response. It is not simply a question of a small (nominal) flip angle $\alpha_0$. Consider, for example, a rectangular soft pulse with $\alpha_0 = \pi/2$ radians. Near resonance the NMR response is appreciably nonlinear; witness the considerable deviation of the response calculated by Fourier transformation compared with that obtained via the Bloch equations (Figure 5.4). However, at appreciable offsets the two curves are virtually indistinguishable, implying that the linear approximation is justified. In

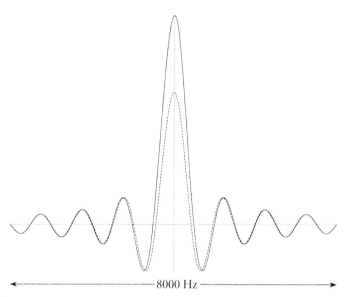

── 8000 Hz ──

**Figure 5.4** The absorption-mode excitation pattern calculated for a 1 ms soft rectangular pulse assuming a linear response (full line) and the Bloch equations (dotted line). Note that at large offsets the two curves are virtually indistinguishable. The intensities at exact resonance are in the ratio $\pi/2:1$.

terms of magnetization trajectories, starting with equilibrium magnetization aligned along $+z$, linearity requires that the subsequent motion be confined to regions near the top of the unit sphere (within, say, the Arctic circle). When magnetization vectors stray nearer to the equator, significant nonlinearities occur, and near the south pole the response is very far from linear.

Linearity also implies that two or more soft pulses applied simultaneously produce an overall effect that is simply the superposition of the actions of the individual pulses; they do not interfere with each other (see § 5.3.7). This is important for considering the polychromatic pulses described in § 5.5.1. One of the first composite pulses for solvent suppression was Redfield's 2–1–4 sequence[9] which implicitly invoked this superposition principle to achieve a null response at the water frequency.

In general the Fourier relationship should be treated only as a rough guide to problems of soft pulse design. Indeed, we shall see below that the nonlinear terms can be of crucial importance when we need a pulse with special properties, for example, one that produces pure absorption-mode signals right across the operating band. The $3\pi/2$ Gaussian soft pulse[10] involves large deviations from linearity, allowing it to refocus some magnetization vectors close to resonance. The pure phase pulses[11] (§ 5.2.6) operate in a nonlinear regime, and achieve both refocusing and a uniform response.

### 5.2.5 **Theoretical calculations**

The most general theory for calculating the effect of a shaped pulse employs the Liouville–von Neumann equation to describe the time evolution of the density operator $\sigma$:

$$\frac{d\sigma}{dt} = -\mathrm{i}[\mathcal{H}, \sigma] \tag{5.8}$$

where the Hamiltonian $\mathcal{H}$ is expressed in angular frequency units. For the special case of a time-independent Hamiltonian, the solution is

$$\sigma(t) = \exp(-\mathrm{i}\mathcal{H}t)\,\sigma(0)\,\exp(+\mathrm{i}\mathcal{H}t) \tag{5.9}$$

We can apply this solution if the amplitude and phase remain constant throughout the pulse, that is to say if the motion is described by the Hamiltonian

$$\mathcal{H} = \Delta\omega I_z + \omega_1(I_x \cos\psi + I_y \sin\psi) \tag{5.10}$$

where $\omega_1 = \gamma B_1$ is the intensity of the radiofrequency field, and $\Delta\omega = \gamma\Delta B$ is the offset from exact resonance. For this purpose the soft pulse envelope is broken down into a histogram of very short periods ($\tau$) of constant radiofrequency phase ($\psi$) and intensity ($\omega_1$). Each step may be described by a propagator, such as

$$U = \exp(-\mathrm{i}\mathcal{H}\tau) \tag{5.11}$$

which is independent of the initial state of the system, so successive propagators may be cascaded to calculate the overall effect. A general propagator causes a rotation about the effective radiofrequency field ($B_{\text{eff}}$) through an angle $\alpha$ about an axis $n_0$:

$$U = \exp[-\mathrm{i}\alpha(I \cdot n_0)] \tag{5.12}$$

In terms of the unit vectors $i, j$ and $k$, the rotation axis is given by

$$n_0 = [\omega_1\,(i\cos\psi + j\sin\psi) + k\Delta\omega]\,/\omega_{\text{eff}} \tag{5.13}$$

where $\omega_{\text{eff}} = \gamma B_{\text{eff}}$. For the entire soft pulse, modulated in amplitude and phase, we can write an overall propagator which is a product of these unitary transformations:

$$U_0 = U_N\,U_{N-1}\ldots U_2\,U_1 \tag{5.14}$$

This is used to predict the net effect on the density operator over the duration $T$ of the soft pulse:

$$\sigma(T) = U_0\sigma(0)\,U_0^{-1} \tag{5.15}$$

Repetition of these calculations at all resonance offsets of interest gives the frequency-domain excitation profile. Alternative treatments employ the concept of the average Hamiltonian (possibly approximated by the Magnus expansion[12]). The simplest treatment is by way of the Bloch equations, but we must bear in mind that they are strictly applicable only to the case of isolated spin-$\frac{1}{2}$ nuclei; spin–spin coupling is not taken into account.

What is actually needed is the reverse calculation—the prediction of a suitable pulse envelope that will give the desired target excitation profile—and this has to be achieved by indirect methods. Usually this involves an iterative scheme in which the pulse shape parameters are varied to give the best fit of a trial excitation profile to the target profile. Many different optimization routines have been explored, including simulated annealing,[13] artificial neural networks,[14] and evolutionary algorithms.[15] A crucial first step is to find an efficient method for defining the pulse shape, since ideally this should be a continuous, smooth function and be defined by a relatively small number of variable parameters. A finite Fourier series satisfies these requirements rather well, for example

$$\gamma B_1(t) = \omega \left\{ A_0 + \sum_{n=1} [A_n \cos(n\omega t) + B_n \sin(n\omega t)] \right\} \qquad \textbf{5.16}$$

where $\omega = 2\pi/T$ and $T$ is the pulse duration. The higher-order Fourier coefficients can be introduced one at a time so as to keep the shape as simple as possible. For practical reasons, very convoluted shapes should be avoided since they tend to impose unreasonable demands on the pulse shaping hardware and the radiofrequency amplifiers.

## 5.2.6 Pure-phase pulses

Once the sidelobe problem has been solved it is evident that soft pulse schemes have a second serious drawback—the phase gradient mentioned in § 5.2.1. It arises because magnetization vectors at different offsets rotate about different tilted effective fields and, instead of reaching the $+y$ axis (pure absorption), spread out in the $xy$ plane. It is as if they had originated from a virtual focus somewhere within the pulse width. To a fairly crude approximation, a soft pulse may be factored into a hard pulse followed by a period of free precession during which this phase error accumulates. When we are only concerned with excitation we may describe the frequency-domain profile in terms of the absolute-value mode, assuming that a linear phase gradient will be corrected by the spectrometer phase setting procedure after data acquisition. For most other applications the phase error is a serious problem. In magnetic resonance imaging applications it can be refocused by reversing the applied field gradient, but in high resolution work this remedy is not available. A $\pi$ pulse could be used for refocusing, but the effects of spin–spin coupling would still remain.

Pulses which correct this phase dispersion are called pure-phase pulses and they rely for their success on the nonlinearity of the spin response. Instead of allowing the magnetization trajectories to diverge, they bring them to a focus along the $+y$ axis by a process analogous to that of spin-echo formation. Of course, this self-focusing property can only be effective over a certain specified operating bandwidth. It is useful to have one further feature, a top-hat excitation profile. More precisely, we aim for an excitation pattern that follows the form of a trapezium—a central region ($\Delta f$) of uniform excitation, flanked by two narrow transition regions, and negligible excitation everywhere else. Rather surprisingly, all these properties

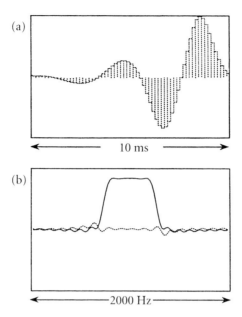

(a)

10 ms

(b)

2000 Hz

**Figure 5.5** (a) The 64-step histogram describing the radiofrequency envelope of the soft E-BURP-2 self-focusing pulse. (b) The corresponding frequency domain excitation profiles for absorption (full curve) and dispersion (dotted curve). The appropriate Fourier coefficients are set out in Table 5.1. The bandwidth scales as the inverse of the pulse duration.

can be achieved by suitable amplitude modulation of the pulse intensity. An entire family of these BURP pulses[11] has been designed by the simulated annealing algorithm and they have proved their worth in many soft pulse applications where a uniform response and pure phase spectra are important. Figure 5.5 shows the pulse envelope and frequency-domain excitation profile of the E-BURP-2 pulse, used for selective excitation. This has been calculated for a 10 ms pulse, giving an effective bandwidth of approximately 400 Hz, but the frequency range can be scaled simply by changing the pulse duration $T$. The relevant Fourier coefficients are given in Table 5.1.

We can gain some insight into the focusing process if we imagine the soft E-BURP-2 pulse as being broken down into a repeated sequence of small angle rotations about the $+x$ axis, interspersed with small angle precessions about the $+z$ axis. When the magnetization trajectory is carried into the hemisphere where $y$ is negative, free precession occurs and causes divergence, but there is a compensating precession in the opposite sense when the magnetization vector is in the $+y$ hemisphere. We can illustrate this repeated divergence and convergence by following a bundle of trajectories covering a small range of offsets (Figure 5.6). The ribbon expands and contracts several times, reaching a well-defined focus at the end of the trajectory. The beauty of the E-BURP scheme is that the refocusing occurs over an appreciable range of offsets.

Outside the operating bandwidth, the trajectories return to points close to the

**Table 5.1** Fourier coefficients[a] for the
E-BURP-2 band-selective excitation pulse

| | | | |
|---|---|---|---|
| $A_0$ | +0.26 | | |
| $A_1$ | +0.91 | $B_1$ | −0.12 |
| $A_2$ | +0.45 | $B_2$ | −1.79 |
| $A_3$ | −1.31 | $B_3$ | +0.01 |
| $A_4$ | −0.12 | $B_4$ | +0.41 |
| $A_5$ | +0.03 | $B_5$ | +0.08 |
| $A_6$ | +0.01 | $B_6$ | +0.07 |
| $A_7$ | +0.06 | $B_7$ | +0.01 |
| $A_8$ | +0.01 | $B_8$ | −0.04 |
| $A_9$ | −0.02 | $B_9$ | −0.01 |
| $A_{10}$ | −0.01 | $B_{10}$ | −0.00 |

[a] To be used in conjunction with the expression:

$$\gamma B_1(t) = \omega\{A_0 + \sum_{n=1}^{10} [A_n\cos(n\omega t) + B_n\sin(n\omega t)]\}$$

where $\omega = 2\pi/T$ and $T$ is the soft pulse duration.

$+z$ axis and the overall excitation is negligible, although the excursions during the pulse can still be appreciable. In many applications, a high degree of out-of-band suppression is just as important as uniformity within the excitation band. Recent refinements of BURP pulses have improved both the uniformity[16] and the degree of suppression.[17]

Note that a pulse designed to incorporate this relatively wideband focusing feature is fairly sensitive to the setting of the nominal flip angle and careful calibration may be required. Otherwise the performance degrades significantly with respect to phase purity, uniformity of excitation and suppression of off-resonance signals. The design of the E-BURP pulses assumes an initial condition with magnetization along the $+z$ axis (Boltzmann equilibrium) or the $-z$ axis (population inversion). A time-reversed version of the E-BURP pulse takes magnetization from the $+y$ axis back to the $+z$ axis,[18] and can be used as the second soft pulse in band-selective correlation spectroscopy (soft-COSY).

When we require a $\pi/2$ rotation from an arbitrary initial condition a different form of pure-phase pulse is required—the universal-rotation pulse (U-BURP).[11] Other members of the BURP family rotate magnetization vectors through $\pi$ radians. The simplest is a spin inversion pulse, I-BURP, which is designed to act on a magnetization vector initially aligned along the $\pm z$ axis. If the pulse is required for refocusing vectors in the $xy$ plane to give a spin echo, it should be a general refocusing pulse (RE-BURP).[11]

The BURP pulses are employed in multiplet-selective or band-selective experiments. The product of the effective bandwidth $\Delta f$ Hz and the pulse duration $T$ s is a constant, approximately equal to four, so the bandwidth can be scaled up by decreasing the pulse duration accordingly. Naturally the transition regions also

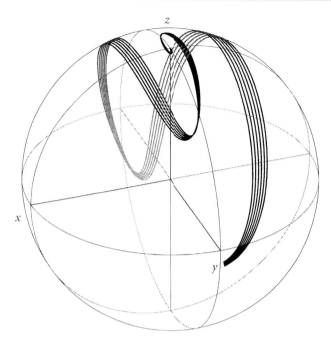

**Figure 5.6** A family of magnetization vectors calculated for a 1 ms E-BURP-2 pulse with a flip angle slightly greater than $\pi/2$ radians. The resonance offsets run from 500 Hz to 700 Hz, and bunching indicates a refocusing effect. The lightly shaded trajectories are on the far side of the unit sphere.

increase in proportion to the bandwidth. When a BURP pulse is used to excite a spin multiplet, all the component lines remain in pure absorption; there is negligible distortion attributable to spin–spin coupling effects during the soft pulse. However if two coupled sites are excited simultaneously with E-BURP pulses, several new effects ensue, including considerable phase dispersion and the excitation of double-quantum coherence (§ 5.3.6).

In general soft pulses should be as short as possible. If there is appreciable spin–spin relaxation during the pulse there is an inevitable loss of signal, and, more seriously, a distortion of the excitation profile.[19] However the design of the BURP pulses may be reoptimized[20–22] to allow for relaxation and retain the desired top-hat pattern and the pure absorption-mode excitation. Recently BURP pulses have been redesigned for applications where an extremely flat excitation profile is required[16] and for extremely low out-of-band excitation.[17]

## 5.3 **Practical considerations**

### 5.3.1 **DANTE sequence**

At the time of writing, not all high resolution spectrometers are suitably equipped for generating shaped soft pulses. It may not be feasible to switch the radiofrequency level from the hard pulse to the soft pulse setting. Sometimes the transmitter frequency cannot be adjusted finely enough for certain selective experiments, or the frequency does not settle rapidly enough after being switched. Finally, waveform generators and linear amplifiers suitable for pulse shaping may not be available. The DANTE sequence[23–25] was devised to circumvent these problems.

It is made up of a sequence of $N$ hard radiofrequency pulses of very short duration separated by regular periods ($\tau$) for free precession. At exact resonance no free precession occurs and the $N$ pulses have a cumulative effect on the magnetization vector, carrying it from the $+z$ to the $+y$ axis in exactly the same manner as a single pulse with the same total flip angle. Off-resonance, the magnetization executes a zig-zag path made up of alternating periods of rotation about the $x$ axis and rotation about the $z$ axis (§ 2.5.1). This trajectory approximates that obtained with a soft pulse of the same duration. If $N$ is made sufficiently large, the individual rotation and precession angles of the DANTE sequence are very small and the zig-zag trajectory deviates very little from the smooth curve obtained with a single soft pulse. If the DANTE sequence has the same overall duration ($N\tau$) as the soft pulse, they both have the same frequency selectivity. Figure 5.7 illustrates the trajectory of a DANTE sequence of 11 pulses for a tilt angle $\theta \approx 1$ radian. It has an overall effect very similar to that of the corresponding soft-pulse trajectory for an offset of 400 Hz shown in Figure 5.1.

One may think of the DANTE sequence as being derived from the equivalent soft pulse (represented as a histogram of time segments) by converting each soft pulse segment into a hard pulse plus a free precession interval $\tau$. The correct way to do this is to place the hard pulse in the centre, flanked by two intervals of $\tau/2$. Otherwise the excitation profile differs slightly from that of the equivalent soft pulse, having more prominent undulations and a baseline offset.[26] (Alternatively the first and last pulses of the DANTE sequence should be halved in pulsewidth.) These distortions are more serious the smaller the number of pulses in the DANTE sequence. Note that the trajectory in Figure 5.7 has been calculated for the case that the first and last pulses are halved.

Since it employs hard pulses, a DANTE sequence avoids having to switch the transmitter level. In experiments that employ both hard and soft pulses, DANTE has the advantage that its pulses are phase coherent with the hard pulses (since they use the same transmitter). The pulse envelope can be shaped simply by modulating the pulse widths, keeping the transmitter level constant. Some care is required with these modulated sequences to ensure that individual hard pulses do not become too short, say less than a microsecond, since switching transients can cause a phase glitch.[27] A trick which largely cancels the phase glitch is to make up a pulse of very small flip angle $\varepsilon$ as a composite of two wider forward and backward pulses differing in flip angle by $\varepsilon$ radians.

The excitation profile of DANTE differs in one important respect from that of the equivalent soft pulse; there are sideband responses at multiples of the pulse repetition rate ($1/\tau$). These arise because free precession through a small angle $\phi$ radians gives the same result as free precession through ($\phi \pm 2n\pi$) radians. Conse-

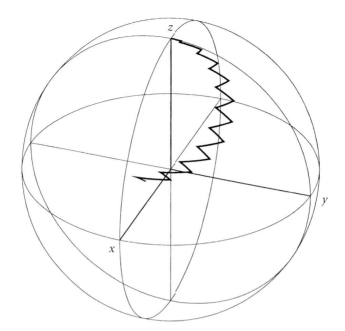

**Figure 5.7** Magnetization trajectories calculated for a 1 ms DANTE sequence consisting of eleven pulses (*x* rotations) separated by 10 equal intervals of free precession (*z* rotations). The first and last pulses are of half intensity. The result is very closely equivalent to that of a 1 ms soft pulse with $\gamma B_1/2\pi = 250$ Hz at an offset of 400 Hz (the fifth trajectory in Figure 5.1).

quently the first sidebands appear at resonance offsets of $\pm 1/\tau$ Hz. It is therefore important to work with a sufficiently high pulse repetition rate that the undesirable sideband responses are well outside the spectrum of interest, or at least fall in empty regions.

By working with the first sideband rather than the centreband response we can use the repetition rate $1/\tau$ to fine-tune the frequency of excitation. This is useful if the transmitter frequency itself cannot be switched rapidly enough or if it has only coarse tuning. It was the sideband response that gave DANTE its name. In Dante's *Purgatorio* the lost souls are condemned to a mountain which comprises seven ledges corresponding to the seven deadly sins. A complete circumnavigation of each ledge is required before promotion to the next highest level and eventual escape. With a trivial time-reversal this is essentially the same as the trajectory of a magnetization vector in a DANTE sequence at the first sideband condition.

The DANTE sequence lends itself quite readily to simultaneous excitation at two arbitrary frequencies.[28] Two DANTE sequences with the same repetition rate $1/\tau$ Hz can be interleaved without any significant interaction between them. One sequence has all the radiofrequency pulses of the same phase, the other has the pulse phase incremented linearly with time in steps of $\Delta\phi$ radians (Figure 5.8) and behaves as if the transmitter frequency had been shifted by $\Delta\phi/(2\pi\tau)$ Hz. The steps

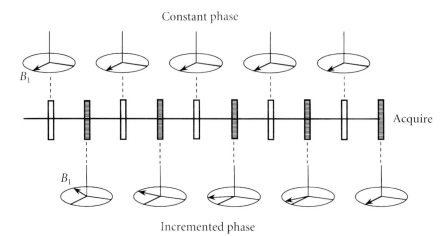

Constant phase

$B_1$

Acquire

$B_1$

Incremented phase

**Figure 5.8** A sequence for simultaneous soft pulse excitation at two arbitrary frequencies. There are two interleaved DANTE sequences, one of which (shaded) has the radiofrequency phase incremented linearly with time, creating the effect of a frequency shift. Note that both sequences are in phase just prior to signal acquisition.

$\Delta\phi$ must not be too large that aliasing occurs. We may think of each separate DANTE sequence as operating in its own moving reference frame, one rotating $\Delta\phi/\tau$ radians per second faster than the other. To ensure that both excitations are pure absorption, we must arrange that the two frames bring their $y$ axes into registration at the end of the double-DANTE sequence. Multiple excitation can be achieved with several interleaved DANTE sequences with different rates of phase ramping (see § 5.3.5).

The DANTE sequence offers one other useful feature—the NMR signal may be monitored (one data acquisition point at a time) in the $\tau$ intervals between the hard pulses, thus allowing magnetization trajectories to be traced out during the selective excitation process itself. For example, BURP pulse trajectories similar to those in the simulation of Figure 5.6 can be derived experimentally, the $z$ component of magnetization being deduced from measurements of the $x$ and $y$ components, using the fact that $M_x^2 + M_y^2 + M_z^2 = M_0^2$ if there is negligible relaxation.

## 5.3.2 Spin pinging

An alternative way to tackle the problem of frequency-dependent phase gradients is to eliminate all dispersion-mode contributions by difference spectroscopy. This can be achieved by the scheme known as spin pinging[29,30] or the related method called DANTE-Z.[31] Spin pinging starts with a hard $\pi/2$ pulse which excites $y$ magnetization. This is followed by a soft $\pi$ pulse applied alternately along the $+x$ and $+y$ axes of the rotating frame:

$$\text{hard}(\pi/2) \; \text{soft}(\pi)_x \; \text{acquire } (+)$$
$$\text{hard}(\pi/2) \; \text{soft}(\pi)_y \; \text{acquire } (-) \hspace{2cm} \textbf{5.17}$$

With the definitions of tilt angle $\theta$, flip angle $\alpha$ and effective field $B_{\text{eff}}$ introduced above, simple trigonometry shows that the dispersion-mode signal is identical after each scan, being given by

$$M_x = M_0 \sin\theta \, \sin\alpha \qquad\qquad \textbf{5.18}$$

where $M_0$ is the initial equilibrium magnetization. Consequently all dispersion-mode contributions vanish in the difference mode at all resonance offsets. Furthermore, this result is valid for all possible choices of nominal flip angle $\alpha_0$ of the soft pulse. The difference-mode absorption signal is given by

$$\Delta M_y = M_0 \cos^2\theta(1 - \cos\alpha) \qquad\qquad \textbf{5.19}$$

The detected signal at resonance (where $\alpha = \alpha_0$) is therefore a maximum for a soft pulse flip angle $\alpha_0 = \pi$ radians. The expression for the frequency-domain excitation profile is

$$M_y = \tfrac{1}{2}M_0\alpha_0{}^2\left[\frac{\sin(\alpha/2)}{\alpha/2}\right]^2 \qquad\qquad \textbf{5.20}$$

The argument $\alpha/2$ is approximately proportional to the resonance offset $\Delta B$ at large offsets but not at small offsets. Consequently Equation 5.20 represents a distorted $\mathrm{sinc}^2$ function. It has much weaker sidelobe responses than a (similarly distorted) sinc function and is therefore much better suited to selective excitation than a rectangular soft $\pi/2$ pulse (Figure 5.9(a)). If the $\pi$ pulse in the spin pinging experiment is shaped as an isosceles triangle, the excitation profile becomes

$$M_y = \tfrac{1}{2}M_0\alpha_0{}^2\left[\frac{\sin(\alpha/2)}{\alpha/2}\right]^4 \qquad\qquad \textbf{5.21}$$

and the sidelobe responses virtually disappear (Figure 5.9(b)). These expressions for the shape of the excitation profile are relatively independent of the nominal flip angle $\alpha_0$. Consequently, careful soft pulse calibration may not be necessary and even a severe spatial inhomogeneity of the radiofrequency field $B_1$ does not degrade the performance. These features make spin pinging a user-friendly selective excitation technique, although it must always be remembered that its effectiveness relies on difference spectroscopy.

### 5.3.3 Pulse shaping methods

Pulse envelopes may be defined as analytical functions, such as a Gaussian or a hyperbolic secant, or by numerical calculation from, for example, the coefficients appropriate to a particular Fourier series expansion. Several of the common pulse shapes are usually preprogrammed in the NMR spectrometer. Alternatively, one may enter an experimental shape by hand as a table of ordinates (and phases) or write a macro-routine to derive these from the Fourier coefficients. This data table is used to drive a digitally controlled waveform generator and modulator, followed by a linear radiofrequency amplifier. A digitally controlled attenuator sets

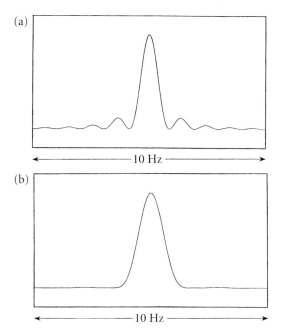

**Figure 5.9** Frequency-domain excitation profiles calculated for the spin pinging pulse sequence with (a) a rectangular soft $\pi$ pulse and (b) a triangular soft $\pi$ pulse. Note the virtual absence of sidelobes in (b). The soft pulse had a duration of 1 s.

the overall radiofrequency level and is used in the calibration procedure. Alternatively a DANTE sequence may be used, with pulse-width modulation of the hard pulses to give the equivalent of amplitude modulation of the soft pulse. This has the advantage of not relying on the linearity of the radiofrequency amplifiers.

### 5.3.4  **Choice of pulse shape**

There is no single shaped pulse that is superior to all others. Many different features must be considered before making a choice for a particular application. Magnetic resonance imaging sets a high store on an excitation profile with sharp edges, whereas high resolution experiments might be more concerned with effective suppression of off-resonance signals. One should never use a very complex pulse shape when a simpler form will suffice, since it is likely that the very complexity of the pulse envelope will entail some other drawbacks. For example, the BURP pulses have practical advantages *vis-à-vis* uniformity of excitation, relatively sharp transition regions, and lack of dispersion contributions, making them attractive for many applications, but this is at the expense of some sensitivity to pulse-width miscalibration. Simple pulse shapes are naturally more tolerant of any nonlinearities that might exist in the pulse generation equipment. A robust behaviour with respect to variation in $B_1$ (whether through poor calibration or spatial inhomogeneity) is a valuable attribute. This is one of the strengths of the spin pinging and DANTE-Z sequences.[30]

One important parameter is the dimensionless quantity $T\Delta f$, the product of the pulse duration and the effective excitation bandwidth. It should be as small as possible. While it is a simple matter to increase the bandwidth by reducing the pulse duration, we are normally interested in obtaining high selectivity without allowing the pulse to become so long that relaxation losses become appreciable. Some soft pulse applications are tolerant of frequency-dependent phase shifts, particularly if the induced phase gradient is a linear function of offset and can therefore be easily corrected; others demand pure phase, suggesting the use of E-BURP, spin pinging or a purging scheme. Many of these desirable features are far more readily achieved for $\pi$ spin inversion pulses than for $\pi/2$ excitation pulses. Sideband responses can be a distinct liability in some applications of the DANTE sequence and even a single soft pulse generates weak sidebands if the histogram used as the pulse envelope has coarse frequency increments.

### 5.3.5 Simultaneous soft pulses

A growing number of high resolution experiments demand the application of two or more soft pulses at the same time. There are several possible practical approaches to this problem. A symmetrical array of pulses in the frequency domain can be achieved by cosine modulation of the soft pulse amplitude at the appropriate frequencies. For example, if the transmitter (at frequency $f_0$) is modulated in amplitude according to the expression

$$A_t = A_0\cos(2\pi f_0 t)\ \cos(2\pi f_1 t)\ \cos(2\pi f_2 t) \qquad\qquad \textbf{5.22}$$

then the excitation spectrum contains four components at frequencies $(f_0 - f_1 - f_2)$, $(f_0 - f_1 + f_2)$, $(f_0 + f_1 - f_2)$, and $(f_0 + f_1 + f_2)$ Hz. Such a scheme may be used for triple irradiation if the fourth sideband is positioned in an empty region of the high resolution spectrum.

A second approach is to interleave $k$ individual DANTE sequences all with the same repetition rate $(1/\tau)$. We have seen in § 5.3.1 that if one DANTE sequence has pulses of constant phase while another has linear phase incrementation at a rate $\Delta\phi/\tau$ radians per second, then their excitation frequencies are displaced by $\Delta\phi/(2\pi\tau)$ Hz. A large number $k$ of soft pulses may be combined in this manner and it is often convenient to adjust the individual duty cycles so that the combined DANTE sequence contains no intervals for free precession; the windows are completely filled. Since the individual duty cycles fall off as $k$ increases, there must be a compensating increase in transmitter level to maintain the same flip angle for the individual soft pulses. One advantage of phase modulation is that the shaping of the pulse envelope is then an entirely independent process. Precautions must be taken to keep the pulse repetition rate $(1/\tau)$ sufficiently high so that DANTE sidebands do not intrude into the spectrum of interest. A limit on the number of soft pulses that can be combined is eventually set by the minimum practical radiofrequency pulse width.

The most general scheme for combining many soft pulses[32] is based on the phase-ramping procedure. Each soft pulse is broken down into a sequence of narrow time segments and its frequency shift with respect to the transmitter is determined by the rate of phase incrementation between one segment and the next. It is therefore important to have a large number of segments and relatively small phase jumps. Each segment is also modulated in amplitude for the purposes of pulse shaping. Consequently the soft pulse may be represented as an array of vectors defining the phase and amplitude of each time segment.

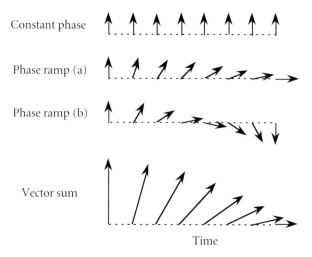

Combination of several soft pulses simply involves vector summation, segment by segment, throughout the duration of the soft pulse. Provided that the dynamic range of the waveform generator is adequate, very large numbers of soft pulses (at arbitrary frequencies) can be combined in this manner. There is no digitization error on the amplitudes of the primitive vectors, only on their resultant. Individual pulses may have different shapes and different selectivities.

### 5.3.6 Spin–spin coupling during a soft pulse

In some of the situations treated above it has been tacitly assumed that magnetization trajectories may be calculated through the Bloch equations rather than from the Liouville–von Neumann equation. This assumption that the component lines of a spin multiplet behave independently under the influence of a radiofrequency pulse is justifiable for a hard pulse because of its short duration, but it is not necessarily correct for a soft pulse that might persist for hundreds of milliseconds. Experience shows that it is still a good approximation for excitation by a pure phase pulse—the components of a spin multiplet are refocused just as if they were independent magnetization vectors, and they can all be recorded in the pure absorption mode.

However, complications arise if two or more groups are excited simultane-

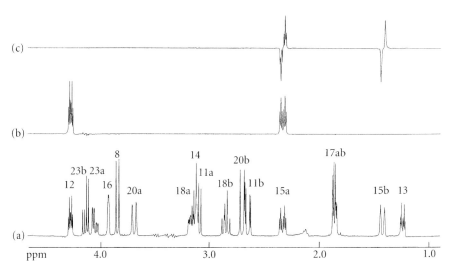

**Figure 5.10** Simultaneous excitation of two proton sites in strychnine with soft E-BURP-2 pulses. (b) For two protons H12 and H15a that are not coupled, the two signals are in absorption and have the same phase. (c) For two coupled protons (H15a and H15b) with the condition $J_{IS}T = 1.5$, the signals are in antiphase dispersion. For clarity, they are shown here as antiphase absorption after a 90° receiver phase shift.

ously.[33–35] Consider the simple case of two weakly-coupled spins $I$ and $S$, each excited by an E-BURP pulse. If $I$ and $S$ have zero mutual coupling (but are coupled to other spins) the two multiplets are excited quite normally in the pure absorption mode (Figure 5.10(b)). However if there is an appreciable mutual coupling $J_{IS}$, we find that both doublets exhibit a phase distortion. When the pulse durations $T$ are set near the condition $J_{IS}T = 1.53$, the two doublets are excited in antiphase dispersion, and an overall phase shift of 90° puts them into antiphase absorption (Figure 5.10(c)). For longer pulse durations the phase distortion increases and multiple-quantum coherence is generated.

The phenomenon has been called the double-resonance two-spin effect (TSETSE).[35] It can be visualized in terms of a decoupling effect.[36] The pure-phase pulse is designed to take the two components ($\alpha$ and $\beta$) of the $S$-spin doublet from the $+z$ axis to the $+y$ axis. If the $I$ spins are simultaneously irradiated, this induces transitions between the $\alpha$ and $\beta$ states at a rate determined by the instantaneous value of $B_1$. The result is that the original $S$-spin offsets ($\pm\frac{1}{2}J_{IS}$) are overcompensated by the pure-phase pulse, and the trajectories overshoot the $+y$ axis by an amount determined by the pulse duration $T$. Trajectories calculated for the condition $J_{IS}T = 1.53$ are illustrated in Figure 5.11. They are seen to terminate near the $\pm x$ axes, but if the calculation is repeated for the single resonance case, the normal focusing on the $+y$ axis occurs.

These coupling effects were first observed with spin inversion pulses[33,34] and were regarded as an unfortunate complication for soft pulse experiments. They can,

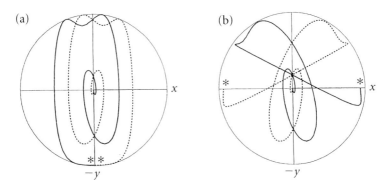

**Figure 5.11** Trajectories calculated for pure-phase E-BURP pulses with $J_{IS}T = 1.53$, shown as projections onto the $xy$ plane. (a) Single resonance, showing focusing along the $-y$ axis; (b) double resonance, showing an increased divergence to the $\pm x$ axes. Trajectories are shown for offsets of $+\frac{1}{2}J_{IS}$ (dotted curves) and $-\frac{1}{2}J_{IS}$ (full curves); they start at the origin ($+z$) and terminate as indicated by the asterisks.

however, be put to good use as a correlation method—a simple test for spin coupling between two designated groups.[35] As can be appreciated from Figure 5.10(c), the spectra are very clean and free from extraneous responses from the other protons in the spectrum, and the antiphase nature of the two multiplets is an easily recognized feature.

### 5.3.7 **Close encounters between soft pulses**

When two soft pulses approach one another in the frequency domain there is an inevitable interaction between them. This is particularly important for pure-phase pulses, since the delicate refocusing scheme is rather easily upset by a second alien radiofrequency pulse in the close vicinity.[37] It is useful to think of the frequency domain as being divided into four regions—the excitation band, transition regions, penumbra and umbra.

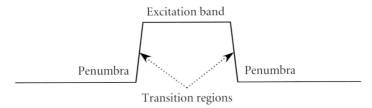

The umbra are the distant regions where the offset is so large that the effective radiofrequency field is essentially along $+z$ and has very little effect. In the penumbra the magnetization excursions can be quite large, but all the trajectories terminate close to $+z$ so the overall excitation is rather small.

(b)

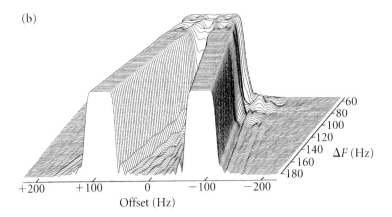

(a)

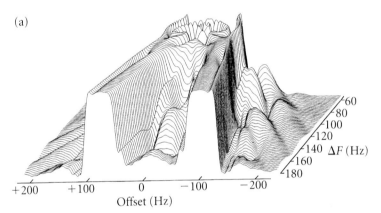

**Figure 5.12** Close encounters between two simultaneous pure-phase pulses, showing the frequency-domain excitation patterns in the absolute-value mode. (a) There is a strong interaction if the separation between centres ($\Delta F$) is less than four times the operating bandwidth ($\Delta f = 45$ Hz). (b) If one pulse is time-reversed, this interaction is considerably reduced.

If one soft pulse encroaches on the penumbra of its neighbour, then there is an appreciable interaction. Figure 5.12(a) shows simulations of this effect as two pure-phase pulses[17] are brought closer and closer together in the frequency domain. Note how their excitation profiles become distorted and how strong waves of signals appear in regions that should have negligible excitation.

This undesirable interaction may be minimized by a simple trick. Since the pure-phase pulses use only a very low level of $B_1$ during the first half of the pulse interval, time-reversal of one of the two pulses[37] greatly reduces the mutual interaction and the consequent distortion of the excitation profiles (Figure 5.12(b)). However, this does induce a frequency-dependent phase shift in the signals excited by the time-reversed pulse.

## 5.4  Applications of soft pulses

### 5.4.1  Pseudo-two-dimensional spectroscopy

NMR information it not normally distributed uniformly throughout the frequency space under investigation; indeed, the useful information is often contained in very specific spectral regions, and may be required to be recorded with high definition along the frequency axis. In much the same way as a microscope zooms in on an interesting feature, the spectroscopist often needs to focus his attention on a particular detail in a spectrum, largely ignoring the rest. Multidimensional spectroscopy (§ 8.6) lacks this zoom facility, treating all frequency components equally. If very fine detail is required then the only recourse is to employ vast data tables and protracted experiments; all the data space is finely digitized but most of it may be wasted for the purpose at hand. Limitations on spectrometer time and data storage often force two-dimensional experiments to terminate at a value of $t_1(max)$ which is too short to provide optimum resolution. This problem is aggravated if long phase cycles are used, and is compounded for three-dimensional spectroscopy.

This was the incentive for developing pseudo-two-dimensional[38] and pseudo-three-dimensional spectroscopy.[39] Instead of programming periods of free evolution followed by Fourier transformation, pseudo-multidimensional spectroscopy scans the corresponding frequency dimensions directly, by moving the frequency of a soft radiofrequency pulse one small step at a time (§ 5.4.1). There is no longer any restriction to uniform frequency increments and no necessity to satisfy the Nyquist condition. The soft pulse selectivity should be comparable with the frequency increments; usually this implies line-selective pulses. This sacrifices the multiplex advantage enjoyed by multidimensional spectroscopy and would be incredibly tedious if wide frequency ranges had to be examined. However for many proton NMR experiments sensitivity is not a critical factor and multiscan averaging is not required. Although the pseudo-multidimensional experiment has inherently poorer signal-to-noise, it is often completed in a much shorter time. It is ideal for focusing attention on a small area (or volume) of frequency space with high definition. A true zoom operation is feasible if the experiment is repeated with finer and finer frequency increments covering smaller and smaller ranges.

Line-selective excitation has another application. Suppose we are interested in a spin multiplet that is not fully resolved from adjacent resonances, but which does present one transition that is free of overlap. A soft $\pi$ pulse can be used to induce a population inversion for this line. A subsequent nonselective $\pi/2$ pulse has the effect of spreading this population disturbance throughout all the lines of the multiplet and then converting this into observable magnetization, appropriately diluted.[40,41] No transfer to neighbouring multiplets occurs. We shall see below in § 5.4.4 that this line-selective technique can also be useful in a two-dimensional context for separating overlapping cross-peaks.

## 5.4.2 **Reduced dimensionality**

The realization that multidimensional spectroscopy was too profligate with spectrometer time and data storage capacity has engendered an entire family of new soft pulse applications in which a two-dimensional experiment is reduced to its one-dimensional counterpart by the introduction of a soft pulse and the freezing of the evolution period into a fixed delay. Thus the well-known homonuclear correlation (COSY) experiment[42] may be reduced to a one-dimensional analog by using a soft pulse to excite a chosen multiplet (the $I$ spin) followed by a fixed delay $\Delta = 1/(2J_{IS})$ to allow the development of antiphase magnetization. A subsequent hard $\pi/2$ pulse causes magnetization transfer, generating an antiphase signal at the $S$-spin site. When the coupling constant is small, there can be an appreciable loss of intensity through relaxation or field inhomogeneity effects. In the more general case of several coupled spins, coherence is transferred to all sites that are coupled to $I$, but now the $\Delta$ delay must be a compromise and the observed intensities depend on the respective coupling constants.[2] This technique will be recognized as giving the same kind of information as a selective spin decoupling experiment with irradiation of the $I$ spins. Note that many molecular structure problems hinge on solving only one or two pieces of spectroscopic information, not the entire connectivity matrix.

Similar simplifications can be introduced into the other standard two-dimensional experiments by soft pulse excitation.[43,44] If we wish to identify all the protons in a given coupling network, we can excite just one of these and then allow the coherence to spread throughout the system by isotropic mixing[45,46] giving the one-dimensional version[6] of total correlation spectroscopy (TOCSY). Two consecutive stages of coherence transfer can be engineered with soft pulses, establishing two correlations $A \rightarrow B$ and $B \rightarrow C$ with a common coupling partner $B$ in the middle.[47] There is a one-dimensional version[6] of the nuclear Overhauser experiment (NOESY).

There are many schemes for conversion of two-dimensional experiments into their one-dimensional equivalents, but it is not always just a matter of starting with a soft excitation pulse. For example, the widely used heteronuclear multiple-quantum correlation (HMQC) sequence[48] (§ 8.3.4) can be made selective with respect to a chosen $^{13}C$ site by substituting a soft Gaussian pulse for the usual hard $\pi/2$ coherence transfer pulse at the end of the sequence.[49] One obtains a spectrum of all protons directly attached to that particular carbon site. A further elaboration[50] adds homonuclear Hartmann–Hahn coherence transfer[45,46] so that the coherence is passed on to all protons in that particular coupled spin system.

In general, for a given site, the sensitivity of these one-dimensional offspring is roughly comparable to that of the two-dimensional parents. The exact sensitivity relationship is complicated, since it depends on at least three factors—the comparison between relaxation losses during the soft pulse and those during the mean evolution time, the difference between a constant NMR signal and the mean value of its sinusoidally modulated counterpart, and the fineness of $F_1$ digitization required in the two-dimensional experiment. However, if a series of one-dimensional experiments has to be fitted into the same total instrument time, the relative sensitivity is reduced proportionately (but see Hadamard spectroscopy in § 5.5.4). The practical advantage of performing a single soft pulse scan is that the overall duration of the experiment can be quite short.

### 5.4.3 Band-selective two-dimensional spectroscopy

In the routine application of two-dimensional spectroscopy it is seldom that the limiting resolution is achieved in the $F_1$ dimension since this would entail a very large number of $t_1$ increments. It is quite common to accept broader linewidths in the $F_1$ dimension even though this may result in a loss of sensitivity by self-cancellation of antiphase signals. One way to reduce the $F_1$ spectral width is by band-selective or multiplet-selective excitation with a soft pulse. These are still two-dimensional experiments since the evolution period $t_1$ is retained. One example is band-selective (double-quantum filtered) correlation spectroscopy (COSY).[51] In one version of this experiment only the excitation pulse is frequency-selective (see § 8.2.3). Instead of the full data matrix in the frequency domain, we collect only a narrow strip in $F_1$ running the full length in the $F_2$ dimension. The aim is to resolve ambiguities in the assignment of cross-peaks and to facilitate the examination of the fine structure.

The term soft-COSY has been pre-empted[52] by an elaboration in which not only is the excitation pulse selective, but the hard mixing pulse is replaced by a cascade of two soft pulses applied to $I$ and $S$ in turn, thereby restricting the experiment to a single coherence transfer $I \rightarrow S$. Selectivity is achieved in both frequency dimensions.[17] In a certain sense this homonuclear technique is analogous to a heteronuclear correlation experiment. Since the soft pulses have no effect on passive spins, they remain mere spectators, giving the cross-peaks a simplified structure similar to those in E-COSY[53] but with higher sensitivity.

Economy of time and storage space is all the more important in multidimensional spectroscopy, where phase cycles tend to become more complicated and where the data table can easily exceed $10^9$ words. Suppose the three-dimensional correlation spectrum represented schematically in Figure 5.13 were excited by a soft pulse in the $F_1$ dimension followed by two soft mixing pulses to limit the coherence transfer into the $F_2$ dimension. We may use non-selective coherence transfer into the $F_3$ dimension, since there is no penalty in using a long acquisition time $t_3$. The appropriate band of signals in $F_3$ is simply obtained by filtration or by selection, discarding the rest.[54] In this manner it is possible to focus on any desired small volume element (shown here as a small cube) with very fine digitization in all three dimensions, and yet with a reasonably restricted data table and a short data-gathering time. In the new very-high-field NMR spectrometers a certain restriction of the frequency bands along these lines may well be necessary to keep multidimensional data matrices and instrument time within reasonable bounds.

### 5.4.4 Overlapping cross-peaks

One of the arguments for the introduction of three-dimensional NMR methods is that two-dimensional correlation spectra often contain overlapping cross-peaks,

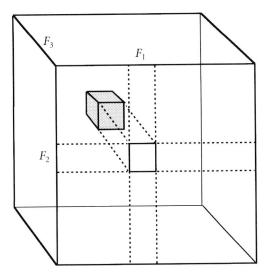

**Figure 5.13** Reduction of the size of the data table and the data gathering time for a three-dimensional experiment by applying band-selective pulses in the $F_1$ and $F_2$ dimensions and by keeping only a small part of the $F_3$ data. Only information within the shaded cube is retained.

rendering the assignment ambiguous and aggravating the difficulties in deciphering the fine structure. The idea is that the new (third) dimension could serve to disentangle two overlapping multiplets $A$ and $B$, but this imposes heavy demands on instrument time and data storage. Alternatively, a soft radiofrequency pulse, like a delicate scalpel, can be used to slice away one piece of $A$ without disturbing $B$. We simply search for a particular excitation frequency that cuts through the $A$ lines but avoids the $B$ lines.

The principle is illustrated with the 400 MHz proton correlation spectrum of a copolymer that was generated biosynthetically. If the monomers are thought of as P and Q, we might expect that there would be four slightly different P cross-peaks in the correlation spectrum from polymer species PPP, PPQ, QPP and QPQ, assuming that only nearest-neighbour interactions perturb the chemical shift of P. Figure 5.14(a) shows such a double-quantum filtered COSY cross-peak, indicating that there is no clear evidence of the underlying structure. Soft-pulse excitation with the spin pinging sequence is sufficiently selective that the four overlapping components can be separated essentially completely.[55] The pulse sequence may be written

$$\text{hard } (\pi/2) \text{ soft } (\pi) \text{ hard } (\pi/2) - t_1 - \text{hard } (\pi/2) \text{ acquisition} \qquad \textbf{5.23}$$

with the appropriate phase cycling. Although the soft pulse is line-selective, the second hard $\pi/2$ pulse serves to distribute this excitation uniformly throughout that particular multiplet without affecting the other three overlapping multiplets. Four individual cross-peaks are illustrated in Figures 5.14(c)–(f). These have been superimposed in Figure 5.14(b) for comparison with Figure 5.14(a).

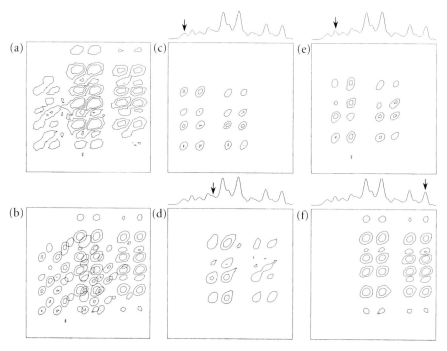

**Figure 5.14** (a) Four severely overlapping cross-peaks in the correlation spectrum of a biopolymer. The individual cross-peaks are separated in (c), (d), (e) and (f) by soft pulse excitation at the frequencies indicated by the arrows. These four cross-peaks have been superimposed in (b) for comparison with (a).

## 5.4.5 **Spatial localization**

Soft pulses are widely used in medical applications of magnetic resonance. Indeed a large part of the methodology of soft pulse excitation has been directed towards imaging and *in vivo* spectroscopy. In the presence of an applied magnetic field gradient, a soft pulse excites only a restricted region of the NMR sample. Although this appears to be self-evident from arguments based on magnetization trajectories, there is an important complication that does not arise when dealing with discrete resonances. The sample is an essentially continuous distribution of nuclear spins along the direction of the applied field gradient. Each small volume element (isochromat) can be assigned a small magnetization vector which experiences a characteristic value of the offset field $\Delta B$. After a soft pulse, these isochromats spread out in such a fashion that their resultant after the pulse is almost zero. The frequency-selectivity actually comes about only because the NMR response is non-linear. To counter this problem it is customary to refocus this divergence before detection, usually by reversing the direction of the pulsed field gradient to form a spin echo (§ 4.1.4).

High resolution spectroscopy can also benefit from these ideas by using spatial selectivity to enhance resolution. If the experimental linewidth is dominated by the inhomogeneity of the applied magnetic field, then in principle we could improve resolution simply by using smaller samples. In practice this is extremely difficult (if not impossible) to achieve because the smaller the sample the greater the influence of the discontinuities of magnetic susceptibility caused by the walls of the sample container. The alternative is to reduce the effective volume of the sample by a spatial localization technique. In high resolution spectrometers the critical gradients are those along the $z$ axis, since they are unaffected by sample spinning. By applying an intense pulsed $z$ gradient during the soft pulse, the excitation may be confined to a small disc-shaped volume. When this imposed gradient is extinguished, the spins precess freely in the natural field gradients, but since the signal arises from a region that is limited in the $z$ dimension, the natural $z$ gradients have little effect, and the resolution is significantly enhanced.[56] Figure 5.15 shows an example of a spin multiplet from furan-2-aldehyde recorded after preparation in increasingly intense

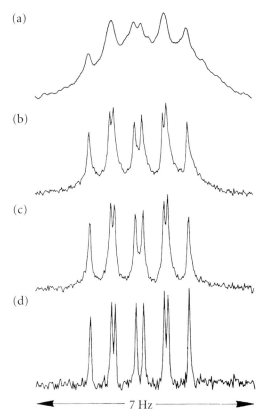

(a)

(b)

(c)

(d)

7 Hz

**Figure 5.15** Spin multiplet from the low-field ring proton of furan-2-aldehyde excited by a soft pulse in increasingly intense applied field gradients which were extinguished before acquisition of the free induction decay. By progressively reducing the effective sample volume, this increases the resolution.

applied $z$ gradients. The resolution improves markedly as the effective volume is decreased.

## 5.4.6 Nuclear Overhauser effect

One of the most powerful structural tools is the nuclear Overhauser effect, the change in the intensity of the resonance of one spin species upon saturation of an adjacent spin with which it has an appreciable dipole–dipole interaction (§ 1.2.1 and § 9.1). For molecules of chemical or biochemical interest, several simplifying assumptions must be made before the measured nuclear Overhauser enhancement can be converted into relative distances. The molecule is assumed to be rigid, the rotational correlation time is usually taken to be about the same for all parts of the molecule, multiple-spin effects are neglected, and the leakage relaxation paths are assumed to be the same for all spins. Although these approximations are hard to justify, they turn out not to be too restrictive in practice because there are usually so many proton–proton interactions in a large molecule that the problem is overdetermined. In general the enhancement can be taken to be inversely proportional to the sixth power of the internuclear distance. In any case, the nuclear Overhauser information is normally used qualitatively, to establish a distance constraint by testing whether two protons are close enough (say, less than 0.6 nm) for a significant enhancement to be evident.

Most work on the Overhauser effect in macromolecules is performed by two- or three-dimensional spectroscopy, the NOESY technique.[57,58] The existence of a NOESY cross-peak indicates that two protons must be in close proximity and this often means that an amino acid chain loops back on itself—a crucial piece of structural information.

For more quantitative determination of internuclear distances, the build-up of the Overhauser enhancement may be studied as a function of the duration of the presaturation pulse in a one-dimensional experiment. This is known as the truncated driven nuclear Overhauser effect.[59] Alternatively, population inversion of one spin with a soft $\pi$ pulse is followed by a variable period $\tau$ to allow cross-relaxation effects to develop—the transient nuclear Overhauser experiment. There is an important precaution that must be taken. To avoid complications from spin–lattice relaxation of the observed spin, and interactions with more distant spins, only the initial section of the build-up curve is monitored (the initial rate approximation). With the same caveats concerning correlation times, molecular rigidity, and the absence of other interacting spins, the measured cross-relaxation rate is again inversely proportional to the sixth power of the internuclear distance. Since the correlation time $\tau_c$ is usually not known, the distance is normally measured with respect to a known reference distance in the same molecule with the same correlation time.[60]

These nuclear Overhauser effect techniques are closely related to the double resonance saturation transfer experiment first proposed by Forsén and Hoffman[61] (§ 8.2.7). When an atom moves between two different molecular species or between two conformers of a given structure, nuclear spin magnetization is carried with it. The perturbation of the nuclear spin populations (saturation or inversion) serves as an innocuous label with which to monitor the slow chemical exchange process. Nowadays this is usually studied by a two-dimensional technique (EXSY)[57] analogous to NOESY, in which the frequencies of the chemical sites are identified during the evolution period, magnetization transfer occurs during the mixing period, and the modified intensities are observed during the detection period. This two-dimensional technique has proved very successful, its only drawback being some minor difficulties in measuring the cross-peak volumes for accurate quantitative work. The alternative is a one-dimensional experiment initiated with a soft population inversion pulse. The need to perturb a band of frequencies as uniformly as possible without disturbing the neighbouring multiplets suggests the use of a shaped soft pulse.

Both nuclear Overhauser and chemical exchange measurements are often carried out by difference spectroscopy. This renders the intensity changes much more prominent although it does nothing to improve the quality of the information. When we need proportional changes in intensity, recourse must be made to the unperturbed spectrum for the reference intensities.

### 5.4.7 **Relaxation measurements**

Before the advent of Fourier transform methods, the natural way to measure individual spin–spin and spin–lattice relaxation times in multi-line spectra was through the use of soft pulses.[62] Inversion-recovery (§ 2.3.1), spin-echo (§ 4.1.7) and spin-locking experiments (§ 2.4.3) all lend themselves to soft pulse methods if the relaxation times are relatively long (in the range of 100 ms to 10 s). However, the relaxation properties are measured only for one chemical site at a time. The Fourier transform revolution has made it possible to perform most of these relaxation studies with hard radiofrequency pulses, with the added advantage that all chemical sites can now be monitored in a single run. These studies are particularly popular in $^{13}$C spectroscopy.

However, there are some more subtle aspects of relaxation where soft pulse experiments are still preferable. One is the investigation of correlation effects through the measurement of the nonexponential relaxation within a strongly coupled spin system.[63,64] For example, if the relaxation is by an external paramagnetic species, the degree of correlation of the fluctuating fields at two different proton sites gives a measure of the distance of closest approach of the relaxation agent. For large distances there is a high degree of correlation, but with close encounters, the degree of correlation is reduced.

Another application is the comparison of spin–lattice relaxation rates after spin inversions—first by a soft pulse and then by a hard pulse. This gives information about cross-relaxation,[65] essentially equivalent to nuclear Overhauser effect measurements. Alternatively cross-relaxation may be studied by synchronized nutation at two adjacent sites,[66] providing a particularly clean approach to the determination of the internuclear distance, unaffected by neighbouring spins.

In coupled spin systems, the measurement of spin–spin relaxation times is complicated by the echo-modulation effect (§ 4.2).[67,68] The problem arises because, in a homonuclear spin system, a hard $\pi$ pulse not only refocuses the $I$-spin magnetization but also flips the spin states of the $S$ spins (§ 4.2.3). One effective remedy is to use a soft $\pi$ pulse (for example, RE-BURP) so that the echo modulation vanishes, and the spin echoes decay exponentially with a time constant equal to the spin–spin relaxation time.

## 5.5 Multiple-pulse experiments

### 5.5.1 Polychromatic pulses

The ability to create a large number of simultaneous soft pulses offers a new approach to the tailoring of the excitation profile for specific applications. The concept is most easily introduced by making the assumption of a linear NMR response since this implies that the pulses act independently of one another, but this is by no means essential to the method. Suppose that two soft rectangular pulses are applied simultaneously and with a frequency separation

$$\Delta f = 1/(2T) \qquad\qquad \textbf{5.24}$$

where $T$ is the pulse duration. Although the individual excitation profiles have the usual sinc function oscillations in the wings, the peaks of one fall in the troughs of the other and mutual cancellation occurs.[18] This cancellation is not quite exact but it improves if a 1:2:1 combination of three soft pulses is employed and is virtually complete if a 1:3:3:1 combination of four soft pulses is used (Figure 5.16).

A polychromatic pulse[18] is an assembly of a large number of soft pulses at regular frequency intervals, usually with $\Delta f$ given by Equation 5.24. The sidelobe problem is merely an edge effect and is countered by halving the intensity of the outermost pulses compared with the rest. For example the polychromatic pulse called PC(9) consists of seven soft pulses of unit intensity flanked by half-intensity pulses at each edge. Not only does it possess an excitation profile that is uniform over the central region but also the phase gradient is approximately linear across the same frequency range and may thus be corrected with the standard phasing routine. There is only a minimal oscillatory response in the wings of the excitation profile.

The beauty of this approach is that we may design composite soft pulses in the frequency domain[69] simply by assembling a suitable number of elements with the appropriate intensities. When the nonlinearity of the NMR response causes a distortion of the desired profile, the intensities of the various elements are simply readjusted to compensate. One attractive application is for broadband excitation with high uniformity across the band. In this scheme the induced phase gradient is essentially linear, and is readily corrected in the resulting spectrum.

Another application is the design of a frequency-domain profile suitable for water suppression (§ 11.2.3), with a rejection notch at the water frequency but essentially uniform excitation elsewhere. The polychromatic pulse PC(54) was

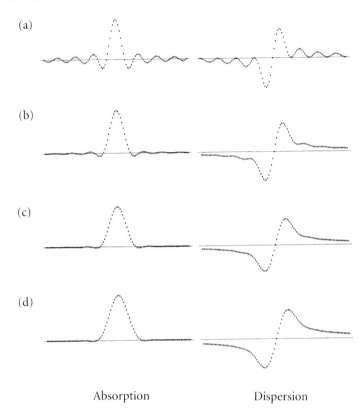

(a)

(b)

(c)

(d)

Absorption                    Dispersion

**Figure 5.16** Frequency-domain excitation profiles calculated for (a) a single soft rectangular pulse, (b) two equal soft rectangular pulses, (c) three soft pulses with intensities 1:2:1, (d) four soft pulses with intensities 1:3:3:1. To minimize the sidelobes, the frequency separations were all set to the condition $\Delta f = 1/(2T)$ where $T$ is the pulse duration.

specifically constructed for this application, with a pulse flip angle of $\pi/4$ radians to ensure an approximately linear response. It may be expressed as

$$PC(54) = 1:(2)_{31}:1:0:0:0:1:(2)_{16}:1 \qquad \textbf{5.25}$$

where $(2)_{31}$ stands for 31 pulses of relative intensity 2. The 54 elements are arrayed in the frequency domain at regular intervals so as to cover the appropriate spectral width. The elements outside the rejection notch are of uniform intensity, whereas in practice the three elements with nominally zero intensity are adjusted slightly to provide fine control of the degree of suppression. A practical example of the application of this pulse to water suppression is shown in § 11.2.3.

## 5.5.2 Janus pulses

Coherence transfer or multiple-quantum experiments normally start by exciting antiphase magnetization ($I_y \rightarrow 2I_xS_z$). The standard method is to apply a hard pulse followed by a free precession delay $\tau = 1/(2J_{IS})$, usually with refocusing of chemical shifts. It would be much more convenient to have an excitation scheme that tolerated a broad range of $J_{IS}$ values since this would provide optimum coherence transfer for several different couplings (§ 2.5.7). Soft pulses designed for this purpose[69] have been named after the Roman god Janus, depicted as facing in two opposite directions at the same time.

Polychromatic pulses can be readily tailored for this purpose. A $\pi/4$ radian soft Janus pulse is constructed as the antisymmetrical array

$$\mathrm{PC}(9{:}0{:}\bar{9}) = 1{:}2{:}2{:}2{:}2{:}2{:}2{:}2{:}1{:}0{:}\bar{1}{:}\bar{2}{:}\bar{2}{:}\bar{2}{:}\bar{2}{:}\bar{2}{:}\bar{2}{:}\bar{1} \qquad \textbf{5.26}$$

where the overbar denotes a $\pi$ radian phase shift. The simulated excitation profile and the experimental offset dependence are illustrated in Figure 5.17 and show excellent agreement.[69] When centred on a $J$ doublet the composite soft pulse rotates one component towards the $+x$ axis and the other towards $-x$. With a pulse duration of 50 ms this polychromatic pulse generates an array of frequency-domain elements at 10 Hz intervals, covering a range of about 180 Hz. It creates antiphase magnetization for spin–spin couplings spanning the range from 40 Hz to 155 Hz.

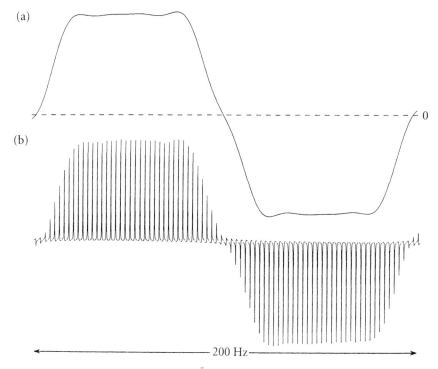

**Figure 5.17** The Janus soft pulse PC(9:0:9̄) designed to create antiphase magnetization from a $J$ doublet over a range of $J$ values: (a) simulated for a 50 ms soft pulse; (b) experimental offset dependence after purging the dispersion-mode contributions. The bandwidth scales as the inverse of the pulse duration.

Janus pulses are useful for initiating magnetization transfer experiments designed to measure long-range carbon–proton or carbon–carbon coupling constants.[70]

### 5.5.3 **Template excitation**

When a large number of simultaneous soft pulses are combined, the data table that defines the resultant pulse consists of a linear array of amplitudes $A_j$ and phases $\phi_j$. This may readily be converted into the equivalent array of real and imaginary components, $A_j \cos\phi_j$ and $A_j \sin\phi_j$. Now this is just the form in which a free induction signal is stored in the spectrometer, so we might consider what would happen if the waveform generator were to be fed with an experimental free induction decay $S(t)$. In practice the array is first reversed in time and then fed to the waveform generator so that all the pulse components come into phase at the end of the soft pulse. Polychromatic pulses generated in this manner possess an excitation profile that closely matches the experimental spectrum $S(f)$, the Fourier transform of $S(t)$. The spectrum has produced its own template excitation pattern.[71] In practice it makes sense to set all experimental intensities to unity so that all the individual soft-pulse flip angles are equal, thus driving all the resonances equally hard. If necessary, we may also remove noise or certain undesirable signals from the experimental data before creating the template.

The concept is in fact much more general; any desired excitation pattern may be constructed as a frequency-domain histogram and then converted to a set of soft pulses for template excitation. This was essentially the goal of the Tomlinson–Hill experiment,[7] although their method of implementation was quite different. Any frequency-domain pattern may be specified. Simulated spectra may be used; band limited regions of excitation may be defined; even noise excitation could be considered.

One application of template excitation is the separation of spectra from two components of a mixture. Each pure component A and B produces its characteristic free induction decay which is subsequently Fourier transformed and converted into the appropriate soft-pulse template. Applied to the spectrum of the mixture, template A excites only spectrum A, template B only spectrum B. Even when there is an overlap of the two spectra the method can be adapted by deleting those parts of the two templates that fall in the overlap region. The resulting partial integrals of spectra A and B are still directly related to the full integrals by known factors.

Template excitation has been used to separate the $^{13}$C spectra of the $\alpha$ and $\beta$ anomers of D-glucose by excitation with tailored soft pulse arrays derived from the spectra of the pure components. This allows the mixture to be analysed as it slowly converts from the pure $\alpha$ form to an equilibrium mixture containing approximately 36% $\alpha$ and 64% $\beta$ (Figure 5.18).

### 5.5.4 **Hadamard spectroscopy**

For a given amount of instrument time, a two-dimensional experiment (§ 8) is more efficient than the equivalent set of one-dimensional experiments using soft

(a)

(b)

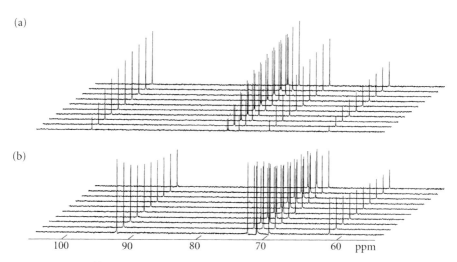

**Figure 5.18** The $^{13}C$ spectra of the $\alpha$ and $\beta$ anomers of glucose monitored independently as a function of time (30 min intervals) after dissolution of pure $\alpha$-D-glucose in heavy water. The separation of subspectra was achieved by excitation with soft-pulse templates derived from the spectra of the pure components. The $\beta$ form (a) grows with time, whereas the $\alpha$ form (b) decays towards its equilibrium concentration (36%). Note that the $\alpha$ and $\beta$ resonances near 70 ppm are only 0.05 ppm apart, but are well separated.

pulse excitation. This reflects the much-vaunted multiplex advantage of two-dimensional spectroscopy—it gathers all the signals in the spectrum while a soft pulse experiment only acquires the signal from one chosen chemical shift. Now, in terms of sensitivity, there is little to choose between gathering several identical time-domain signals (one-dimensional spectroscopy with time averaging) or an equal number of time-varying signals (two-dimensional spectroscopy). Compared with a two-dimensional experiment with a total duration $t_{exp}$, the equivalent sequence of $N$ different one-dimensional measurements can only allocate a time $t_{exp}/N$ to each, thus sacrificing a factor of $N^{\frac{1}{2}}$ in sensitivity. This appears to be a serious limitation on soft pulse methods.

Fortunately this sensitivity loss is not inevitable. It may be retrieved by a trick that was first employed in infrared spectroscopy[72] and which has been exploited in magnetic resonance *in vivo* spectroscopy[73,74] and more recently in high resolution NMR.[75,76] It is perhaps most easily explained in terms of a weighing analogy. Suppose we wish to weigh four different objects A, B, C and D on a balance. Let us make the reasonable assumptions that the balance makes a small random error that is independent of the weight on the pans and is uncorrelated from one weighing to the next. We could of course take each of the four objects and weigh them one at a time. However, a higher accuracy can be achieved by putting all four objects on the balance together, but partitioned between the two pans according to the scheme

|               | A | B | C | D |
|---------------|---|---|---|---|
| weighing (1)  | + | + | + | + |
| weighing (2)  | + | + | − | − |
| weighing (3)  | + | − | + | − |
| weighing (4)  | + | − | − | + |

Here a + signifies that the object is on the left-hand pan and a − means it is on the right-hand pan (together with the standard weights). By suitably combining the four different weighings, we can deduce the weight of each object separately. For example, addition of all four weighings gives four times the weight of object A. Subtraction of weighings (2) and (3) from weighings (1) and (4) gives four times the weight of object D; all the other contributions cancel identically. For random, uncorrelated errors, the overall error from four weighings merely doubles, so the accuracy of the measurement is doubled in this case. In the general case it improves by the square root of $N$, where $N$ is the number of experiments.

The scheme shown above is a Hadamard matrix[77] of order four. It is believed that such matrices exist for all values $N=4n$, where $n$ is an integer, at least up to $N=264$. The moral of the story is clear—soft pulse experiments carried out in sets of $N$ at a time can achieve much higher sensitivity than the same experiments performed individually, one after the other. The trick is to encode the simultaneous excitations with a + or − phase based on the Hadamard scheme. In the first stage, $N$ soft pulses are applied simultaneously, with phase inversions that follow the signs in the top row of the appropriate Hadamard matrix. This is repeated a further $(N-1)$ times with different phase combinations, according to the recipes in subsequent rows of the matrix. The result is a set of $N$ different composite spectra, each containing all the desired information, but in far too complex a form for easy analysis. We extract the individual subspectra by combining these $N$ results together, adding or subtracting according to the signs in one of the columns of the Hadamard matrix. After $N$ different combinations, we derive $N$ simple subspectra, each equivalent to a spectrum we would have obtained with just a single soft pulse. However, the sensitivity is increased by the square root of $N$. For the optimal cases $N = 4, 8, 12, 16, 20$ etc. the improved sensitivity matches that of the equivalent two-dimensional experiment performed in the same total time. We have won back the multiplex advantage. Indeed the performance may well be better than that of the two-dimensional scheme because each soft-pulse experiment may be optimized independently.

The most satisfactory applications of the Hadamard stratagem are multiple soft-pulse experiments that involve excitation of $^{13}C$ spins since, for natural abundance material, this avoids the possibility that the simultaneous excitations might interact. This may be illustrated with the long-range C–H shift correlation spectra of strychnine (Figure 5.19). The eight carbon → proton correlation experiments employ eight simultaneous half-Gaussian pulses applied to the $^{13}C$ sites. Eight experiments are performed, each with a different combination of soft-pulse phases, following the prescription of the Hadamard matrix of order eight. The individual subspectra are separated by recombining the eight scans according to the columns of the same matrix.

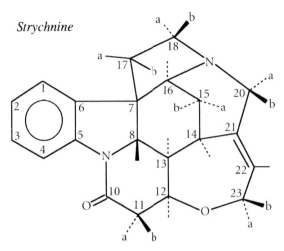

**Figure 5.19** The molecular structure of strychnine showing the labelling scheme used in Figures 5.10 and 5.20.

These subspectra are clean and unaffected by cross-talk between the various correlations (Figure 5.20), giving reliable values for the long-range C–H coupling constants.

### 5.5.5 Phase encoding

This multiplexing concept can be extended to cover any arbitrary number $N$ of soft pulse scans, not necessarily those represented by a Hadamard matrix, by invoking a phase-encoding scheme[78] instead of phase inversion. Consider for example the case of $N=5$ which, by the Hadamard method, would have entailed the inefficient use of a matrix of order eight.

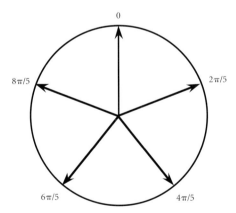

Five soft pulses at the appropriate frequencies excite five distinct chemical sites A, B, C, D, E simultaneously. Initially all five soft pulses have the same phase, but in four

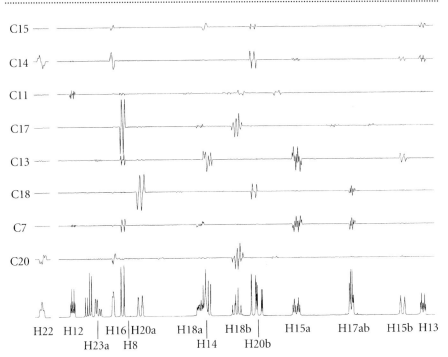

C15

C14

C11

C17

C13

C18

C7

C20

H22   H12   | H16|H20a   H18a|   H18b|   H15a   H17ab   H15b H13
          H23a   H8          H14       H20b

**Figure 5.20** Long-range C–H (antiphase) splittings in strychnine observed by soft-pulse coherence transfer from selected $^{13}$C sites. All eight correlations were performed simultaneously, based on a Hadamard matrix coding scheme, and then separated by decoding according to the same matrix. This improves the sensitivity almost threefold.

subsequent scans the phases are advanced in units of $2\pi/5$ radians according to the matrix:

| Scan | A | B | C | D | E |
|------|---|---|---|---|---|
| 1 | 0 | 0 | 0 | 0 | 0 |
| 2 | 0 | $2\pi/5$ | $4\pi/5$ | $6\pi/5$ | $8\pi/5$ |
| 3 | 0 | $4\pi/5$ | $8\pi/5$ | $2\pi/5$ | $6\pi/5$ |
| 4 | 0 | $6\pi/5$ | $2\pi/5$ | $8\pi/5$ | $4\pi/5$ |
| 5 | 0 | $8\pi/5$ | $6\pi/5$ | $4\pi/5$ | $2\pi/5$ |

Because of the cyclic nature of the phase incrementation, $12\pi/5$ is the same as $2\pi/5$ etc. The results from the five scans are stored in separate locations so that they may be recombined in different ways. For example, if all five results are added, this gives the spectrum excited by the soft pulse at site A, all the other signals cancel since they may be represented by five vectors uniformly distributed around a circle. Modern NMR spectrometers usually have a facility for resetting the phase angle of a free induction signal before Fourier transformation (the zero-order phase correction). If the five results are now reprocessed with the phase angle incremented in steps of $2\pi/5$, according to the second column of the matrix, then we only obtain spectra arising from excitation of site B. Analogous combinations gives the results for sites C, D, and E. Appropriate phase-encoding matrices may be devised for any value of $N$ and, by analogy with the Hadamard scheme, the signal-to-noise ratio should be improved by a factor $N^{\frac{1}{2}}$. Incidentally this technique also demonstrates the generality of the concept of phase cycling; it is by no means limited to the well-known four-fold $0°, 90°, 180°, 270°$ cycle.

These two multiplexing schemes promise to be quite useful for reducing the complexity of multidimensional spectroscopy (§ 8.6.1). Take, as an example, the case of proton–proton Overhauser spectroscopy (NOESY) of large molecules, where the custom is to reduce the spectral complexity by isotopic enrichment in $^{15}N$. Normally this would be handled by a protracted three-dimensional experiment involving round-trip coherence transfer H $\rightarrow$ $^{15}N$ $\rightarrow$ H, spreading out the information into the third ($^{15}N$) dimension in order to remove any anbiguities created by overlap of proton responses. Usually the molecules of interest have a limited number of $^{15}N$ sites, and it would seem logical to pick these out with soft pulses applied at the frequencies of the $^{15}N$ shifts. A separate two-dimensional proton–proton NOESY subspectrum is recorded for each $^{15}N$ site. Furthermore, phase-encoding allows all these sites to be excited simultaneously and the individual NOESY sub-spectra to be separated by the appropriate decoding scheme (Figure 5.21). The total

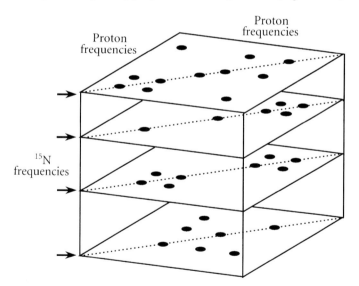

**Figure 5.21** Schematic representation of an experiment designed to record a two-dimensional proton–proton NOESY subspectrum for each of a set of $^{15}N$ sites excited by a soft radio-frequency pulse. This would represent a significant time-saving compared with the conventional three-dimensional spectroscopy method, where a second evolution period is introduced to allow precession at the $^{15}N$ chemical shifts. The sensitivity of the soft pulse experiment would be improved by the Hadamard trick (see text).

elapsed time would be far shorter than that required for the commonly-used three-dimensional experiment, but the sensitivity would be the same or better.

### 5.5.6  Hartmann–Hahn coherence transfer

In a classic paper, Hartmann and Hahn[79] describe a revolutionary solid-state experiment which exploits the dipole–dipole interaction to enhance the sensitivity of a low-abundance nuclear spin species by bringing it into contact with an abundant species prepared at a very low spin temperature in the rotating reference frame. This is what we normally recognize as the Hartmann–Hahn experiment (§ 1.4.5). Almost as an afterthought, the same paper outlines an analogous liquid-phase experiment which relies on an entirely different mechanism (spin–spin coupling) and where the spin temperature concept is not applicable.

Two (abundant) coupled spins $I$ and $S$ are spin locked by radiofrequency fields $B_1$ and $B_2$ that satisfy the matching condition for equal precession rates in their respective rotating frames:

$$\gamma_I B_1 = \gamma_S B_2 \qquad\qquad \textbf{5.27}$$

If the $I$-spin magnetization is initially aligned along $+y$ and the $S$ spin magnetization along $-y$, then a cyclic interchange of coherence ensues at a frequency $\frac{1}{2}J_{IS}$ Hz and persists until damped by spin–spin relaxation. This is illustrated for the case of two protons in uracil in Figure 5.22. If the duration $\tau$ of spin locking is set to the condition

$$\tau = 1/(2J_{IS}) \qquad\qquad \textbf{5.28}$$

then the maximum interchange of coherence is recorded.

The scalar-coupled Hartmann–Hahn effect is the basis of the well-known two-dimensional TOCSY[45] or HOHAHA[46] spectroscopy, used to indicate all possible proton-proton correlations at the same time. The soft-pulse analogue of this technique has some interesting possibilities.[80] It is doubly selective; both the source and the target spins have to be designated. This means that not only can the optimum condition for transfer be satisfied (Equation 5.28) but also that there is no dissipation of coherence through coupling paths to other spins. It is easy to suppress all signals save those generated by Hartmann–Hahn transfer, so the spectra are free of artifacts and overlapping resonances. The signals detected at the target spin are pure absorption and in-phase. Several such selective Hartmann–Hahn transfers may be implemented simultaneously by choosing different $B_1$ levels for each transfer to ensure that there is no cross-talk during the spin-locking period. Coding and decoding may be accomplished by the Hadamard method or the related phase-encoding scheme described above.

### 5.5.7  Concatenated Hartmann–Hahn transfers

Sometimes a molecular structure problem may be solved simply by demonstrating that a set of protons is scalar coupled in a linear chain or in a closed loop. This

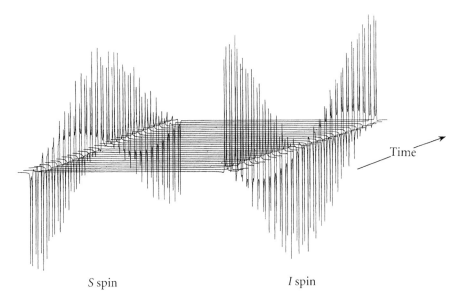

**Figure 5.22** The cyclic exchange of coherence between the *I* and *S* sites of an *IS* proton spin system (uracil) during a Hartmann–Hahn experiment with equal weak radiofrequency fields at the two sites. The *S*-spin signal was initially aligned along the −*y* axis by a soft E-BURP pulse, whereas the *I*-spin signal was aligned along the +*y* axis. The time increments are 10 ms and $J_{IS}$ = 7.5 Hz.

information can of course be extracted from conventional two-dimensional correlation spectroscopy but there is always a danger of ambiguity in crowded spectra. We can attack this problem in a more direct manner by linking together several successive Hartmann–Hahn transfers.[81–87] This reverses the normal philosophy in which we first obtain an excess of correlation information from a single two-dimensional spectrum and then analyse it all *post mortem*. Instead we design an experiment to test a specific hypothesis about the structure; if this is done wisely the experiment gives a yes-or-no answer.

Since Hartmann–Hahn transfer delivers in-phase magnetization from the source to the target site, it is then in the right form to be passed on to a third site, and then to a fourth, etc.

$$A \rightarrow B \rightarrow C \rightarrow D$$

Naturally we need reasonable estimates of the coupling constants $J_{AB}$, $J_{BC}$, $J_{CD}$, but the optima are quite broad and a small mistiming of the spin locking intervals is readily forgiven. The aggregate spin locking time must be short enough that the signal loss through spin–spin relaxation is acceptable. In practice as many as six successive transfers have been implemented in high resolution proton NMR while still retaining a strong signal at the end of the chain. If any link in the chain has a vanishingly small coupling, no coherence reaches the last site.

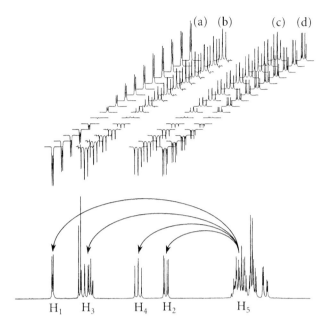

**Figure 5.23** Spin–lattice relaxation rates of protons in sucrose octaacetate after a non-selective population inversion followed by one, two, three and four successive Hartmann–Hahn transfers from the hidden proton H5. All the observed relaxation times are equal (0.88 s) and correspond to that of the source proton H5. The target protons H1, H2, H3 and H4, when measured directly, are found to relax significantly more slowly than H5.

The transferred signals are surprisingly clean and free from overlapping resonances that are not directly involved in the Hartmann–Hahn effect. The experiment thus lends itself to studies of crowded spectra. For example one can measure spin–lattice relaxation times of protons that are completely obscured in the conventional high resolution spectrum. Figure 5.23 shows spin–lattice relaxation rates of protons in sucrose octaacetate after one, two, three and four successive Hartmann–Hahn transfer from proton H5. In each case the relaxation rate is the same, that of the source proton H5 rather than that of the various target protons.[85]

Another application provides an alternative[86] to the familiar INADEQUATE technique[88] but with far higher sensitivity. A series of successive proton–proton transfers is initiated along a coupled chain, and the final coherence is passed to the directly-bound $^{13}C$ site by a nonselective INEPT magnetization transfer.[89] By gradually increasing the number of proton–proton steps, we can identify the carbon sites one by one.

## 5.6 Discussion

The development of soft pulse experiments has been a rather gradual process, dating from a pioneering experiment by Alexander.[1] The nuclear Overhauser effect, relaxation measurements, and the study of slow chemical exchange revived interest in selective pulses for high resolution NMR spectroscopy, but the Fourier transform revolution soon overshadowed this kind of investigation, because it gave the results for all chemical sites in a single experiment. The pendulum has begun to swing back only rather slowly, as it is realized that some chemical studies do not require global information and can therefore be carried out more effectively one site at a time. Hadamard spectroscopy is important here, since it offers the chance to retrieve the vaunted multiplex advantage of two-dimensional spectroscopy. Practical limitations on multidimensional spectroscopy have reinforced the idea that it may be advantageous to reduce the dimensionality by means of band-selective pulses. This chapter has set out to review the applications of soft pulse excitation and to point out some possibilities for future research.

# References

1. S. Alexander, *Rev. Sci. Instrum.* **32**, 1066 (1961).
2. C. J. Bauer, R. Freeman, T. Frenkiel, J. Keeler and A. J. Shaka, *J. Magn. Reson.* **58**, 442 (1984).
3. R. N. Bracewell, *The Fourier Transform and its Applications*, McGraw-Hill, New York, 1978.
4. W. S. Warren, *J. Chem. Phys.* **81**, 5437 (1984).
5. J. Friedrich, S. Davies and R. Freeman, *J. Magn. Reson.* **75**, 390 (1987).
6. H. Kessler, U. Anders, G. Gemmecker and S. Steuernagel, *J. Magn. Reson.* **85**, 1 (1989).
7. B. L. Tomlinson and H. D. W. Hill, *J. Chem. Phys.* **59**, 1775 (1973).
8. R. Freeman, H. D. W. Hill, B. L. Tomlinson and L. D. Hall, *J. Chem. Phys.* **61**, 4466 (1974).
9. A. G. Redfield, S. D. Kunz and E. K. Ralph *J. Magn. Reson.* **19**, 114 (1975).
10. L. Emsley and G. Bodenhausen, *J. Magn. Reson.* **82**, 221 (1989).
11. H. Geen and R. Freeman, *J. Magn. Reson.* **93**, 93 (1991).
12. W. Magnus, *Commun. Pure Appl. Math.* **7**, 649 (1954).
13. H. Geen, S. Wimperis and R. Freeman, *J. Magn. Reson.* **85**, 620 (1989).
14. J. D. Gezelter and R. Freeman, *J. Magn. Reson.* **90**, 397 (1990).
15. X. L. Wu and R. Freeman, *J. Magn. Reson.* **85**, 414 (1989).
16. J. M. Nuzillard and R. Freeman, *J. Magn. Reson. A* **110**, 252 (1994).
17. Ē. Kupče and R. Freeman, *J. Magn. Reson. A* **112**, 134 (1995).
18. Ē. Kupče and R. Freeman, *J. Magn. Reson. A* **102**, 122 (1993).
19. D. A. Horita, P. J. Hajduk and L. E. Lerner, *J. Magn. Reson. A* **103**, 40 (1993).
20. J. M. Nuzillard and R. Freeman, *J. Magn. Reson. A* **107**, 113 (1994).
21. J. Shen and L. E. Lerner, *J. Magn. Reson. A.* **112**, 265 (1995).
22. Ē. Kupče, J. Boyd and I. D. Campbell, *J. Magn. Reson. B* **106**, 300 (1995).
23. R. Freeman, G. A. Morris and M. J. T. Robinson, *J. Chem. Soc. Chem. Comm.* 754 (1976).
24. G. Bodenhausen, R. Freeman and G. A. Morris, *J. Magn. Reson.* **23**, 171 (1976).
25. G. A. Morris and R. Freeman, *J. Magn. Reson.* **29**, 433 (1978).
26. X. L. Wu, P. Xu and R. Freeman, *J. Magn. Reson.* **81**, 206 (1989).
27. M. Mehring and J. S. Waugh, *Rev. Sci. Instr.* **43**, 649 (1972).
28. H. Geen, X. L. Wu, P. Xu, J. Friedrich and R. Freeman, *J. Magn. Reson.* **81**, 646 (1989).
29. X. L. Wu, P. Xu, and R. Freeman, *J. Magn. Reson.* **83**, 404 (1989).
30. X. L. Wu, P. Xu, and R. Freeman, *J. Magn. Reson.* **99**, 308 (1992).

31. D. Boudot, D. Canet, J. Brondeau and J. C. Boubel, *J. Magn. Reson.* **83**, 428 (1989).

32. Ē. Kupče and R. Freeman, *J. Magn. Reson. A* **105**, 234 (1993).

33. B. Ewing, S. J. Glaser and G. P. Drobny, *Chem. Phys.* **147**, 121 (1990).

34. L. Emsley, I. Burghardt and G. Bodenhausen, *J. Magn. Reson.* **90**, 214 (1990).

35. Ē. Kupče, J. M. Nuzillard, V. S. Dimitrov and R. Freeman, *J. Magn. Reson. A* **107**, 246 (1994).

36. J. M. Nuzillard and R. Freeman, *J. Magn. Reson. A.* **112**, 72 (1995).

37. Ē. Kupče and R. Freeman, *J. Magn. Reson. A* **112**, 261 (1995).

38. S. Davies, J. Friedrich and R. Freeman, *J. Magn. Reson.* **75**, 540 (1987).

39. S. Davies, J. Friedrich and R. Freeman, *J. Magn. Reson.* **76**, 555 (1988).

40. S. V. Ley, A. J. Whittle and G. E. Hawkes, *J. Chem. Res. (S)* **8**, 210 (1983).

41. C. Bauer and R. Freeman, *J. Magn. Reson.* **61**, 376 (1985).

42. W. P. Aue, E. Bartholdi and R. R. Ernst, *J. Chem. Phys.* **64**, 2229 (1976).

43. H. Kessler, S. Mronga and G. Gemmecker, *Magn. Reson. Chem.* **29**, 527 (1991).

44. R. Freeman, *Chem. Rev.* **91**, 1397 (1991).

45. L. Braunschweiler and R. R. Ernst, *J. Magn. Reson.* **53**, 521 (1983).

46. A. Bax and D. G. Davis, *J. Magn. Reson.* **65**, 355 (1985).

47. H. Kessler, H. Oschkinat, C. Griesinger and W. Bermel, *J. Magn. Reson.* **70**, 106 (1986).

48. A. Bax, R. H. Griffey and B. L. Hawkins, *J. Magn. Reson.* **55**, 81 (1983).

49. S. Berger, *J. Magn. Reson.* **81**, 561 (1989).

50. R. C. Crouch, J. P. Shockar and G. E. Martin, *Tetrahedron Lett.* **31**, 5273 (1990).

51. J. Cavanagh, J. P. Waltho and J. Keeler, *J. Magn. Reson.* **74**, 386 (1987).

52. R. Brüschwieler, J. C. Madsen, C. Griesinger, O. W. Sørensen and R. R. Ernst, *J. Magn. Reson.* **73**, 380 (1987).

53. C. Griesinger, O. W. Sørensen and R. R. Ernst, *J. Chem. Phys.* **85**, 6837 (1986).

54. H. Oschkinat, C. Griesinger, P. J. Kraulis, O. W. Sørensen, R. R. Ernst, A. M. Gronenborn and G. M. Clore, *Nature* **332**, 374 (1988).

55. P. Xu, X. L. Wu and R. Freeman, *J. Magn. Reson.* **89**, 198 (1990).

56. A. Bax and R. Freeman, *J. Magn. Reson.* **37**, 177 (1980).

57. J. Jeener, B. H. Meier, P. Bachmann and R. R. Ernst, *J. Chem. Phys.* **71**, 4546 (1979).

58. R. Richarz and K. Wüthrich, *J. Magn. Reson.* **30**, 147 (1978).

59. G. Wagner and K. Wüthrich, *J. Magn. Reson.* **33**, 675 (1979).

60. D. Neuhaus and M. Williamson, *The Nuclear Overhauser Effect in Structural and Conformational Analysis*, VCH, New York, 1989.

61. S. Forsén and R. A. Hoffman, *J. Chem. Phys.* **39**, 2892 (1963).

62. R. Freeman and S. Wittekoek, *J. Magn. Reson.* **1**, 238 (1969).

63. A. Abragam, *The Principles of Nuclear Magnetism*, Clarendon Press, Oxford, 1961.

64. R. Freeman, S. Wittekoek and R. R. Ernst, *J. Chem. Phys.* **52**, 1529 (1970).

65. I. D. Campbell and R. Freeman, *J. Chem. Phys.* **58**, 2666 (1973).

66. I. Burghardt, R. Konrat, B. Boulat, S. J. Vincent and G. Bodenhausen, *J. Chem. Phys.* **98**, 1721 (1993).

67. E. L. Hahn and D. E. Maxwell, *Phys. Rev.* **84**, 1246 (1951).

68. R. Freeman and H. D. W. Hill, *Dynamic Nuclear Magnetic Resonance Spectroscopy*, Eds: F. A. Cotton and L. M. Jackman, Academic Press, New York, 1975.

69. Ē. Kupče and R. Freeman, *J. Magn. Reson. A* **103**, 358 (1993).

70. Ē. Kupče and R. Freeman, *J. Magn. Reson. A* **104**, 234 (1993).

71. Ē. Kupče and R. Freeman, *J. Magn. Reson. A* **106**, 135 (1994).

72. M. J. E. Golay, *J. Opt. Soc. Am.* **39**, 437 (1949).

73. R. J. Ordidge, A. Connelly and J. B. Lohman, *J. Magn. Reson.* **66**, 285 (1986).

74. L. Bolinger and J. S. Leigh, *J. Magn. Reson.* **80**, 162 (1988).

75. C. Müller and P. Bigler, *J. Magn. Reson. A* **102**, 42 (1993).
76. V. Blechta and R. Freeman, *Chem. Phys. Lett.* **215**, 341 (1993).
77. J. Hadamard, *Bull. Sci. Math.* **17**, 240 (1893).
78. Ē. Kupče and R. Freeman, *J. Magn. Reson. A* **105**, 310 (1993).
79. S. R. Hartmann and E. L. Hahn, *Phys. Rev.* **128**, 2042 (1962).
80. R. Konrat, I. Burghardt and G. Bodenhausen, *J. Am. Chem. Soc.* **113**, 9135 (1991).
81. S. J. Glaser and G. P. Drobny, *Chem. Phys. Lett.* **164**, 456 (1989).
82. S. J. Glaser and G. P. Drobny, *Chem. Phys. Lett.* **184**, 553 (1991).
83. Ē. Kupče and R. Freeman, *J. Magn. Reson. A* **100**, 208 (1992).
84. Ē. Kupče and R. Freeman, *J. Am. Chem. Soc.* **114**, 10671 (1992).
85. Ē. Kupče and R. Freeman, *J. Magn. Reson. A* **101**, 225 (1993).
86. Ē. Kupče and R. Freeman, *Chem. Phys. Lett.* **204**, 524 (1993).
87. S. J. Glaser, *J. Magn. Reson. A* **104**, 283 (1993).
88. A. Bax, R. Freeman and T. Frenkiel, *J. Am. Chem. Soc.* **103**, 2102 (1981).
89. G. A. Morris and R. Freeman, *J. Am. Chem. Soc.* **101**, 760 (1979).

# 6

# Separating the wheat from the chaff

A high resolution NMR spectrum contains two kinds of information—signals that will help us solve the chemical problem in hand, and the remaining signals that are of no real interest. In many cases the latter are more numerous or more intense and may mask the former. So we must find some specific property that allows us to discriminate between them. It may be based on differences of chemical shift, spin–spin coupling, relaxation times, the number or topology of coupled spins, or on symmetry properties. Some of these techniques are so important that they are in everyday use, others are more specific and perhaps less well-known.

## 6.1 Phase cycling

The most commonly used separation procedure is the phase cycle. In general terms, this arranges for the undesirable artifact to acquire some systematic phase shift $\phi$ where $\phi = 360°/N$, whereas the signal of interest is always accumulated with the same phase. After $N$ scans are collected with the receiver reference phase programmed to follow the phase of the signal, the responses from the artifact are cancelled. The case $N=2$ constitutes difference spectroscopy, where the artifact is alternated in sense while the signal keeps the same sense in both scans.

The most common example of a phase cycle has $N=4$ with $\phi = 0°$, $90°$, $180°$ and $270°$, since these phase shifts are easy to implement on most spectrometers. Other schemes with different phase shifts can be employed. Although it is convenient to think in terms of an actual cycle with a progressive shift in phase, in fact the four scans can be taken in any order, indeed in experiments where the relaxation delays are rather short, there may be a particular ordering that minimizes the undesirable steady-state effects. We consider first of all how the idea of phase cycling first emerged.

## 6.1.1 CYCLOPS

Quadrature detection is employed to determine the sense of the NMR precession in a reference frame rotating at the transmitter frequency. We need to be able to recognize whether the magnetization vector is moving in a clockwise sense or in a counterclockwise sense in the $xy$ plane. This allows the transmitter to be set near the centre of the NMR spectrum where it is most effective, instead of at the edge, allowing the radiofrequency power to be reduced by a factor of four. This also provides a sensitivity improvement of the square root of two because noise is now restricted to a bandwidth that has been reduced by a half. (If the transmitter is at the extreme edge of the spectral width, the receiver accepts noise from a range of frequencies on the far side of the transmitter frequency, where there are no signals.)

Quadrature detection is accomplished by simultaneously recording the components $M\sin\omega t$ and $M\cos\omega t$ of a precessing vector $M$, whereas the single-channel phase detector only records $M\cos\omega t$, leaving an ambiguity about the sense of precession (the sign of $\omega$). For simplicity of notation we might call the two orthogonal signals absorption *abs* and dispersion *dis*. For effective operation the two channels must be well balanced in gain, and the phase shift between them must be 90°. If this is not the case, a weak version of the true spectrum is reflected about the transmitter frequency and is superimposed on the normal spectrum. Contamination by this quadrature image could prove very misleading in complex high resolution spectra.

These shortcomings may be compensated by an ingenious scheme devised by Hoult and Richards.[1,2] This is the first example of a four-step phase cycle— CYCLOPS. It corrects for imbalance between the two channels and for deviations from exact orthogonality. Consider, first of all, the question of amplitude imbalance, where (say) the A channel amplifies correctly but the B channel has a lower gain. A precessing vector $M$ would then appear to trace out an ellipse rather than a circle; this may be represented as the combination of a strong vector $M$-$\varepsilon$ precessing clockwise and a weak vector $\varepsilon$ moving counterclockwise.

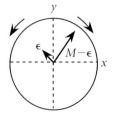
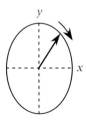

The normal signal arises from $M\text{-}\varepsilon$ while the quadrature image is generated by $\varepsilon$. In the limit that the B channel is completely disabled ($\varepsilon = 0.5\,M$) the scheme reverts to conventional single-phase detection, where there is no discrimination of the sense of precession. Half the intensity of the spectrum is then reflected about the transmitter frequency and superimposed on the remaining half, giving a symmetrical result. Aliasing of this kind is to be avoided at all costs.

The Hoult and Richards scheme continuously exchanges the roles of the two channels A and B so that the imperfections are shared equally. The first scan, with transmitter phase 0°, feeds the absorption signal into the A channel and the dispersion signal (which is weaker than it should be) into the B channel.

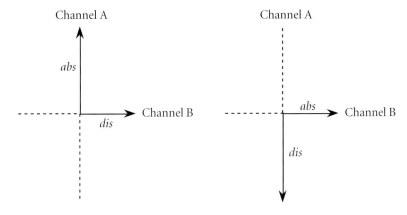

In scan (2) with the transmitter phase 90°, both signal components are advanced by 90°, so now it is the absorption signal that is weaker and the dispersion signal that is stronger, and inverted. The two absorption signals $A_1 + B_2$ are collected in a computer storage location $S$, and the two dispersion signals $B_1 - A_2$ in another location $T$. Both $S$ and $T$ will have signals weaker than the ideal but they are weaker by exactly the same factor $(M - \varepsilon)/M$. The amplitude imbalance has been corrected exactly, and there is no quadrature image in the spectrum.

The second kind of imperfection arises when the two channels are not exactly orthogonal. Suppose that one channel is skewed out of the correct phase alignment by a small angle $\phi$ radians and the other by $-\phi$ radians. These are exaggerated here for the purpose of illustration.

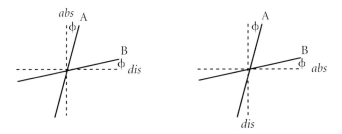

The A channel picks up a slightly reduced absorption signal but also a small fraction of dispersion. The B channel receives a slightly reduced dispersion signal plus a slight leakage of absorption.

$$A_1 = +abs\ \cos\phi + dis\ \sin\phi \qquad\qquad \textbf{6.1}$$

$$B_1 = +dis\ \cos\phi + abs\ \sin\phi \qquad\qquad \textbf{6.2}$$

The parent signal has an amplitude $M\cos\phi$, which is almost the full signal, since $\phi$ is small. However, the cross-talk between the two channels induces some dispersion in channel A and some absorption in channel B.

After a second scan with a 90° increment of the transmitter phase, the two channels receive the signals:

$$A_2 = -dis\ \cos\phi + abs\ \sin\phi \qquad\qquad \textbf{6.3}$$

$$B_2 = +abs\ \cos\phi - dis\ \sin\phi \qquad\qquad \textbf{6.4}$$

Combination of these according to the Richards and Hoult recipe gives identical signals in the storage locations $S$ and $T$, just as if the quadrature detector had been operating ideally:

$$A_1 + B_2 = +2abs\ \cos\phi \qquad\qquad \textbf{6.5}$$

$$B_1 - A_2 = +2dis\ \cos\phi \qquad\qquad \textbf{6.6}$$

The amplitude is slightly reduced, but the relative phase, arctan $(abs/dis)$, is correct.

Although the CYCLOPS scheme compensates exactly for amplitude or phase error if only one kind of imperfection is involved, there are some second-order errors when both occur simultaneously.[2] Present-day spectrometers have pairs of phase-sensitive detectors that are well matched, reducing the necessity for CYCLOPS, but it is still widely used. Another strategy is to double the sampling rate and use the same phase detector for both channels, shifting phase by 90° at each step; this circumvents the problem of amplitude imbalance.

So far this procedure is not strictly a phase cycle, but in practice the phase incrementation of the transmitter is continued (0°, 90°, 180°, 270°) with the appropriate rerouting of the signals to complete a full cycle of four steps, although this is not essential to the CYCLOPS compensation scheme.

| transmitter | storage $S$ | storage $T$ |
|:-----------:|:-----------:|:-----------:|
| 0°   | $+A$ | $+B$ |
| 90°  | $+B$ | $-A$ |
| 180° | $-A$ | $-B$ |
| 270° | $-B$ | $+A$ |

This has the further advantage of compensating another type of artifact, the so-called quadrature glitch. This arises from a small dc offset of the incoming free induction signal, and after Fourier transformation creates a spike in the centre of the spectrum (at zero frequency). It is cancelled by the CYCLOPS scheme.

## 6.1.2  **EXORCYCLE**

A much more general practical problem arises once we set our minds to inventing pulse sequences to manipulate nuclear spins. Ideally the pulses should provide exact $\pi/2$ or $\pi$ rotations about the $x$ or $y$ axes of the rotating frame. In practice, resonance offset and $B_0$ and $B_1$ inhomogeneity conspire to upset the calibration and induce a tilt of the effective field. A phase cycle is a simple way to compensate some of these shortcomings.

Early two-dimensional experiments in heteronuclear ($^{13}$C–H) $J$ spectroscopy showed spurious responses called ghosts and phantoms that were weak replicas of the main $J$ spectrum but displaced in the $F_1$ frequency dimension.[3] Their frequency locations indicate that they arise from magnetization components that are not properly rotated by the radiofrequency pulses. Phantoms originate in transverse magnetization excited by an imperfect $\pi$ refocusing pulse applied to $^{13}$C and they change sign if this pulse is inverted in phase. Working on the principle of difference spectroscopy (inverting the artifacts without inverting the desired signal) we can eliminate the phantom responses by alternating the phase of the $\pi$ pulses (0° and 180°). Ghosts arise from magnetization that is unaffected by the $\pi$ pulse. This allows us to employ the Meiboom–Gill trick,[4] shifting the phase of the $\pi$ pulse by 90° so as to invert the sense of the $^{13}$C spin echo, but leaving the ghost signals unchanged (§ 4.1.6). If the receiver reference phase is also alternated, the echoes accumulate but the ghost signals cancel. A combination of the two modes of compensation leads to the EXORCYCLE scheme[5]:

| Phase of $\pi$ pulse | Receiver reference phase |
| :---: | :---: |
| 0° | 0° |
| 90° | 180° |
| 180° | 0° |
| 270° | 180° |

As with CYCLOPS, it is convenient to write this as a progressive shift of the radiofrequency phase, but in fact the ordering is immaterial. Phase cycles of this type are widely used to compensate for pulse imperfections in multipulse experiments. For example, selective (soft) radiofrequency pulses (§ 5.2) are inherently imperfect when they act on spins that have an appreciable offset, but the resulting artifacts may usually be cancelled by a scheme based on EXORCYCLE.

## 6.1.3  **Coherence transfer pathways**

The CYCLOPS and EXORCYCLE schemes may be thought of as elaborations of difference spectroscopy, but they paved the way to a more general acceptance of the concept of a phase cycle, where the receiver follows the phase of the desired signal component in equal phase increments around a circle, while the artifacts follow a different phase evolution and are cancelled. The key to designing such phase cycles is the study of the appropriate coherence transfer pathways.[6] Consider the energy level diagram for three coupled spin-$\frac{1}{2}$ nuclei. The possible transitions are classified as zero-quantum ($\Delta m = 0$), single-quantum ($\Delta m = \pm 1$), double-quantum ($\Delta m = \pm 2$) and triple-quantum ($\Delta m = \pm 3$). Not all these categories induce observable signals in the receiver coil, but all can be excited by a suitable pulse sequence. Multiple-quantum transitions are discussed in more detail in § 1.3, § 3.4, and § 8.4.2.

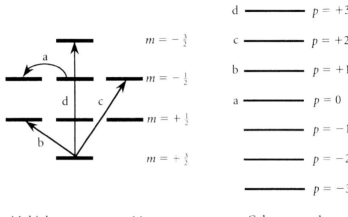

Multiple-quantum transitions                    Coherence order

In analysing multiple pulse experiments, it is useful to distinguish the sign of the change in magnetic quantum number and sketch out a diagram in terms of orders of coherence. In this case there would be seven orders. It is convenient to include $z$ magnetization in the zero-order coherence level. Free precession always maintains the same order of coherence, but radiofrequency pulses cause jumps between various orders. Boltzmann equilibrium corresponds to $p = 0$, and most experiments start with this initial condition. For an observable response, the experiment must end with single-quantum coherence. If quadrature phase detection is used, it selects either $p = +1$ or the counter-rotating component $p = -1$; it is conventional to choose $p = -1$. All pathways that do not terminate on $p = -1$ may therefore be neglected.

Radiofrequency pulses induce changes in the order of coherence. To evaluate what happens, we may need to represent a hard pulse as a cascade of selective pulses. In general, each hard radiofrequency pulse (or sequence of pulses) causes a branching of the coherence transfer paths into several possible orders; we only draw those paths that eventually lead to observable magnetization ($p = -1$). Most experiments start with the creation of transverse magnetization from a system initially at Boltzmann equilibrium.

$$+I_z \xrightarrow{\ (\pi/2)\, \bar{I}_x\ } -I_y$$

$$p = 0 \qquad\qquad p = \pm 1 \qquad\qquad \textbf{6.7}$$

We have already seen (§ 3.4.1) how homonuclear double quantum coherence can be created from antiphase transverse magnetization:

$$+2I_yS_z + 2I_zS_y \xrightarrow{\ (\pi/2)\ \tilde{I}_y\ } +2I_yS_z + 2I_xS_y \xrightarrow{\ (\pi/2)\ \tilde{S}_y\ } +2I_yS_x + 2I_xS_y$$

$$(p = \pm 1) \qquad\qquad\qquad\qquad\qquad (p = \pm 2) \qquad\qquad \textbf{6.8}$$

A $\pi$ pulse interchanges homonuclear zero- and double-quantum coherence:

$$+2I_xS_x - 2I_yS_y \xrightarrow{\ (\pi)\ \tilde{I}_y\ } -(2I_xS_x + 2I_yS_y)$$

$$(p = \pm 2) \qquad\qquad\qquad\qquad (p = 0) \qquad\qquad \textbf{6.9}$$

$$+2I_xS_x + 2I_yS_y \xrightarrow{\ (\pi)\ \tilde{I}_y\ } -(2I_xS_x - 2I_yS_y)$$

$$(p = 0) \qquad\qquad\qquad\qquad (p = \pm 2) \qquad\qquad \textbf{6.10}$$

We shall see below several instances where this interchange is used.

The different orders of coherence have different properties. As we have seen in § 3.4.3, the extent of free precession about the $z$ axis depends on the coherence order $p$ and on the resonance offsets in the appropriate rotating frame. Thus zero-quantum coherence precesses at $(\omega_I - \omega_S)$, whereas double-quantum coherence precesses at $(\omega_I + \omega_S)$, where $\omega_I$ and $\omega_S$ are measured with respect to the transmitter frequency. Consequently the spectra derived from the various orders of coherence can be separated by a suitable choice of the transmitter frequency.

For the present purpose, a more important property is the effect of a phase shift of the propagator representing the radiofrequency pulse (or sequence of pulses). If this propagator undergoes a phase shift $\phi$, and causes a change $\Delta p$ in coherence order, then the resulting coherence suffers a phase shift $\phi \Delta p$ (§ 3.4.4). When a sequence of propagators is used, the phase shifts are cumulative. The accepted convention is that the final coherence observed in the receiver has the order $p = -1$. If the detected signal is derived from coherence of order $+p$, then the change in order $\Delta p = -(p+1)$.

This is the key to separation based on coherence order, and it provides the recipe for designing the appropriate phase cycle. We perform a set of $N$ scans with the radiofrequency pulse phase advanced in steps of $\phi = 360°/N$. For each new scan, the receiver phase is stepped by $-\phi \Delta p$. (This negative sign is a consequence of having chosen $p = -1$ as the coherence level accepted by the quadrature phase detector.) In this manner, the receiver follows the phase of the desired signal component for all $N$ steps, rejecting signals with different values of $\Delta p$. Alternatively, all the radiofrequency pulses in the sequence may be incremented by $\Delta p\phi$, and the detected signals simply added, with the receiver phase held constant.

Note that there is always the possibility of aliasing when we are dealing with phase increments around a circle, since a jump of $\phi$ has the same effect as a jump of $(360°n + \phi)$, where $n$ is an integer. Consequently $N$ must be sufficiently large (and the phase increments $\phi$ correspondingly small) to avoid accepting undesirable aliased components. In many cases $N = 4$ and the phase increments are 90°.

Diagrams of coherence pathways usually indicate only those routes that lead to the desired final coherence order. Sometimes there may be two (or more) such routes. An important example is when pure absorption-mode responses are required. This requires two mirror-image pathways up to the point where the read pulse is applied. An example is shown below.

### 6.1.4 **INADEQUATE**

As an illustration of the design of a practical phase cycle, consider the case of the INADEQUATE experiment,[7] which exploits the fact that double-quantum coherence is twice as sensitive to radiofrequency phase shifts as single-quantum coherence. This permits the separation of signals arising from pairs of coupled $^{13}$C spins from the much more intense signals from isolated $^{13}$C spins (§ 8.4.1). The pulse sequence for the (one-dimensional) INADEQUATE experiment may be expressed as

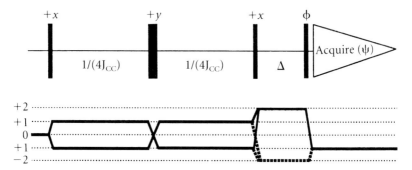

For coupled spin pairs, homonuclear double quantum coherence is created by the first three pulses and precesses during the short period $\Delta$. The coherence transfer pathways are drawn for this case, two pathways leading to coherence level $p = -1$ at the end.

We need to discover how the receiver reference phase ($\psi$) should be cycled to accumulate signals that have passed through double-quantum coherence, while rejecting the rest. The first point to note is that, in the absence of pulse imperfections, the first three pulses act on isolated $^{13}$C spins to give an overall $\pi$ rotation, creating $-z$ magnetization in the $\Delta$ interval. Consequently, if the fourth pulse is cycled clockwise ($+x$, $-y$, $-x$, $+y$) the undesirable but very intense signal from isolated $^{13}$C spins follows in a clockwise sense:

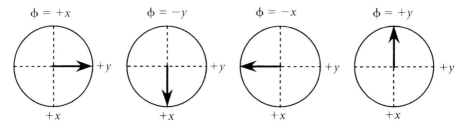

The coupled $^{13}C$–$^{13}C$ systems are affected differently. If we represent the magnetization of the two components of the $J$ doublet by a solid arrow (fast) and a dashed
arrow (slow), they first diverge through 90°, are reflected and interchanged by the
$(\pi)_y$ pulse, and then diverge through a further 90° to the antiparallel configuration.
The third pulse $(\pi/2)_x$ generates double-quantum coherence during the interval $\Delta$.

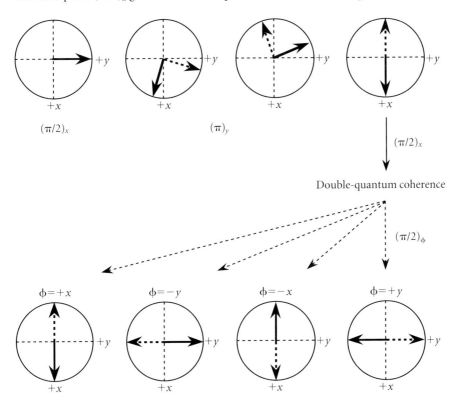

Double-quantum coherence

The fourth pulse $(\pi/2)_\phi$ reconverts the double-quantum coherence into observable
(antiphase) magnetization. Since $\Delta p = +1$ or $-3$, a 90° phase shift of the read pulse
generates a counter-clockwise shift of 90° (a clockwise shift of 270°) for the detected
signal (an antiphase doublet). This is illustrated in the bottom row. Consequently
we must cycle the receiver reference phase ($\psi = +x, +y, -x, -y$) to pick up the
desired response from coupled $^{13}C$–$^{13}C$ systems and suppress the response from
isolated $^{13}C$ spins. In practice further phase cycling (including CYCLOPS and
EXORCYCLE) is usually incorporated to correct for spectrometer imperfections of
various kinds.[8]

## 6.1.5 Less-common phase cycles

Practical considerations have favoured the choice of four-step phase cycles, but
other schemes are possible. The simplest cycle where the sense of rotation can be

defined is $N=3$ and this has been used by Bodenhausen *et al.*[6] Phase cycles with any arbitrary number $N$ have been proposed[9] as a method of phase-encoding information in multiple selective excitation experiments (§ 5.5.5). Instead of performing $N$ separate soft-pulse experiments one after the other, all $N$ selective excitations are carried out simultaneously. The experiment is repeated $N$ times, and in each new scan the pulse phases are coded differently, using phase increments of $360°/N$. If the signals are recorded with the appropriate combinations of the receiver phase, a response can be obtained from each of the $N$ sites in turn, while the remaining responses cancel, being represented by vectors distributed uniformly around a circle. Sensitivity is improved according to the square root of $N$, bringing it into line with that of the corresponding two-dimensional technique.

## 6.2  Pulsed field gradients

The application of field gradients in NMR is almost as old as NMR itself. For example, Hahn[10] demonstrated the effects of field gradients on spin echoes in the presence of molecular diffusion, and Carr and Purcell[11] improved the methodology by deliberately introducing an intense pulsed field gradient into this experiment. Later Anderson *et al.*[12] suggested the idea of suppressing undesirable signals by bracketing a $\pi$ refocusing pulse with two identical gradient pulses, and also proposed a scheme for filtering stimulated echoes by the application of a gradient while the magnetization of interest was stored along the $z$ axis. Here, already, we have most of the basic ideas for the use of pulsed gradients as a selection device in high resolution spectroscopy.

Without doubt, medical imaging and *in vivo* spectroscopy have appropriated the lion's share of pulsed field gradient applications since this is the essence of the imaging method.[13] This section narrows the focus and concentrates on the use of pulsed gradients as an alternative to phase cycling. Phase cycling spreads out the various NMR components in phase space whereas pulsed gradients spread them out in physical space—along a gradient of the magnetic field—then recovers them by a suitable recall gradient. The principal advantage of the gradient method is that a one-shot experiment is possible. Only a single free induction decay is acquired and there is no need to rely on the vagaries of difference spectroscopy, where spectrometer instabilities lead to imperfect cancellation. In situations where the signal-to-noise ratio is relatively high, gradient methods can greatly shorten the overall duration of the measurement by eliminating the repetition inherent in a phase cycle. Finally, for aqueous solutions of large molecules, the relatively fast diffusion of water molecules can be exploited to dissipate the intense water signal in an applied gradient whereas the macromolecules are largely unaffected because they move too slowly (§ 11.2.1). It is now clear that pulsed field gradient experiments will gradually replace phase cycling for many applications, once the appropriate hardware is available on all spectrometers.

Pulsed gradients are not without their own technical problems. One quite serious difficulty stems from the eddy currents created in surrounding conductors and their subsequent slow decay as a function of time. Magnetic resonance imaging technology[13] has provided solutions. One is pre-emphasis, the imposition of a short pulse of reversed polarity just before the main gradient. Another is active shielding, where a secondary coil outside the primary is fed with current of the opposite polarity to protect the surrounding conductors from the primary gradient. A gradient coil in an intense applied field is essentially a loud-speaker, and may vibrate quite strongly and noisily; it should therefore be embedded in a suitable insulator such as epoxy resin, in order to limit the movement. If the field gradient profile needs to be shaped, then the amplifiers used to generate these fields should preferably be linear, must have adequate dynamic range, and the current in the gradient-off condition must be very small indeed.

In a typical high resolution spectrometer, pulsed gradients interact adversely with the deuterium lock signal, temporarily broadening this signal. The effect can be minimized by using short pulses and by blanking the lock channel during the gradient pulse. Finally, we must be careful about spinning the sample if gradient pulses are employed, and avoid it altogether if $x$ or $y$ gradients are used, otherwise intense modulation sidebands are generated. It was probably concern about all these technical difficulties (which are by no means insurmountable) that slowed the acceptance of field gradient methods in high resolution spectroscopy.

Pulse field gradients are normally applied along the $z$ axis, since this avoids interaction with sample spinning (although many experiments are now run without spinning). A linear gradient in the $z$ direction disperses magnetization vectors along a helix, with a pitch inversely proportional to the intensity of the gradient.

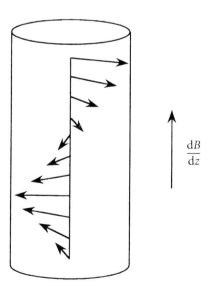

We may distinguish two categories of field gradient experiments. In the first, a single gradient pulse is used to disperse or purge undesirable magnetization components;

in the second, two or more matched gradients are used to defocus and then refocus some particular signal component.

### 6.2.1 **Purging gradients**

The simplest application of an applied field gradient is to disperse transverse components of magnetization in situations where only the $z$-component is of interest, as in Hahn's early experiments with stimulated echoes.[10] The idea was revived for high resolution measurements of spin-lattice relaxation times where imperfections of the $\pi$ inversion pulse were spoiled with a suitable $z$ gradient.[14] If the $\pi$ pulse creates some transverse magnetization, it can have a quite serious effect if the sample has a high spin density, because a radiation damping signal is induced on the coil and rotates the magnetization vector further from the $z$ axis (§ 11.3.2).

A similar pulsed field gradient strategy can be used to disperse transverse components of magnetization in nuclear Overhauser spectroscopy (§ 9.2.5) where the magnetization of interest is stored along the $z$ axis for a short mixing period. These rather primitive applications are often known as homospoil experiments since they were initially implemented by grossly missetting the homogeneity correction currents. If the direction of the spoiling gradient corresponds to the direction of an appreciable naturally occurring gradient, some caution should be exercised to ensure that subsequent radiofrequency pulses do not cause a partial refocusing of the dispersed magnetization. If repeated homospoil pulses are used, they should be of random length or intensity, to avoid accidental refocusing.

### 6.2.2 **Matched field gradients**

Wokaun and Ernst[15] showed theoretically that the sensitivity of multiple-quantum transitions to field gradients is proportional to the order of the coherence (§ 3.4.3). They demonstrated experimentally that all orders of coherence save zero-quantum were purged by a field gradient pulse. Later Maudsley et al.[16] showed that the same phenomenon causes coherence transfer echoes to occur at different delay times for different orders of coherence. During an evolution interval $\tau$, multiple-quantum coherences unwind, spreading out in phase space according to the local intensity of the field due to the gradient. If the coherence order is then switched to $-1$ by means of a mixing pulse, the coherences from various regions take a time $p\tau$ to rewind and form an echo at time $p\tau + \tau$. This permits separation of different orders at different times.

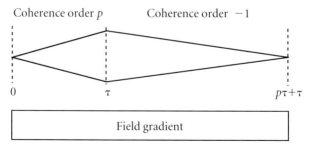

The collection of signals from different orders of coherence at different times is an inconvenient procedure, but an extension of the basic idea circumvents this problem. Bax *et al.*[17] demonstrated experimentally that the different orders of multiple-quantum coherence could be separated by using short gradient pulses of carefully matched durations. Consider the case of $p$-quantum coherence converted to single-quantum coherence ($p = -1$) by a mixing pulse. A gradient of duration $t_g$ is applied just before the mixing pulse, with a gradient of equal intensity but duration $pt_g$ just after the mixing pulse. This selects the coherence transfer echo arising from $p$-quantum coherence, but rapidly disperses magnetization derived from all other orders. If the coherence orders have the same signs then the two gradients must be applied in opposite senses.

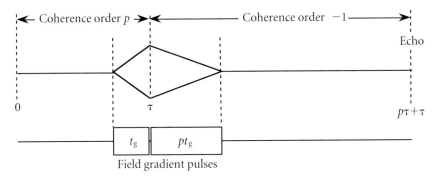

Field gradient pulses

Note that, in addition to the applied field gradient pulses, there are also naturally occurring field gradients operating in this experiment, and they cause the echo to occur at time $p\tau + \tau$, because the single-quantum coherence refocuses $p$ times more slowly than the $p$-quantum coherence defocuses.

All such matched field gradient experiments assume that macroscopic diffusion effects are small enough to be neglected over the duration of the gradient pulse, otherwise the refocusing would be blurred or even nonexistent. Note that in gradient experiments magnetization is not destroyed, but is spread out as a function of the spatial coordinate (usually the $z$ axis); subsequent pulses may have the effect of refocusing this magnetization if they are applied before diffusion or spin–spin relaxation has had time to destroy the signal irreversibly.

The condition for matched gradients is that a given isochromat accumulates equal but opposite phase excursions during the two consecutive gradient pulses, requiring that the gradient intensities, $G_1$ and $G_2$, the pulse durations $\tau_1$ and $\tau_2$, the coherence orders $p_1$ and $p_2$, and (in the case of coherence transfer between different species) the magnetogyric ratios $\gamma_1$ and $\gamma_2$, obey the equality

$$G_1\tau_1 p_1\gamma_1 = -G_2\tau_2 p_2\gamma_2 \qquad\qquad \textbf{6.11}$$

Note that $G$, $p$ and $\gamma$ all carry intrinsic signs.

The matched gradient scheme is the basis for many present-day separation experiments that were previously accomplished by phase cycling. Again it is the analysis of the coherence transfer pathway that is the guide for determining which gradient pulses to use. The familiar homonuclear shift correlation experiment (COSY) may be carried out in a one-shot mode by application of two equal gradients just before and just after the mixing pulse.[15] The advantage is that the duration of the experiment can be greatly curtailed by eliminating the phase cycling, provided that the signal-to-noise ratio does not require appreciable signal averaging.

However, restricting the experiment to a single coherence transfer path ($+1$ to $-1$) introduces phase-modulation in the evolution period, and thus precludes the usual phase-sensitive mode of display unless special tricks are used. The first pulsed-gradient COSY experiments[18] simply use pseudo-echo weighting[19] to extract the pure absorption mode. A more recent extension[20] retains both the ($+1$ to $-1$) and ($-1$ to $-1$) coherence pathways (in double-quantum filtered COSY) by avoiding the application of a gradient pulse during the evolution period. Gradients are only imposed just before and just after the read pulse:

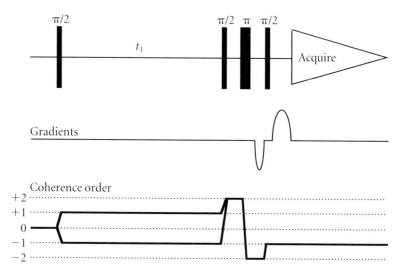

Precautions must be taken to minimize the effects of the phase evolution that occurs during the intervals used for applying pulsed gradients. A $\pi$ pulse is used to refocus the precession of double-quantum coherence, and the acquisition is delayed until the end of the second gradient pulse. Since the mirror-image pathway is not used, there is a 50% loss in signal intensity, but some of this loss can be retrieved by using a $\pi/3$ read pulse instead of $\pi/2$.

Several authors have reluctantly accepted some of the shortcomings of difference spectroscopy and opted for a two-shot pulsed field gradient scheme where the two mirror-image coherence transfer pathways are explored independently and the detected signals are recombined later. For example, Tolman et al.[21] have used this technique to obtain phase-sensitive $^{13}C$–H correlation (HMQC) spectra with the sequence

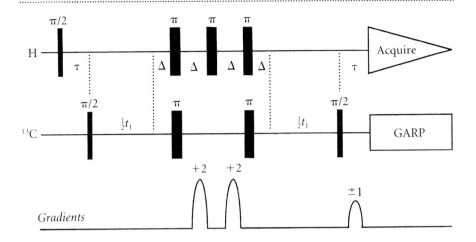

*Gradients*

(The broadband decoupling sequence GARP is treated in § 7.4.3.) The two coherence pathways involve heteronuclear zero- and double-quantum coherences (interchanged by the central $\pi$ pulse) during evolution. Signals acquired *via* these two mirror image pathways are collected separately for each $t_1$ increment by alternating the sense of the final field gradient pulse. Subtraction of these signals gives the real part of the signal at a particular $t_1$ increment, whereas addition gives the corresponding imaginary part.

This scheme derives from a technically more difficult idea proposed by Hurd *et al.*[22] Twofold oversampling in the acquisition dimension with gradient pulses between digitizer sampling points permits both the echo ($p = +1$ to $p = -1$) and antiecho ($p = +1$ to $p = +1$) components to be collected alternately during a single free induction decay. This switched acquisition time technique puts very severe demands on the rate of gradient switching but achieves phase-sensitive two-dimensional spectra in a single shot per $t_1$ increment.

For experiments where it is necessary to refocus chemical shift (resonance offset) effects and yet still employ a field gradient, the gradient pulse may be incorporated into a spin echo module, either (a) before or after the refocusing pulse or (b) split into two equal pulses of opposite polarity:

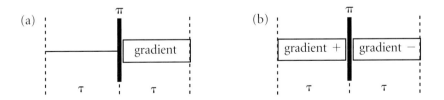

Although in mode (b) the gradient pulses are matched and opposed, they act in the same sense and constitute a purging pulse of duration $2\tau$.

## 6.3 **Filters**

Many of the items in the spectroscopist's toolkit are simple devices that can be slipped into an existing experimental procedure to remove undesirable components and leave a cleaner final spectrum. In some cases they are not an essential part of the manipulation. We group them under the general heading of filters.

### 6.3.1 **Multiple-quantum filters**

We have seen above that either phase cycling or gradient methods can be used to separate signals by temporarily generating a particular order of coherence ($p$) and then reconverting this into observable magnetization. The procedure may be condensed into a simple building block or module that can be inserted into an existing pulse sequence to act as a multiple-quantum filter. Typically it would consist of a pair of hard radiofrequency pulses following the requisite phase cycle (as in the INADEQUATE experiment) or with the appropriate pulsed gradients. The module is inserted into the preparation or mixing periods of a two-dimensional experiment.

A $p$-quantum filter rejects all responses arising from systems of less than $p$ coupled spins. For example, a four-quantum filter has been used to select the spectrum of the four coupled protons of $m$-bromonitrobenzene from a mixture with a three-spin system (2-furancarboxylic acid) and a two-spin system (2,3-dibromothiophene).[23] Elaborations are possible. For example, a filter can be designed to pass only $p$-quantum coherence (a band-pass rather than high-pass filter) and it may even be made to favour a particular topology of the coupling network, although this is seldom 100% efficient. This can be useful in dealing with the spectra of peptides and proteins. A good example is the simplification of the proton correlation spectrum of bovine pancreatic trypsin inhibitor (BPTI) obtained with a four-quantum filter specifically designed to pass the responses from $AX_3$ spin systems in order to identify the alanine residues.[24]

Double-quantum filtration has been widely adopted, notably in conjunction with homonuclear correlation spectroscopy in the famous DQ-COSY technique.[25] The pulse sequence employs three $\pi/2$ pulses, with the phases

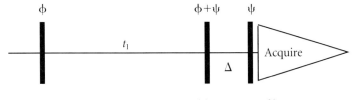

Double-quantum filter

The second pulse creates double-quantum coherence. After a delay $\Delta$ of a few microseconds to allow phase switching, this is reconverted into observable magnetization by the third pulse as in the INADEQUATE experiment described above. The phase $\phi$ is cycled 0°, 90°, 180° and 270° while the receiver reference phase is alter-

nated. Quadrature detection in the evolution dimension is achieved by setting $\psi$ to
0° and 90°. Double-quantum filtration offers two important advantages. The diagonal
peaks (like the cross-peaks) are now principally in antiphase absorption (unlike
those in the COSY technique which are dispersive). This permits the detection of
cross-peaks closer to the diagonal. Secondly, the resonances from isolated spins, in
particular solvent peaks, are suppressed since they cannot support double-quantum
coherence. Although there is a 50% loss in intensity of the cross-peaks compared
with conventional COSY spectra, this is a small price to pay for the improved clarity.

   The general acceptance of pulsed field gradient methods in the early 1990s was
undoubtedly triggered by the excellent spectra obtained by Hurd[26,27] for aqueous
solutions examined by double-quantum filtered correlation spectroscopy (DQ-
COSY). The crucial factor seems to have been the high degree of rejection of $t_1$ noise
in these spectra. This arises because a single-shot experiment gives far less opportunity
for spectrometer instabilities to affect the result, whereas phase-cycling is particularly sensitive to these shortcomings, as in all measurements by difference spectroscopy. The suppression of the enormous water signal is facilitated by the
double-quantum filtration but it is also considerably improved by the fact that
water molecules diffuse significantly through the applied gradient and cannot then
be refocused by mistake. Hurd[26] was able to emphasize the virtual absence of
artifacts by plotting contours at such a low level that they picked up the thermal
baseplane noise (Figure 6.1). Previously, such temerity would have been rewarded
by the appearance of enormous bands of $t_1$ noise. It is not clear whether the use of
actively shielded gradient coils was essential to the success of these experiments, but
the availability of commercial pulsed gradient hardware certainly opened up the
field to many spectroscopists, leading to a veritable explosion of publications on
gradient-enhanced spectroscopy.[27]

## 6.3.2 Purging by the radiofrequency field

The radiofrequency field $B_1$ is proportionately far more inhomogeneous than the
static field $B_0$, and this can be put to good use as a technique for dispersing undesirable signal components. A short spin-lock pulse is applied with the radiofrequency
phase set so that $B_1$ lies along the signal to be retained ($M_y$), whereas signals at right-
angles fan out in all directions around a circle in the $xz$ plane.

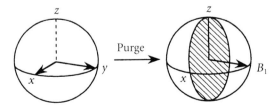

An example of purging in a homonuclear system[28] is the removal of dispersion-
mode contributions after a half-Gaussian selective pulse.[29] This is useful because

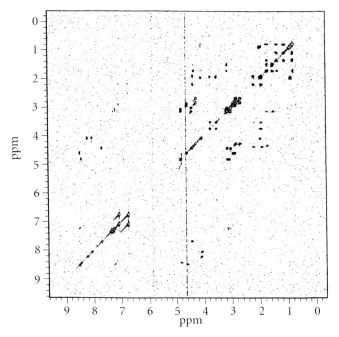

**Figure 6.1** Double-quantum filtered COSY spectrum (400 MHz) of protons in 8mM angiotensin II in water obtained by using pulsed field gradients, no phase cycling and a nonspinning sample. The lowest contour level has been set to pick up baseplane noise in order to emphasize the absence of the water signal (4.6–6.0 ppm) and its associated $t_1$ noise.

the frequency domain excitation profile for the absorption-mode ($M_y$) has a narrow bandwidth whereas the dispersion mode profile ($M_x$) is very broad.

    In heteronuclear systems, undesirable antiphase $S$-spin components (for example $2I_zS_y$) may be purged by a ($\pi/2$) pulse applied to the $I$ spins, thereby converting them into unobservable heteronuclear multiple-quantum coherence. This has been used in the refocused INEPT experiment to remove dispersion contributions and to restore the usual binomial intensities to the spin multiplets.[30] As always, precautions must be taken to avoid reconversion of multiple-quantum coherence into observable single-quantum coherence by any subsequent pulses. Creation of unobservable multiple-quantum coherence is a general method for suppressing unwanted signal components.

### 6.3.3 Z filters

This idea of providing a safe haven for desired components of magnetization while scrambling the phases of all others is also exploited in the z filter.[31] Suppose we wish to retain $I_y$ but discard other transverse components. A ($\pi/2$)$_x$ pulse creates longitudinal magnetization which can be safely stored along the z axis for a short time $\tau$ until returned to the $xy$ plane by a second $(\pi/2)$ pulse. It is customary to cycle the phase of either the first pulse (and all preceding pulses) or the second pulse (and all subsequent pulses together with the receiver reference phase) according to the CYCLOPS procedure.[1] A possible cuckoo in the nest is two-spin order $(2I_zS_z)$ which would also be retained during $\tau$, but the second pulse converts this into multiple-quantum coherence, and we endeavour to locate the $z$-filter immediately before acquisition or just before a period of free precession so that this multiple-quantum coherence is not reconverted into observable magnetization. Zero-quantum coherence will also be retained during $\tau$, but may be suppressed by collecting several scans while varying $\tau$, in steps comparable with the reciprocal of the chemical shift difference. With these caveats, a $z$ filter can be effective in eliminating phase and intensity distortions that can arise in spin echo experiments where the timing parameter cannot be correctly matched to the relevant coupling constants. It is a useful module, provided care is taken with its positioning within the pulse sequence. Typically $\tau$ would be several tens of milliseconds. An analogous scheme, albeit with considerably longer variable delays, is used to eliminate coherent effects during the cross-relaxation interval in transient nuclear Overhauser experiments (NOESY).[32]

### 6.3.4 Isotope filters

Providence has decreed that the interesting atoms of carbon and nitrogen should have rather low abundances of spin-$\frac{1}{2}$ nuclei. Consequently we are often faced with the task of discriminating between protons attached to $^{13}$C or $^{15}$N spins and those attached to $^{12}$C or $^{14}$N. This is accomplished with a device known as an isotope filter. The basic principle is that proton spin echoes are modulated by the heteronuclear coupling if the heteronucleus is flipped by a $\pi$ pulse synchronized with the $\pi$ refocusing pulse applied to the protons (§ 4.2.3). If the time of the echo is chosen to be $1/(J_{CH})$ or $1/(J_{NH})$, the proton vectors attached to $^{13}$C or $^{15}$N are brought into alignment along the $-y$ axis whereas the proton vectors attached to $^{12}$C or $^{14}$N spins are carried to the $+y$ axis[33]. For the $^{13}$C case the vector diagram may be drawn as

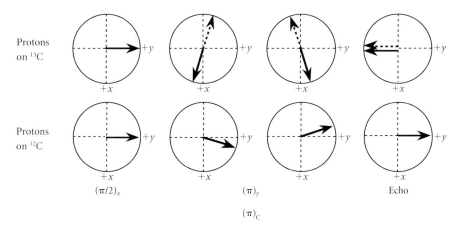

Protons on $^{13}$C

Protons on $^{12}$C

$(\pi/2)_x$       $(\pi)_y$       Echo

$(\pi)_C$

On every second scan the $(\pi)_C$ pulse is omitted, and all vectors are carried to the $+y$ axis; thus the $^{13}$C satellite spectrum is obtained by difference spectroscopy. The same cluster of pulses is widely used in the bilinear rotation decoupling (BIRD) module[34] which can invert directly-bonded $^{13}$C protons but not $^{12}$C protons (§ 4.2.8). There is also a related excitation module[35] called TANGO (§ 4.2.8).

For work on biological macromolecules the isotopes $^{13}$C and $^{15}$N are employed in a different manner. Here the main concern is simplification of the proton spectra, accomplished by studying isotopically labelled proteins, often enriched in both $^{13}$C and $^{15}$N. Multidimensional spectroscopy is used (§ 8.6.2), with the information spread out in the $^{13}$C and $^{15}$N dimensions as well as two proton dimensions.[36] Alternatively, soft pulse excitation schemes can be employed (§ 5.5.5).

## 6.3.5 J filters

The BIRD and TANGO modules (§ 4.2.8) also discriminate between direct and long range $^{13}$C–H couplings, since the former are much larger in magnitude and give a $\pm 180°$ divergence of proton vectors in a short interval $1/(^1J_{CH})$, during which time the weak long-range couplings induce very little divergence. An analogous separation module has been employed in relayed heteronuclear correlation spectroscopy

$$H_{remote} \rightarrow H_{adjacent} \rightarrow {}^{13}C \qquad \textbf{6.12}$$

where it may be advantageous to suppress the direct-connectivity signals, leaving only the remote-connectivity responses.[37] Filtration is achieved by introducing a short delay $\tau$ into the sequence to allow antiphase proton magnetization vectors to diverge through appreciable angles due to $^1J_{CH}$, while the corresponding vectors controlled by $^nJ_{CH}$ are scarcely affected. The antiphase magnetization is then converted into unobservable heteronuclear multiple-quantum coherence with a $\pi/2$ pulse applied to the $^{13}$C spins. Repetition with different lengths of the delay $\tau$ extends the effectiveness over a range of coupling constants. This can be thought of as a low-pass $J$ filter.

## 6.3.6 **Chemical shift filters**

Similar principles are involved when we wish to select a given signal on the basis of its chemical shift while suppressing the rest by difference spectroscopy. The first application of this idea was an experiment by Hore *et al.*[38] to select a desired double-quantum frequency in a proton–proton version of the INADEQUATE technique. What would normally have been the evolution period of the two-dimensional experiment $(t_1)$ was varied in suitable steps (2 ms) and the corresponding spectra added together. These signals add coherently only when the transmitter is adjusted so that the double-quantum coherence does not precess in a reference frame rotating in synchronism with the transmitter frequency. The same principle applies to selection of single-quantum coherence by an analogous filter, although such selectivity would probably be more cleanly achieved by soft pulse experiments. Note that precession through integer multiples of 360° also gives rise to coherent addition, unless the delay is suitably randomized. A refinement[39] based on the constant time spin echo technique removes the effect of homonuclear spin–spin coupling and thus behaves as a true chemical shift filter.

### 6.3.7 **Binomial filters**

Suppression techniques that rely on random delays are inherently inefficient because it is difficult to achieve a truly isotropic distribution of vectors around a circle. It is usually preferable to find a deterministic scheme for signal cancellation. For exact cancellation at a particular offset $\Delta f$, we may simply introduce a fixed delay $\tau$ before acquisition of the free induction decay, thus creating a phase shift $2\beta = 2\pi\Delta f\tau$ for a signal component at that offset.[40] If the delay $\tau$ is set equal to $1/(2\Delta f)$, the combination with the corresponding free induction decay acquired without a delay cancels the offending signal. There are two quite distinct ways to perform this experiment. The first is a post-acquisition data processing scheme where a free induction decay is added to a delayed version of the same data set. The second involves adding the signals from two consecutive experimental scans, one with the requisite delay before acquisition. This mode is subject to perturbation by instrumental instabilities analogous to those that create $t_1$ noise. In both cases the rejection profile of the sum of the two signals follows a cosine curve:

$$S(1{:}1) = \sin(2\pi\Delta ft) + \sin(2\pi\Delta ft + 2\beta) = 2\sin(2\pi\Delta ft + \beta)\cos\beta \qquad \textbf{6.13}$$

The exact null condition is quite sharp, and occurs for $\tau = 1/(2\Delta f)$ where $2\beta = \pi$ radians, and the signal changes sign as we cross the null (Figure 6.2).

Most suppression experiments work better if there is more tolerance towards small changes in offset. If we think of the basic experiment as analogous to the 1:1

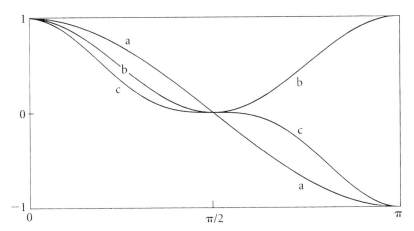

**Figure 6.2** Response curves for binomial filters as a function of $\beta$. (a) The 1:1 filter, which has a sharp null condition at $\beta = \pi/2$. (b) The 1:2:1 filter, which has a broader null condition. (c) The 1:3:3:1 filter which has an even broader null condition.

'jump and return' water suppression sequence (§ 11.2.4), we can imagine higher-order binomial combinations.[41] For example, the null condition can be broadened by the equivalent 1:2:1 combination. This involves collecting three separate free induction decays with phase shifts 0, $2\beta$ and $4\beta$, and then combining them with the relative intensities 1:2:1.

$$S(1:2:1) = \sin(2\pi\Delta ft) + 2\sin(2\pi\Delta ft + 2\beta) + \sin(2\pi\Delta ft + 4\beta)$$
$$= 4\sin(2\pi\Delta ft + 2\beta)\cos^2\beta \qquad\qquad \textbf{6.14}$$

The delays are set at $\tau_1 = 0$, $\tau_2 = 1/(2\Delta f)$ and $\tau_3 = 1/(\Delta f)$, and the null is now broader because of the $\cos^2\beta$ dependence (Figure 6.2). Even higher tolerance at the null condition is achieved with the equivalent 1:3:3:1 combination where the phase shifts are 0, $2\beta$, $4\beta$ and $6\beta$.

$$S(1:3:3:1) = \sin(2\pi\Delta ft) + 3\sin(2\pi\Delta ft + 2\beta) + 3\sin(2\pi\Delta ft + 4\beta) + \sin(2\pi\Delta ft + 6\beta)$$
$$= 8\sin(2\pi\Delta ft + 3\beta)\cos^3\beta \qquad\qquad \textbf{6.15}$$

Note that all three schemes introduce frequency-dependent phase shifts into the spectra.

The same quality of suppression can be translated to zero frequency (the transmitter frequency) by taking the appropriate sums and differences of free induction decays in experiments that may be designated $1{:}\bar{1}$, $1{:}\bar{2}{:}1$ and $1{:}\bar{3}{:}3{:}\bar{1}$. The response profiles are now proportional to $\sin(\beta)$, $\sin^2(\beta)$ and $\sin^3(\beta)$, respectively. Binomial filters may be used as building blocks in various pulse sequences, for example as a form of $J$ filter[42] or to suppress aliasing in two-dimensional spectroscopy.

## 6.4  Spectral editing

In $^{13}C$ spectroscopy we often need to be able to identify $CH_3$, $CH_2$, $CH$ and quaternary carbon sites for assignment purposes. Although this information is present in the proton-coupled $^{13}C$ spectrum (or the off-resonance decoupled spectrum) sensitivity considerations and the need for simple spectra usually demand broadband decoupled $^{13}C$ spectra. Procedures that break down the decoupled $^{13}C$ spectrum into subspectra based on the number of attached protons are known as editing methods. They may be based on the dephasing effect of CH coupling[43] or, more elegantly, on $^{13}C$ spin echo modulation.[44-48] Quaternary $^{13}C$ sites have such small C–H couplings that dephasing can usually be neglected in this context.

### 6.4.1  Spin echo modulation

The spin echo modulation method may take several forms; one implementation can be written as

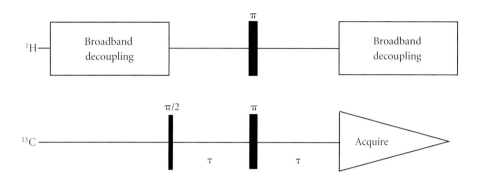

The echo modulation from the four possible sites follows the time development

$$S_{2\tau}(C) = S_0 \qquad\qquad\qquad\qquad\qquad\qquad \textbf{6.16}$$

$$S_{2\tau}(CH) = S_0 \cos(2\pi J_{CH}\tau) \qquad\qquad\qquad \textbf{6.17}$$

$$S_{2\tau}(CH_2) = S_0 \cos^2(2\pi J_{CH}\tau) \qquad\qquad\qquad \textbf{6.18}$$

$$S_{2\tau}(CH_3) = S_0 \cos^3(2\pi J_{CH}\tau) \qquad\qquad\qquad \textbf{6.19}$$

If the interval $2\tau$ is chosen to be $1/(J_{CH})$, the CH and CH$_3$ signals are inverted, whereas the CH$_2$ (and quaternary) signals are upright.[47,48] This serves as a good method for deciding on the parity, that is to say, whether there is an odd or even number of protons directly attached to carbon. Often this information is sufficient to resolve questions of assignment. When complete separation into subspectra is required, four experiments may be performed with four different settings of $\theta = 2\pi J_{CH}\tau$. The optimum setting for separation of CH$_3$ and CH signals[48] occurs for the magic angle where $\cos^2\theta = 1/3$. In practice, the separation is almost as efficient with $\theta = \pi/3$. Four subspectra are recorded with $\theta = 0$, $\pi/3$, $2\pi/3$ and $\pi$, and are then combined in the appropriate linear combinations to give the edited subspectra. Where there is a range of $^1J_{CH}$ values, the separation is imperfect, and we have cross-talk between the different subspectra.

## 6.4.2 DEPT editing

A more popular method of editing is based on the different orders of multiple-quantum coherence that can be excited in the CH, CH$_2$ and CH$_3$ groups, the DEPT technique.[49] The principal advantage over spin echo editing methods is that the separation is achieved by employing different pulse flip angles rather than different timing delays; this renders it less sensitive to the natural variation of $^1J_{CH}$ coupling constants. The mechanism of DEPT editing is complex and is treated in detail by the product operator formalism in § 3.4.6. The pulse sequence may be written:

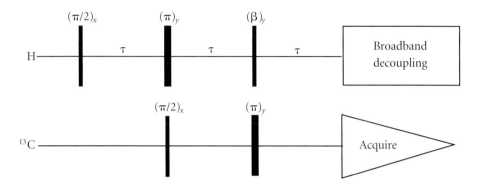

The delay $\tau = 1/(2J_{CH})$. The technique exploits the creation and evolution of heteronuclear multiple-quantum coherence, together with the flip angle effect of the $\beta$ pulse. The treatment in § 3.4.6 demonstrates that the observed signals have the following time dependences:

$$S_t(CH) = S_0 \sin\beta \qquad\qquad\qquad \mathbf{6.20}$$

$$S_t(CH_2) = S_0 \sin 2\beta \qquad\qquad\qquad \mathbf{6.21}$$

$$S_t(CH_3) = \tfrac{3}{4} S_0 (\sin\beta + \sin 3\beta) \qquad\qquad\qquad \mathbf{6.22}$$

(The quaternary carbon sites are not excited in the DEPT technique.) When flip angles $\beta = \pi/4$, $\pi/2$ and $3\pi/4$ are employed, appropriate linear combinations of these spectra give a clear separation into $^{13}$C subspectra based on the number of directly attached protons. If DEPT spectra are recorded without proton broadband decoupling, the natural spread of $J_{CH}$ values cause phase and intensity anomalies within the spin multiplets; they can be suppressed by elaborations of the sequence, for example the DEPT$^{++}$ scheme.[50]

A simpler and more effective scheme for editing $^{13}$C spectra is the SEMUT technique.[51,52] The pulse sequence may be written:

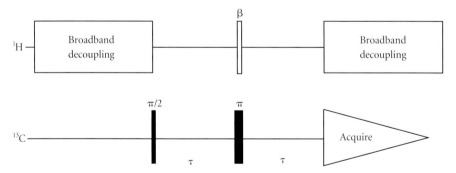

The delay $\tau$ is set equal to $1/(2J_{CH})$. The proton $\beta$ pulse converts $^{13}$C magnetization into unobservable heteronuclear multiple-quantum coherences with an efficiency which depends on the flip angle $\beta$. The dependence of the observed $^{13}C$ intensities for $CH_3$, $CH_2$, CH groups follows the same expressions as the spin echo editing method ($\cos^n\beta$) if $2\pi J_{CH}\tau$ is replaced by $\beta$. Quaternary carbon sites are excited by the SEMUT technique but are essentially unmodulated. As outlined above for the spin echo method, four spectra with $\beta = 0$, $\pi/3$, $2\pi/3$ and $\pi$, are recorded and the appropriate linear combinations give the required subspectra.

Although editing by DEPT or SEMUT relies on changing the flip angle $\beta$, the delays $\tau$ must nevertheless be correctly adjusted to the condition $\tau = 1/(2J_{CH})$. In practice there is a natural spread of $^1J_{CH}$ values making it impossible to satisfy this condition exactly. The result is a phenomenon known as cross-talk, where weak spurious signals originating in (say) a $CH_3$ group appear in the subspectrum of $CH_2$ groups, leading to possible false assignments. Cross-talk only occurs in a downward direction, for example from $CH_3$ to $CH_2$ or CH. Refinements[52] called SEMUT-GL and DEPT-GL introduce a purging sandwich that greatly attenuates cross-talk and thus allows for the natural variation in magnitude of $^1J_{CH}$. These more sophisticated procedures yield edited spectra with the least ambiguity.

# 6.5 Post-acquisition processing

Even after all the spin gymnastics are over, and the NMR signal has been acquired, there are still useful ways to improve the quality of the data before presentation. With the increasing sophistication of laboratory computers there is growing emphasis on data processing schemes of this type.

## 6.5.1 Maximum entropy

Despite the remarkable success of the Fourier transform algorithm, there are still situations where alternative schemes for converting the time-domain signal into a high resolution spectrum may be preferable. One of these is the maximum entropy method. The advantage of this data processing technique lies in the possibility of removing artifacts caused by some perturbation of the raw experimental data. We shall see below that the maximum entropy reconstruction can do this because it is an inverse method.

We start with an experimental free induction decay consisting of $N_t$ complex numbers $d_k$, which each have an uncertainty measured by $\sigma_k$, the standard deviation of the instrumental noise. Instead of calculating the spectrum directly by Fourier transformation in the normal manner, we seek a trial spectrum (made up of $N_f$ ordinates $x_m$) that is compatible with $d_k$ within the uncertainty represented by $\sigma_k$. Fourier transformation is performed in the reverse sense, converting $x_m$ into a trial time-domain data set $y_k$, which is then compared with the experimental data set $d_k$ according to the usual $\chi^2$ statistic:

$$\chi^2 = \sum_{k=1}^{N_t} |y_k - d_k|^2/\sigma_k^2 \qquad \textbf{6.23}$$

This trial time-domain signal is taken to be consistent with the experimental free induction decay if $\chi^2$ is equal to the number of data points $N_t$, because then the uncertainty at any given ordinate $y_k$ is of the order of the experimental root-mean-square noise.

At this point it is a convenient fiction to imagine a vast array of possible trial spectra. Most of them will not be consistent with the experimental data set, but because the experimental free induction decay is noisy, there will be a small number of trial spectra that *do* satisfy the criterion that $\chi^2 = N_t$. We may call these the acceptable solutions and then try to decide which of them is the best. It is not possible to find the true solution to the problem; it could be any one of the acceptable solutions defined above. We must single out a trial spectrum based on some other criterion. We do this by choosing the one that has the least information content. This may at first seem surprising, but it simply reflects the laudable aim of excluding any features in the solution for which there is no sound evidence in the experimental data. The least information content implies the maximum entropy of the trial spectrum. The entropy may be defined in different ways, but one simple example is

$$S = -\sum_{m}^{N_f}(x_m/b)\ln(x_m/b) \qquad\qquad \textbf{6.24}$$

where $b$ is a normalization parameter, chosen to prevent the program from increasing the entropy merely by raising the baseline of the spectrum. This is the "maximally noncommittal" solution, usually called the maximum entropy reconstruction.[53–60] This definition of entropy only allows positive values of $x_m$, and it is sometimes argued that this is an advantage of the method, since NMR absorption signals should always be positive, whereas noise can have either sign. More sophisticated expressions can be written for the entropy to allow the signal phase to be considered, and these are probably more appropriate to NMR spectra in general.

In practice, of course, there is no question of guessing an entire set of possible trial spectra; we take a single trial spectrum and refine it iteratively. It is conventional to start from a completely featureless trial spectrum, so that no bias is introduced. Fortunately it can be shown that there is a unique maximum entropy solution for a given experimental data set; it does not matter where the iteration starts since there are no false minima in the multidimensional space used for the search.

The power of the method lies in the elimination of instrumental artifacts. If the experimental free induction decay is incomplete or corrupted in some manner, we can impose an identical perturbation on the trial time-domain data so that the comparison algorithm works with both data sets on an equal footing. For example, there may be premature truncation of the time domain signal. Normally we take care to acquire a free induction signal for a sufficiently long period that it has fully decayed (or we impose a weighting function to accelerate the decay). However, in multidimensional spectroscopy it is not always feasible to gather data for a long enough time in the evolution dimensions, and some truncation may have to be tolerated. Direct Fourier transformation of a truncated time-domain signal gives sinc function distortions of the lineshapes. A related situation occurs when the first few points of the free induction decay are corrupted by radiofrequency pulse break-through; it would be better if this section of the response were discarded rather than have it distort the spectrum.

Alternatively, we may consider that any free induction decay is intrinsically incomplete, because there is no signal for $t < 0$, that is, before the excitation pulse. If it were possible to record a signal that is symmetrical in time (as in a whole spin echo), then there would be no dispersion-mode contributions in the spectrum. Maximum entropy reconstruction offers a chance to generate the entire signal (for both positive and negative values of $t$) and thus eliminate the dispersion mode. This would be a decided advantage in certain types of two-dimensional experiment (§ 8.7.1).

Maximum entropy sidesteps these problems by introducing an identical blanking of the trial free induction decay, comparing the raw and trial data only within the appropriate time interval, where the experimental data set contains no discontinuity or corruption.

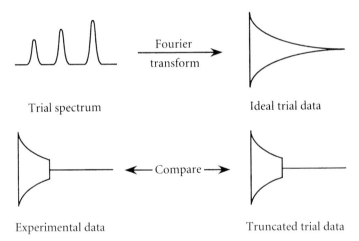

Trial spectrum          Ideal trial data

Experimental data          Truncated trial data

Once the maximum entropy reconstruction has been calculated, we may use the trial spectrum *before* any perturbation is introduced; it serves as an artifact-free version of the experimental data.

The maximum entropy algorithm is nonlinear, and some care must be taken with its application, since the noise level in the reconstructed spectrum is greatly reduced. We must continually remind ourselves that in a nonlinear regime, sup-pression of noise does not necessarily imply improved sensitivity (§ 10.3.1). Signals weaker than the general level of the noise are also suppressed by the algorithm, and it is just such signals that we need to detect if the sensitivity is to be improved. Indeed, the very crux of the maximum entropy reconstruction is the omission of any feature for which there is no good evidence in the raw experimental data. An even more insidious problem is that the noise reduction applies only to baseline noise; the signal peaks still carry the normal level of noise and their intensities remain just as unreliable as ever, although the cleaner appearance of the maximum entropy reconstruction might lead us to think otherwise (§ 10.3.1). Similar caveats apply to attempts to improve resolution by nonlinear data processing; a narrowing of the linewidth does not necessarily imply that two previously unresolved lines can be separated.[55]

### 6.5.2 **Symmetry filters**

High resolution NMR spectra obey certain symmetry rules, and this information can be exploited to distinguish true signals from artifacts. Correlation spectroscopy (COSY) provides one very simple example—if we exclude differences in linewidth in the two dimensions, the two-dimensional spectrum has reflection symmetry with respect to the principal diagonal. The spurious responses known as $t_1$ noise lack this symmetry feature; they consist of bands of interference running parallel to the $F_1$ axis. Baumann *et al.*[61] use a nonlinear data processing algorithm to impose this reflection symmetry on the experimental data, thereby suppressing $t_1$ noise. Each pixel $S(x,y)$ in the two-dimensional spectrum is compared with its symmetry-related counterpart $S(y,x)$ and the lower ordinate placed at both locations; true signals are therefore retained, but artifacts are reduced down to the general level of the noise, except in the unlikely case that two artifacts accidentally satisfy the symmetry relation. This symmetrization operation has proved useful in several other contexts.

The separation of chemical shifts from coupling constants in homonuclear systems is one of the most difficult separation techniques of all, and has remained an unfulfilled challenge to spectroscopists since it was first proposed.[62] While we are well accustomed to dealing with broadband decoupled $^{13}C$ spectra, we have become reconciled to the fact that proton spectra are sometimes so congested with overlapping multiplets that they cannot be assigned and analysed.

Two-dimensional proton $J$ spectroscopy holds out the promise of broadband decoupled proton spectra where all the spin–spin splittings have been eradicated,[63] but this goal has not yet been satisfactorily realized in practice. However, certain display modes can be devised (§ 8.5.3) where the two-dimensional spin multiplets possess $C_4$ symmetry and this feature can be exploited to separate signals from adjacent overlapping multiplets.[63] In this case the centres of symmetry all lie on the line $F_1 = 0$, and the search is therefore one dimensional. At each step the program searches a $50 \times 50$ Hz test zone for responses that satisfy the symmetry criterion. All signals within the test zone are processed with an algorithm that examines the ordinates at eight symmetry-related locations, and replaces them all with the lowest (absolute) value. The result is then subtracted from the experimental data matrix, thus minimizing overlap effects with nearby multiplets. In effect, the algorithm acts as a symmetry filter rejecting all responses that do not conform to the $C_4$ test. (Note,

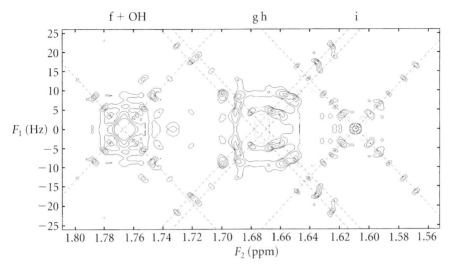

**Figure 6.3** A congested region of the 400 MHz two-dimensional proton *J* spectrum of 1-dehydro-testosterone, after symmetrization. There are four overlapping spin multiplet patterns, with responses that fall on the diagonal lines (dashed), together with a singlet from the OH group that is exactly degenerate in frequency with proton f. The chemical shifts of protons g and h differ by less than 0.01 ppm but they are cleanly separated by the symmetry filter.

however, that if this trick is used to discriminate against noise, it does not improve the sensitivity (§ 10.3).) The power of the method to separate interpenetrating two-dimensional multiplets is illustrated in Figure 6.3 for a crowded section of the two-dimensional proton *J* spectrum of 1-dehydrotestosterone.[63] Two adjacent multiplets (*g* and *h*) that are only 0.01 ppm apart are separated by this symmetry filter. Once all the multiplets have been located, a spectrum of chemical shift frequencies can be constructed (§ 8.5.3).

### 6.5.3 **Pattern recognition**

Two-dimensional correlation spectroscopy supplies an embarrassing wealth of information and it is quite tedious to extract all the chemical shift and spin coupling parameters by inspection. In the absence of passive splittings and degenerate couplings, the basic COSY cross-peak has a well-defined symmetry—the familiar doubly antiphase square array with splittings equal to the active coupling constant (§ 1.4.2 and § 8.2.3). It is easy to imagine a pattern-recognition algorithm[64–67] that searches the experimental data array for this arrangement of four peaks with the requisite sign alternation, a feature that proves very useful in discriminating against spurious responses. The sum of the moduli of the intensities is then placed at the centre of symmetry.

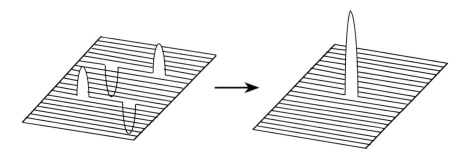

The coalescence of the spin multiplet structure leads to a welcome increase in the signal-to-noise ratio. The existence of a similar cross-peak symmetrically disposed with respect to the diagonal is also taken into account, and the two intensities combined. In the more general case, each passive coupling duplicates the antiphase square pattern, necessitating a higher-dimensional search. Eventually the two-dimensional COSY spectrum is reduced to a set of singlets at the chemical shift coordinates and the coupling constants are stored in a table together with their assignments. All the connectivity and coupling information is retained, but in a much more convenient format. The main drawback of the method is the protracted nature of the search, since every location in the two-dimensional spectrum must be examined as a possible centre for a cross-peak.

The search program can be considerably simplified by supplying prior information about the proton chemical shifts, since cross-peaks must be centred at the chemical shift coordinates.[68] This information can be obtained from the two-dimensional $J$ spectrum, as described in the previous section. These chemical shift values may be used to construct a two-dimensional grid of proton shift coordinates. If such a chemical shift grid is superimposed on the corresponding COSY, TOCSY or NOESY spectrum, each cross-peak must necessarily lie at an intersection on the grid, but of course, not all intersections correspond to cross-peaks.

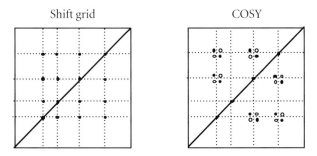

Shift grid                                      COSY

This enormously simplifies any search for cross-peaks, since instead of considering every point in the two-dimensional matrix as a possible centre, we need only examine those lying on intersections on the chemical shift grid. The search program

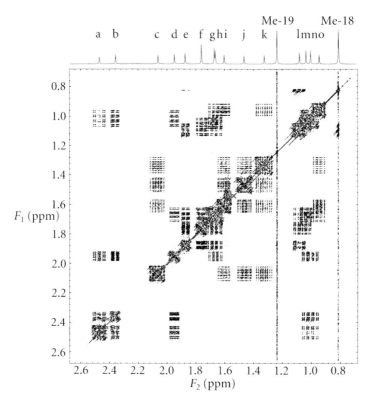

**Figure 6.4** Part of the 400 MHz two-dimensional double-quantum filtered correlation spectrum of 1-dehydrotestosterone with the decoupled proton spectrum shown along the top margin. There is a very congested region (centre right) suspected of harbouring eight overlapping cross peaks. There are also two weak correlations involving the protons of the methyl group (Me-18) and partly obscured by $t_1$ noise from the intense methyl resonance.

uses the symmetry filter described in the previous section, thus suppressing any overlapping responses that do not possess the correct symmetry. The integral over the test zone (ignoring signs) serves as a measure of the confidence that there is a genuine cross-peak at that location. Such automated data reduction procedures relieve the spectroscopist of the task of making subjective judgements about the presence or absence of a correlation peak in a two-dimensional spectrum, and they offer an effective way to analyse very crowded spectra.

The operation of this pattern recognition method is illustrated for the very crowded two-dimensional correlation spectrum (§ 8.2.2) of the protons in 1-dehydrotestosterone (Figure 6.4). This spectrum contains one region of severe overlap where it is suspected that there are eight interpenetrating cross-peaks. The pattern recognition algorithm, operating only in the vicinity of points on the chemical shift grid, locates 55 cross-peaks, and in particular recognizes these eight overlapping responses.[69] The correlations may then be presented as a table, with

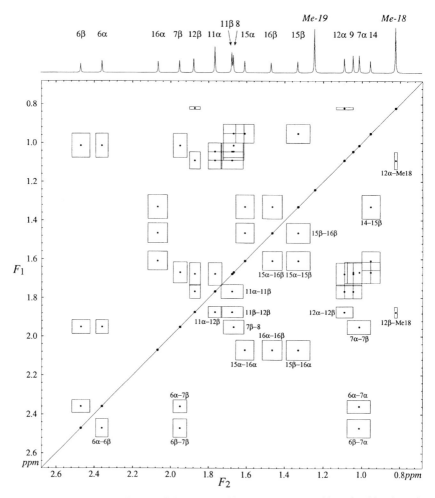

**Figure 6.5** The spectrum of Figure 6.4 processed by a pattern recognition algorithm based on cross-peak antisymmetry with respect to vertical and horizontal axes through the chemical shift coordinates. Correlations are now shown as dots surrounded by rectangles indicating the perimeter of the multiplet structure. This confirms the suspicion that there are eight overlapping cross-peaks (centre right) and two weak correlations between the methyl group (Me-18) and protons 12α and 12β.

estimates of the confidence levels, or as points on a two-dimensional chart with an indication of the perimeter of each cross-peak (Figure 6.5).

# References

1. D. I. Hoult and R. E. Richards, *Proc. Roy. Soc. (London)* A **344**, 311 (1975).
2. E. O. Stejskal and J. Schaefer, *J. Magn. Reson.* **14**, 160 (1974).

3. G. Bodenhausen, R. Freeman, R. Niedermeyer and D. L. Turner, *J. Magn. Reson.* **26**, 133 (1977).

4. S. Meiboom and D. Gill, *Rev. Sci. Instr.* **29**, 688 (1958).

5. G. Bodenhausen, R. Freeman and D. L. Turner, *J. Magn. Reson.* **27**, 511 (1977).

6. G. Bodenhausen, H. Kogler and R. R. Ernst, *J. Magn. Reson.* **58**, 370 (1984).

7. A. Bax, R. Freeman and S. P. Kempsell, *J. Am. Chem. Soc.* **102**, 4849 (1980).

8. A. Bax, *Two-Dimensional Nuclear Magnetic Resonance in Liquids*, Delft University Press, Dordrecht, 1982.

9. Ē. Kupče and R. Freeman, *J. Magn. Reson. A.* **105**, 310 (1993).

10. E. L. Hahn, *Phys. Rev.* **80**, 580 (1950).

11. H. Y. Carr and E. M. Purcell, *Phys. Rev.* **94**, 630 (1954).

12. A. G. Anderson, R. L. Garwin, E. L. Hahn, J. W. Horton, G. L. Tucker and R. M. Walker, *J. Appl. Phys.* **26**, 1324 (1955).

13. P. C. Lauterbur, *Nature*, **242**, 190 (1973).

14. R. L. Vold, J. S. Waugh, M. P, Klein and D. E. Phelps, *J. Chem. Phys.* **48**, 3831 (1968).

15. A. Wokaun and R. R. Ernst, *Chem. Phys. Lett.* **52**, 407 (1977).

16. A. A. Maudsley, A. Wokaun and R. R. Ernst, *Chem. Phys. Lett.* **55**, 9 (1978).

17. A. Bax, P. G. de Jong, A. F. Mehlkopf and J. Smidt, *Chem. Phys. Lett.* **69**, 567 (1980).

18. P. Barker and R. Freeman, *J. Magn. Reson.* **64**, 334 (1985).

19. A. Bax, R. Freeman and G. A. Morris, *J. Magn. Reson.* **43**, 333 (1981).

20. A. L. Davis, E. D. Laue, J. Keeler, D. Moskau and J. Lohman, *J. Magn. Reson.* **94,** 637 (1991).

21. J. R. Tolman, J. Chung and J. H. Prestegard, *J. Magn. Reson.* **98**, 462 (1992).

22. R. E. Hurd, B. K. John and D. Plant, *J. Magn. Reson.* **93**, 666 (1991).

23. A. J. Shaka and R. Freeman, *J. Magn. Reson.* **51**, 169 (1983).

24. M. H. Levitt and R. R. Ernst, *J. Chem. Phys.* **83**, 3297 (1985).

25. U. Piantini, O. W. Sørensen and R. R. Ernst, *J. Am. Chem. Soc.* **104**, 6800 (1982).

26. R. E. Hurd, *J. Magn. Reson.* **87**, 422 (1990).

27. D. M. Freeman and R. E. Hurd, *NMR Basic Principles and Progress*, **27**, 200 (1992).

28. H. Kessler, U. Anders, G. Gemmecker and S. Steuernagel, *J. Magn. Reson.* **85**, 1 (1989).

29. J. Friedrich, S. Davies and R. Freeman, *J. Magn. Reson.* **75**, 390 (1987).

30. O. W. Sørensen and R. R. Ernst, *J. Magn. Reson.* **51**, 477 (1983).

31. O. W. Sørensen, M. Rance and R. R. Ernst, *J. Magn. Reson.* **56**, 527 (1984).

32. S. Macura, Y. Huang, D. Suter and R. R. Ernst, *J. Magn. Reson.* **43**, 259 (1981).

33. R. Freeman, T. H. Mareci and G. A. Morris, *J. Magn. Reson.* **42**, 341 (1981).

34. J. R. Garbow, D. P. Weitekamp and A. Pines, *Chem. Phys. Lett.* **93**, 514 (1982).

35. S. Wimperis and R. Freeman, *J. Magn. Reson.* **58**, 348 (1984).

36. M. Ikura, L. E. Kay and A. Bax, *Biochemistry*, **29**, 4654 (1990).

37. H. Kogler, O. W. Sørensen, G. Bodenhausen and R. R. Ernst, *J. Magn. Reson.* **55**, 157 (1983).

38. P. J. Hore, R. M. Scheek, A. Volbeda and R. Kaptein, *J. Magn. Reson.* **50**, 328 (1982).

39. L. D. Hall and T. J. Norwood, *J. Magn. Reson.* **76**, 548 (1988).

40. K. Roth, B. J. Kimber and J. Feeney, *J. Magn. Reson.* **41**, 302 (1980).

41. Ē. Kupče and R. Freeman, *J. Magn. Reson.* **99**, 644 (1992).

42. D. L. Rabenstein and T. T. Nakashima, *Anal. Chem.* **51**, 1465A (1979).

43. F. A. L. Anet, N. Jaffer and J. Strouse, 21st Experimental NMR Conference, Tallahassee, Florida, unpublished, 1980.

44. M. H. Levitt and R. Freeman, *J. Magn. Reson.* **39**, 533 (1980).

45. C. LeCocq and J. Y. Lallemand, *J. Chem. Soc., Chem. Comm.* 150 (1981).

46. S. L. Patt and J. N. Shoolery, *J. Magn. Reson.* **46**, 535 (1982).

47. F. K. Pei and R. Freeman, *J. Magn. Reson.* **48**, 318 (1982).
48. H. J. Jakobsen, O. W. Sørensen, W. S. Brey and P. Kanya, *J. Magn. Reson.* **48**, 328 (1982).
49. D. M. Doddrell, D. T. Pegg and M. R. Bendall, *J. Magn. Reson.* **48**, 323 (1982).
50. O. W. Sørensen and R. R. Ernst, *J. Magn. Reson.* **51**, 477 (1983).
51. H. Bildsøe, S. Dønstrup, H. K. Jakobsen and O. W. Sørensen, *J. Magn. Reson.* **53**, 154 (1983).
52. O. W. Sørensen, S. Dønstrup, H. Bildsøe and H. J. Jakobsen, *J. Magn. Reson.* **55**, 347 (1983).
53. S. F. Gull and G. J. Daniell, *Nature (London)*, **272**, 686 (1978).
54. E. D. Laue, J. Skilling, J. Staunton, S. Sibisi and R. Brereton, *J. Magn. Reson.* **62**, 437 (1985).
55. M. A. Delsuc and G. C. Levy, *J. Magn. Reson.* **76**, 306 (1988)
56. S. J. Davies, C. Bauer, P. J. Hore and R. Freeman, *J. Magn. Reson.* **76**, 476 (1988).
57. J. C. Hoch, A. S. Stern, D. L. Donoho and I. M. Johnson, *J. Magn. Reson.* **86**, 236 (1990).
58. J. A. Jones and P. J. Hore, *J. Magn. Reson.* **92**, 363 (1991).
59. P. J. Hore, in *Maximum Entropy in Action*, Eds: B. Buck and V. A. Macaulay, Oxford University Press, 1991.
60. J. A. Jones, D. S. Grainger, P. J. Hore and G. J. Daniell, *J. Magn. Reson. A* **101**, 162 (1993).
61. R. Baumann, G. Wider, R. R. Ernst and K. Wüthrich, *J. Magn. Reson.* **44**, 402 (1981).
62. W. P. Aue, J. Karhan and R. R. Ernst, *J. Chem. Phys.* **64**, 4226 (1976).
63. M. Woodley and R. Freeman, *J. Magn. Reson. A* **109**, 103 (1994).
64. G. Wider, S. Macura, A. Kumar, R. R. Ernst and K. Wüthrich, *J. Magn. Reson.* **56**, 207 (1984).
65. B. U. Meier, G. Bodenhausen and R. R. Ernst, *J. Magn. Reson.* **60**, 161 (1984).
66. P. Pfändler, G. Bodenhausen, B. U. Meier, and R. R. Ernst, *Anal. Chem.* **57**, 2510 (1985).
67. S. Glaser and H. K. Kalbitzer *J. Magn. Reson.* **74**, 540 (1987).
68. M. Woodley and R. Freeman, *J. Am. Chem. Soc.,* **117**, 6150 (1995).
69. M. Woodley and R. Freeman, *J. Magn. Reson. A* **118**, 39 (1996).

# 7

# Broadband decoupling

## 7.1 Introduction

### 7.1.1 Definitions

The recording of $^{13}$C spectra without C–H spin–spin splitting is a widespread and routine operation in high resolution spectroscopy, going under the name of broadband decoupling. It is perhaps useful at this stage to clear up some semantic points. There is a distinction between a splitting, which is an observable in the high resolution spectrum, and coupling, which is a physical interaction between two nuclear spins carried through the valence electrons in the molecule. A double resonance experiment may modify the splitting but it certainly does not change the spin–spin coupling constant. Evidence for this assertion is provided by the fact that broadband decoupling sequences are very effective in promoting the liquid-phase Hartmann–Hahn effect,[1–3] where the coupling is responsible for the coherence transfer (§ 1.4.5). Strong continuous irradiation of a particular spin species also causes saturation, but this has nothing to do with decoupling, as we can test by switching off the irradiation—the spin–spin splittings are restored immediately but the saturation persists for times of the order of the spin–lattice relaxation time.

Suppose we wish to observe the high resolution spectrum of a spin $S$ while removing the spin–spin splitting due to coupling to a heteronuclear species, the $I$ spins. To be more precise, we require that the observed spin–spin splittings be reduced to such a small value that they are no longer resolved and, preferably, no longer influence the S-spin peak heights. This should be effective over the entire range of *I*-spin chemical shifts. We know from experiments of the type:

S spins: $\qquad\qquad\qquad$ $(\pi/2)$—$\tau$ — $\qquad$ —$\tau$—observe

I spins: $\qquad\qquad\qquad\qquad\qquad$ $(\pi)$ $\qquad\qquad\qquad\qquad\qquad$ **7.1**

that at time $2\tau$, the normal divergence of vectors due to the coupling $J_{IS}$ is refocused. Consequently, if we continuously invert the *I* spins while observing the *S* spins, the spin–spin splitting should disappear. The rate of spin inversion should be fast compared with $|J_{IS}|$, consequently it is harder to decouple when the coupling constant is large. We can decouple by applying a continuous irradiation field $B_2$ at exact *I*-spin resonance, but this stirring technique[4] rapidly becomes ineffective as $B_2$ is shifted even a few Hz from exact resonance. The task is to increase the effective bandwidth without imposing an excessive radiofrequency power dissipation on the sample, or very high voltages on the decoupling coil.

### 7.1.2 **Figure of merit**

It is useful to define a figure of merit for the various possible schemes for broadband decoupling. First we should standardize on the line broadening that is acceptable for the S-spin spectrum, since the more drastic the broadening, the more easily the residual splittings will be masked. Some applications, for example quantitative studies, are more demanding than others. For a Lorentzian line shape, a residual splitting of one half the full line width reduces the peak height to about 80%. We define a practical criterion for the effective decoupling bandwidth ($\Delta F^\star$) to be the range within which the S-spin signal remains above 80% of the height observed for coherent decoupling at exact resonance. A tacit convention has emerged where the line broadening is set at 1.5 Hz for this kind of test. In general, the effective bandwidth ($\Delta F^\star$) is directly proportional to the intensity of the $B_2$ field in use, so we normalize with respect to $\gamma B_2/2\pi$. The figure of merit[5] is then

$$\Xi = \frac{2\pi\Delta F^\star}{\gamma B_2}$$  **7.2**

Actually we are more concerned about comparing different decoupling schemes at the same radiofrequency power dissipation, so a correction can be made for windowed sequences, if necessary. (Later, in § 7.4.4, we shall see that the bandwidth achieved by adiabatic decoupling schemes increases as the square of the radiofrequency level, so a new figure of merit may be required for these techniques.)

The various broadband decoupling schemes may then be judged principally on the basis of their $\Xi$ values (Table 1), although practical considerations (tolerance to instrumental imperfections, the magnitude of $J_{IS}$) may also be important. A higher figure of merit $\Xi$ means that all the *I* spins within the given bandwidth can be decoupled for a lower mean radiofrequency power.

**Table 7.1** Figures of merit $\Xi = 2\pi\Delta F^*/\gamma B_2$ for various broadband decoupling schemes[a]

| Sequence | $\Xi$ | Reference |
|---|---|---|
| Continuous-wave | 0.0075 | 19 |
| Pseudorandom noise | 0.3 | 6 |
| MLEV-16 | 1.5 | 9 |
| WALTZ-16 | 1.8 | 22 |
| GARP-1 | 4.8 | 28 |
| SUSAN-1 | 6.2 | 29 |
| STUD[b] | 10.9 | 37 |
| MPF-9[b] | 12.3 | 32 |
| WURST[b] | 16.7 | 40 |

[a] Effective bandwidth $\Delta F^*$ measured between 80% points, with a 1.5 Hz line broadening on the observed resonances ($S$).
[b] The figure of merit of adiabatic schemes depends on the $B_2$ level.

## 7.2 Decoupling methods

### 7.2.1 Continuous-wave decoupling

Naturally enough, the first decoupling experiments were performed with a monochromatic, continuous-wave decoupling field since only very simple spin systems were involved. The theory has been outlined in § 1.4.4; the $J$ splitting is scaled down, leaving a residual splitting given by the expression

$$S_{residual} = \frac{J\Delta B}{[(\Delta B)^2 + B_2^2]^{1/2}} \qquad \textbf{7.3}$$

Thus the decoupling is only really effective if the offset $\Delta B$ is very small with respect to the decoupler field intensity $B_2$. Under the condition $\Delta B << B_2$ the residual splitting is given by the approximate expression

$$S_{residual} \approx \frac{J\Delta B}{B_2} \qquad \textbf{7.4}$$

This mode may be used to scale down $^{13}C$ splittings in order to determine the number of protons directly attached to a given $^{13}C$ site without risking overlap of adjacent multiplets. In the early days of $^{13}C$ spectroscopy it was sometimes even employed as a crude method of shift correlation by calculating approximate values for the proton offsets from the residual splittings in the $^{13}C$ spectrum, although there remains a sign ambiguity unless the experiment is repeated with a different decoupler frequency. Off-resonance continuous-wave decoupling is sometimes used as a practical method for calibrating the intensity of the $B_2$ field in order to ensure that composite pulse methods (described below) operate under optimal conditions.

It is clear that continuous-wave decoupling is far too inefficient in situations where an appreciable range of $I$-spin shifts must be covered, as we can appreciate from a simple example. Suppose we have a decoupling field $\gamma B_2/2\pi = 10$ kHz and a coupling constant $J_{IS} = 200$ Hz. From Equation 7.4, coherent continuous-wave decoupling would leave a residual splitting of 0.75 Hz at an offset $\gamma \Delta B/2\pi = 37.5$ Hz. This would reduce the intensity of a $^{13}C$ line of width 1.5 Hz to 80% of the height observed with decoupling at exact resonance, giving a very low figure of merit $\Xi$ of 0.0075. Some improvement in effective bandwidth can be achieved by coherent audiofrequency modulation of the decoupler field, and several schemes have been tried—sinusoidal, square-wave and sawtooth sweep. In this context it can be dangerously misleading to think in terms of the Fourier spectrum of the $I$-spin irradiation, seeking to pack the entire $I$-spin chemical shift range with modulation sidebands in the hope that there will always be at least one sideband close to exact resonance. Unfortunately the various sidebands do not act independently.

### 7.2.2 **Noise decoupling**

For many years noise decoupling[6] was the method of choice. The decoupling radiation was rendered incoherent by pseudorandom phase inversions (§ 10.5). We may visualize this as mimicking the decoupling effect that occurs when the $I$ spins are undergoing fast chemical exchange, for this too is a random process. Suppose that the $I$ spin that is coupled to a particular $S$ spin is in an $\alpha$ state; when an exchange event occurs, this $I$ spin is replaced with another having either an $\alpha$ or $\beta$ state. If the power level of the noise radiation is high enough it has an effect similar to fast chemical exchange and the $S$-spin resonance becomes a singlet. At intermediate levels there is a distinct danger that noisy components are introduced into the $S$-spin spectrum; spin-echo experiments are particularly sensitive to such noisy perturbations.

Noise decoupling is relatively easy to set up and needs no calibration other than a choice of a suitable $B_2$ level. It served well for over a decade but when $^{13}C$ spectroscopy began to be performed at higher magnetic fields, it became apparent that this technique (for which $\Xi \approx 0.3$) involves too high a radiofrequency power dissipation in the sample, particularly for ionic solutions. Broader-band operation is required but at more modest levels of $B_2$ field.

### 7.2.3 **Composite pulse decoupling**

If we regard the decoupling process as a repeated, rapid interchange of $I$-spin states, the problem reduces to one of finding an efficient way to invert the $I$ spins at large offsets, or more precisely, at large ratios $\Delta B/B_2$. There are two complementary ideas for achieving this aim, both suggested by Levitt.[7–9] The first is to employ composite $\pi$ pulses to invert the $I$ spins. The action of composite pulses of the type

$$R = (\pi/2)_x \, (\pi)_y \, (\pi/2)_x \quad \text{and} \quad R = (\pi/2)_x \, (3\pi/2)_y \, (\pi/2)_x \qquad \textbf{7.5}$$

has been described according to the vector picture in § 2.5.4. These sandwich pulses extend the offset range of effective spin inversion by more than an order of magnitude. Near resonance they also improve the tolerance to variations in the flip angle (arising from miscalibration or from the spatial inhomogeneity of the radio-frequency field), but the two types of compensation do not mix well. More sophisticated composite pulses were later proposed that were able to compensate both types of imperfection simultaneously[10] but they have not yet been used for broadband decoupling.

## 7.2.4 Magic cycles and supercycles

The second line of attack is to assemble the composite pulses into a magic cycle.[8] Since composite pulse decoupling involves the repeated application of spin inversion pulses, even very small pulse imperfections can have appreciable cumulative effects in a long sequence. Magic cycles are designed to correct these tiny shortcomings so that the imperfections do not accumulate with time. This takes a leaf out of the book of the solid state NMR spectroscopists[11,12] where similar problems arise. The first such decoupling sequence was MLEV-4 which may be written

$$R\,R\,\bar{R}\,\bar{R} \qquad\qquad \textbf{7.6}$$

where $R$ represents a composite inversion pulse and $\bar{R}$ is its phase-inverted counterpart. It is clear that the combination $RR$ with both pulses in the same sense is almost a perfect cycle as can be appreciated from the magnetization trajectories calculated for two composite pulses $R = (\pi/2)_x\,(3\pi/2)_y\,(\pi/2)_x$ shown in Figure 7.1. Although the tilt angle $\theta$ given by

$$\tan\theta = \Delta B/B_2 \qquad\qquad \textbf{7.7}$$

is appreciable, these trajectories all terminate at a point very close to the $+z$ axis. The small residual deviation is largely compensated by the second part ($\bar{R}\bar{R}$) of the MLEV-4 cycle. Here we use the term cycle in a special sense; an exact cycle returns the magnetization vector to its starting point, and so there can be no cumulative deviation in a prolonged sequence. Formally it is represented by the identity operator. We shall see later that the concept of cyclicity is the key to understanding of the various broadband decoupling schemes.

The decoupling performance of MLEV-4 ($\Xi \approx 1.0$) proved a significant improvement over noise decoupling at the same power level, encouraging further research on the concept of composite pulse decoupling.

The next stage was the realization[9] that the tiny residual imperfections of this magic cycle could be compensated by assembling an even number of such cycles into a supercycle, for example MLEV-16

$$R\,R\,\bar{R}\,\bar{R} \quad \bar{R}\,R\,R\,\bar{R} \quad \bar{R}\,\bar{R}\,R\,R \quad R\,\bar{R}\,\bar{R}\,R \qquad\qquad \textbf{7.8}$$

The overall length is of course four times longer than MLEV-4, but fortunately the criterion for good decoupling requires only that the primitive cycle be repeated fast

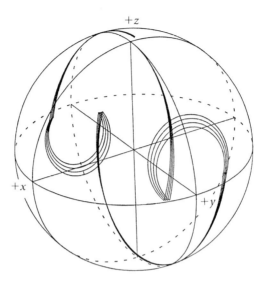

**Figure 7.1** A family of four trajectories for two consecutive composite pulses $RR$, where $R = (\pi/2)_x \, (3\pi/2)_y \, (\pi/2)_x$, near the condition $\Delta B/B_2 = 0.3$. In spite of the appreciable tilt of the effective fields, all the trajectories terminate at a point very close to the $+z$ axis; the sequence is nearly cyclic. This suggests that the combination $RR\bar{R}\bar{R}$ would be an efficient broadband decoupling sequence.

compared with $J_{IS}$. The MLEV-4 unit reduces the splitting from $J_{IS}$ to a much smaller effective splitting $J_{eff}$ and the supercycle only needs to be repeated fast compared with $J_{eff}$. Experimental results from an early $^{13}$C spectrometer are shown in Figure 7.2, where it can be seen that the figure of merit $\Xi \approx 1.5$. This opens the way for more and more complex supercycles[13] such as MLEV-64, which has a figure of merit $\Xi \approx 1.8$. At the time, this expansion procedure was largely intuitive since an exact theory of decoupling was lacking.

## 7.3  Decoupling theory

### 7.3.1  Average Hamiltonian theory

Imagine that an ice skater, rapidly spinning on one spot, is illuminated with a stroboscopic light. If the pulses of the strobe light closely match the spinning frequency, the skater will appear to rotate only very slowly, at the difference between the spinning frequency and the strobe frequency. For exact synchronization the motion will be frozen in the eyes of the observer. Broadband decoupling by pulse methods involves the application of a periodic sequence that induces a motion of $I$-spin vectors that is very nearly cyclic; that is to say, the vectors return to a point very close to their starting point once every period. If the acquisition of the $S$-spin signal only occurs at these points, then the $J_{IS}$ splitting will not be detected.

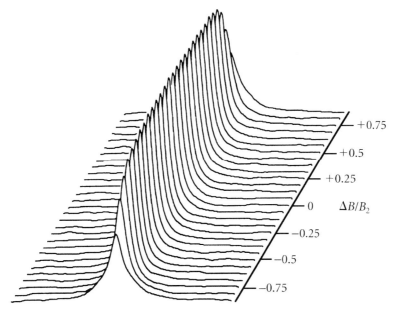

**Figure 7.2** Experimental performance of the MLEV-16 decoupling sequence as a function of the normalized offset parameter $\Delta B/B_2$. The $^{13}C$ peak height remains within 80% of its maximum when the proton offset $\Delta B$ is within the range $\pm 0.75\ B_2$, a figure of merit $\Xi \approx 1.5$.

This idea of applying a periodic perturbation to the $I$-spins while sampling the $S$-spin signal at the end of each period suggests a treatment known as coherent averaging or average Hamiltonian theory.[14,15] Considerable success had been achieved with this theory in the design of complex multipulse sequences for decoupling homonuclear dipolar interactions in the solid state, so it seemed natural to apply it to the problem of broadband decoupling in isotropic liquids. The Hamiltonian $\mathcal{H}$ that describes the $IS$ system under decoupling conditions is of course time-dependent, the $I$ spins undergoing wild evolutions each period, but by transformation into a time-dependent frame of reference (the toggling frame) we can focus on an apparent evolution that is only varying very slowly, described by an average Hamiltonian $\bar{\mathcal{H}}$. Any periodic motion can be made to appear simple if it is sampled only once each period, as the ice skater analogy suggests. The evolution under the average Hamiltonian $\bar{\mathcal{H}}$ is evaluated as a power series (the Magnus expansion) and if the cycle time is short enough, $\bar{\mathcal{H}}$ can be approximated by the zero-order term $\bar{\mathcal{H}}^{(0)}$ which is relatively easy to calculate. Provided that the cyclic condition is satisfied, the effect of $\bar{\mathcal{H}}^{(0)}$ is simply to scale down the splitting in the $S$-spin spectrum to a smaller value $\lambda J_{IS}$ Hz. This approximate theory suggests how broadband decoupling sequences might be constructed. If we accept that the combination $RR$ is not quite cyclic but has a small residual rotation $\varepsilon$, magic cycles and supercycles may be designed by seeking to null $\bar{\mathcal{H}}^{(0)}$ to a certain order in $\varepsilon$. This was how the various MLEV sequences evolved.

## 7.3.2 Exact theory of decoupling

At this point, a timely intervention by Waugh[16,17] provided an alternative theory of decoupling that was exact and did not assume that the decoupling sequence was cyclic. Not only did it explain the MLEV recipe more clearly but it provided a simple way to compute the decoupler performance and hence take account of the myriad possible instrumental shortcomings.

We consider the time evolution of the $I$ and $S$ spins with a view to calculating the $S$-spin spectrum as the $I$ spins are subjected to a periodic perturbation. For simplicity, relaxation effects are neglected; they can be quite complicated to evaluate in double resonance experiments and in practice they have little effect on decoupling performance. Although it is not required, it helps to think about the problem if we assume that the $S$-spin signal is sampled in synchronism with the decoupler period. This stroboscopic sampling was forced on the early experiments by the fact that the same computer controlled the decoupler pulsing and the $S$-spin signal acquisition. Later, when efficient sequences had been discovered and tested, it made sense to build a hard-wired decoupler[18] that was not synchronous with the $S$-spin sampling operations. The main features of the decoupling are unchanged in this variant of the experiment.

In retrospect, a clue to the problem could have been gleaned from straightforward theory of continuous-wave decoupling[19] applied to a two-spin ($IS$) system as outlined in § 1.4.4. We consider the energy level diagram in a frame of reference rotating at the decoupler frequency, offset from the exact $I$-spin resonance by $\Delta$ Hz.

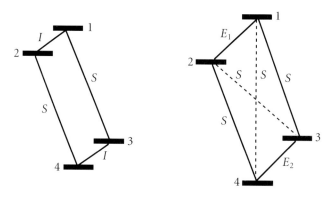

Rotating frame                    Introduction of $B_2$

The two $I$-spin transitions correspond to the two spin states $\alpha$ and $\beta$ of the $S$ nucleus. When the decoupler field $B_2$ is introduced, the $I$-spin transitions experience effective fields $E_1$ and $E_2$ (expressed in frequency units) and given by

$$E_1 = [(\Delta + \tfrac{1}{2}J)^2 + (\gamma B_2/2\pi)^2]^{1/2} \qquad \textbf{7.9}$$

$$E_2 = [(\Delta - \tfrac{1}{2}J)^2 + (\gamma B_2/2\pi)^2]^{1/2} \qquad \textbf{7.10}$$

In the presence of the $B_2$ field, the previously forbidden transitions (1)–(4) and (2)–(3) become partially allowed. Perfect decoupling occurs if the two $I$-spin transitions experience equal effective fields. In practice, acceptable decoupling is achieved if $|E_1 - E_2|$ is smaller than the $S$-spin linewidth; then there is an unresolved doublet at the $S$-spin chemical shift frequency. In the more general case, the two effective fields $E_1$ and $E_2$ are different, resulting in an intense $S$-spin doublet with a splitting $|E_1 - E_2|$ and two weak satellites at a much larger splitting $|E_1 + E_2|$.

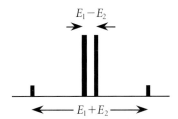

The intensities are given by $\cos^2[(\theta_1 - \theta_2)/2]$ for the central lines and $\sin^2[(\theta_1 - \theta_2)/2]$ for the satellites, where

$$\tan\theta_1 = \frac{\gamma B_2/2\pi}{|\Delta + \frac{1}{2}J|} \tag{7.11}$$

$$\tan\theta_2 = \frac{\gamma B_2/2\pi}{|\Delta - \frac{1}{2}J|} \tag{7.12}$$

where $\theta_1$ and $\theta_2$ define the axes $n_1$ and $n_2$ of the effective fields $E_1$ and $E_2$ in the rotating frame. For continuous-wave coherent decoupling the residual splitting can only be kept small if the $B_2$ field is intense and very close to resonance. These conditions correspond to a near-cyclic motion of the $I$-spin magnetization vectors about axes that lie very close to the $xy$ plane ($\theta_1 \approx \theta_2 \approx \pi/2$).

Waugh[17] showed that multi-pulse broadband decoupling sequences also rely on generating an approximately cyclic motion of the $I$-spin vectors. Instead of free precession where the $I_1$ and $I_2$ vectors have precession frequencies differing by $J_{IS}$, we impose a sequence of forced precessions which bring them back almost to the same point once every period, so that they appear to have moved very little. We present here a simplified version of the Waugh analysis.

Composite-pulse decoupling corresponds to a sequence of pure rotations of the $I$-spin vectors. For example the MLEV-4 sequence is made up of four primitive rotations:

$$R_I = R\,R'\,R''\,R''' \tag{7.13}$$

The resultant $R_I$ is also a pure rotation since the elements $R$ through $R'''$ generate a series of linked arcs on the surface of the unit sphere. It may be written

$$R_I = \exp(-i\phi n \cdot I) \tag{7.14}$$

representing a rotation through an angle $\phi$ about an axis defined by the unit vector $n$.

For the $IS$ spin system, the two $I$-spin lines undergo slightly different net rotations:

$$R_1 = \exp(-i\phi_1 \mathbf{n}_1 \cdot \mathbf{I}) \qquad\qquad \textbf{7.15}$$

$$R_2 = \exp(-i\phi_2 \mathbf{n}_2 \cdot \mathbf{I}) \qquad\qquad \textbf{7.16}$$

If sampling is synchronized with the ends of each full period $\tau$, the $I$-spin vectors appear to have the frequencies

$$F_1 = \phi_1/(2\pi\tau) \qquad\qquad \textbf{7.17}$$

$$F_2 = \phi_2/(2\pi\tau) \qquad\qquad \textbf{7.18}$$

instead of their free precession frequencies $\pm\frac{1}{2}J_{IS}$. This shifts the energy levels in a manner analogous to the Bloch–Siegert shifts[20] considered in § 1.4.3. We consider the motion in a reference frame rotating in synchronism with the $I$-spin chemical shift $(\nu_I)$, arguing that the choice of frame frequency is not critical since the decoupling sequences under investigation can be anticipated to operate effectively over a wide range of frequencies. In this frame, the usual expressions for the energy levels are transformed by lowering level (1) by $\nu_I$ and raising level (4) by the same amount.

$$
\begin{aligned}
(1) & \quad +\tfrac{1}{2}\nu_I + \tfrac{1}{2}\nu_S + \tfrac{1}{4}J_{IS} && \rightarrow && -\tfrac{1}{2}\nu_I + \tfrac{1}{2}\nu_S + \tfrac{1}{4}J_{IS} \\
(2) & \quad -\tfrac{1}{2}\nu_I + \tfrac{1}{2}\nu_S - \tfrac{1}{4}J_{IS} && \rightarrow && -\tfrac{1}{2}\nu_I + \tfrac{1}{2}\nu_S - \tfrac{1}{4}J_{IS} \\
(3) & \quad +\tfrac{1}{2}\nu_I - \tfrac{1}{2}\nu_S - \tfrac{1}{4}J_{IS} && \rightarrow && +\tfrac{1}{2}\nu_I - \tfrac{1}{2}\nu_S - \tfrac{1}{4}J_{IS} \\
(4) & \quad -\tfrac{1}{2}\nu_I - \tfrac{1}{2}\nu_S + \tfrac{1}{4}J_{IS} && \rightarrow && +\tfrac{1}{2}\nu_I - \tfrac{1}{2}\nu_S + \tfrac{1}{4}J_{IS} && \textbf{7.19}
\end{aligned}
$$

This involves a crossover of levels (3) and (4).

We show the transformed energy level diagrams with the splittings $\tfrac{1}{2}J_{IS}$, $F_1$ and $F_2$, grossly exaggerated for the purpose of illustration.

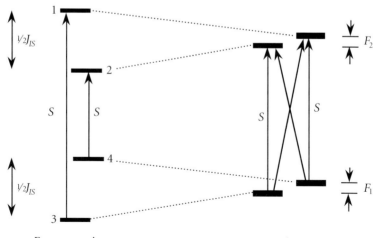

Free precession                 Forced precession

In the absence of any $B_2$ field, the $I$ spins precess freely at $\pm \frac{1}{2} J_{IS}$, but under decoupling conditions, forced precession occurs, and these frequencies are reduced to $\frac{1}{2}F_1$ and $\frac{1}{2}F_2$, lowering levels (1) and (4) and raising levels (2) and (3). This almost returns the energy diagram to that applicable to the uncoupled system, since $F_1$ and $F_2$ are very low frequencies indeed.

$$(1) \quad -\tfrac{1}{2}\nu_I + \tfrac{1}{2}\nu_S + \tfrac{1}{2}F_2$$
$$(2) \quad -\tfrac{1}{2}\nu_I + \tfrac{1}{2}\nu_S - \tfrac{1}{2}F_2$$
$$(3) \quad +\tfrac{1}{2}\nu_I - \tfrac{1}{2}\nu_S - \tfrac{1}{2}F_1$$
$$(4) \quad +\tfrac{1}{2}\nu_I - \tfrac{1}{2}\nu_S + \tfrac{1}{2}F_1 \qquad \textbf{7.20}$$

Note that the two allowed $S$-spin transitions are now (2)–(3) and (1)–(4) and they have almost identical frequencies. When transformed back into the laboratory frame they are:

$$(2)-(3) \quad \nu_S + \tfrac{1}{2}[F_1 - F_2] \qquad \textbf{7.21}$$

$$(1)-(4) \quad \nu_S - \tfrac{1}{2}[F_1 - F_2] \qquad \textbf{7.22}$$

The residual splitting is therefore given by $|F_1 - F_2|$.

This is an important result for it determines the criterion for effective decoupling —the net $I$-spin rotations $\phi_1$ and $\phi_2$ during $\tau$ should be as nearly as possible the same. In fact, since we demand good decoupling over a wide range of $I$-spin chemical shifts, the net rotation angle $\phi$ should vary only very slowly over this entire range. In principle this could be achieved with large deviations from cyclicity provided that the two precession angles $\phi_1$ and $\phi_2$ were always very nearly equal. In practice it is easier to keep $|\phi_1 - \phi_2|$ small by ensuring that both $\phi_1$ and $\phi_2$ remain small, that is to say, that the decoupling sequence be as close as possible to an exact cycle. This is the principal conclusion of the Waugh analysis. Cyclicity guarantees an efficient decoupling sequence; its effective bandwidth will be determined by the offset at which the sequence ceases to be sufficiently cyclic because the component pulses no longer invert the spins properly.

The intensities of the two (strong) components of the decoupled $S$-spin doublet are equal and may be written:

$$A = \tfrac{1}{2}[1 + \boldsymbol{n}_1 \cdot \boldsymbol{n}_2] \qquad \textbf{7.23}$$

If the rotation axes are almost collinear, then A is close to unity (the maximum intensity). This is usually guaranteed in practice if $\gamma B_2/2\pi \gg |J_{IS}|$, since the decoupler field affects both the $I$-spin lines in a very similar fashion.

So far there has been no mention of the two forbidden $S$-spin transitions that are rendered partly allowed by the imposition of the decoupler field $B_2$. Their frequencies are given by:

$$(1)-(3) \quad \nu_S + \tfrac{1}{2}[F_1 + F_2] \qquad \textbf{7.24}$$

$$(2)-(4) \quad \nu_S - \tfrac{1}{2}[F_1 + F_2] \qquad \textbf{7.25}$$

and their intensities by

$$B = \tfrac{1}{2}\left[1 - n_1 \cdot n_2\right] \qquad \qquad \textbf{7.26}$$

These satellite responses are always very weak since $n_1$ and $n_2$ are nearly collinear. In the practical example of coherent decoupling with $\gamma B_2/2\pi = 2$ kHz and $J_{IS} = 200$ Hz, the satellites reach a maximum intensity of only 1% of the intensity of one of the main decoupled lines.

This analysis provides us with a valuable quantitative parameter with which to judge the effectiveness of various decoupling schemes. We define a scaling factor that is the ratio of the differential precession angle under decoupled and coupled conditions:

$$\lambda = \frac{|\phi_1 - \phi_2|}{|2\pi J_{IS}\tau|} \qquad \qquad \textbf{7.27}$$

We set out to ensure that $\lambda \ll 1$ over the entire range of $I$-spin chemical shifts so that the residual splitting of the $S$-spin lines falls well within the experimental line width. Broadband decoupling is simply an extreme case of $J$ scaling.[7]

## 7.3.3 Design of decoupling sequences

Now we can appreciate how a magic cycle like MLEV-4 works. The individual pulse elements $R$ and $R$ link together to give an approximate cycle, whereas the juxtaposition of $\bar{R}\bar{R}$ largely cancels the residual error of $RR$.

We now take advantage of some general properties of pure rotations. If we move one of the elements of a sequence (for example one composite inversion pulse) from the end to the beginning, the net rotation angle $\phi$ of the overall sequence is unchanged.[17] For example,

$$\phi(R\,R\,\bar{R}\,\bar{R} = \phi(\bar{R}\,R\,R\,\bar{R}) \qquad \qquad \textbf{7.28}$$

Furthermore the axis $n$ of the overall rotation is itself rotated by this permutation through $\phi_0$ about $n_0$, where $\phi_0$ and $n_0$ refer to the element that has been permuted. Thus, since the overall rotation axis of $R\,R\,\bar{R}\,\bar{R}$ is approximately along $z$, and since $\bar{R}$ represents an approximate $\pi$ rotation about the $x$ axis, the permutation shown above essentially reverses the direction of the overall rotation axis $n$. It is then clear why the expansion from MLEV-4 to MLEV-8 shows an improvement in decoupling efficiency—the residual rotation of $R\,R\,\bar{R}\,\bar{R}$ about the $z$ axis is largely cancelled by the equal and opposite residual rotation of $\bar{R}\,R\,R\,\bar{R}$.

Incidentally, the invariance of $\phi$ upon cyclic permutation of any part of the sequence justifies the assumption that the sampling of the $S$-spin free induction decay need not be restricted to the end of each decoupler period, provided that it is synchronized and monitors corresponding points in each successive period. We shall see later what happens if this stroboscopic condition is relaxed, so that several samples are taken per period, or even abandoned, so that the sampling is allowed to become asynchronous with respect to the decoupler periodicity.

The second manipulation is phase inversion of all elements in the sequence. Again the overall rotation angle $\phi$ remains invariant but the rotation axis has its $x$ and $y$ components reversed:[17]

$$\phi\{R\,R\,\bar{R}\,\bar{R}\} = \phi\{\bar{R}\,\bar{R}\,R\,R\} \qquad\qquad \textbf{7.29}$$

$$n_x \rightarrow -n_x \text{ and } n_y \rightarrow -n_y \qquad\qquad \textbf{7.30}$$

This accounts for the success of the expansion of MLEV-8 to MLEV-16 since the former has a net rotation about an axis in the $xy$ plane. The residual rotation is cancelled by appending the phase-inverted counterpart. Note that the rotation axis changes its orientation in the transverse plane as a function of decoupler offset but this does not jeopardize the compensation.

These two operations, permutation and phase inversion, are the key to most of the improved decoupling sequences. It is reassuring to see that the mainly intuitive ideas of Levitt, corroborated by experiment, are now vindicated by the exact theory of Waugh.

One might speculate that noise decoupling, for which there has never been any rigorous theory, operates by causing a random walk of trajectories on the unit sphere, with a certain degree of compensation through a periodic pseudorandom reversal of the sense of rotation. Somehow this keeps the difference $|\phi_1 - \phi_2|$ small enough that the decoupling is reasonably effective. Even with rather inefficient decoupling sequences the decoupled resonance tends to dominate the remaining modulation sidebands, particularly when there are many modulation frequencies involved. However, the newer deterministic sequences would be expected to be superior on the basis of the present theoretical analysis, and in practice they have proved far more efficient.

## 7.3.4 General properties

We may now summarize some general rules governing decoupling sequences. Theory predicts that the decoupling performance is determined by trace $(R_s)$, where $R_s$ represents the rotation operator for the sequence, and the trace is invariant with respect to a unitary transformation, such as a rotation. The trace of any rotation operator $R_s$ is equal to $(1 + \cos 2\phi)$ where $\phi$ is the rotation angle of $R_s$.

- Decoupling performance is invariant with respect to a phase shift $(\psi)$ of all the component pulses. Note, however, that the rotation axis of $R_s$ is shifted through the same angle $\psi$. In many practical cases a phase shift $\psi = 180°$ is used in the expansion to the next-highest supercycle.

- Decoupling performance is invariant with respect to cyclic permutation of any part $(R_0)$ of the sequence. The effect is to rotate the axis of $R_s$ by $R_0$. In practice $R_0$ is often a composite $\pi$ pulse or a simple $\pi/2$ pulse.

- Time-reversal of a decoupling sequence reverses the offset dependence.

- A sequence that employs only 180° phase shifts (for example WALTZ decoupling) has a symmetrical offset dependence.

# 7.4 **Practical considerations**

## 7.4.1 **WALTZ decoupling**

When the MLEV sequences were carefully tested experimentally under very high resolution conditions it became clear that instrumental shortcomings could degrade the performance to some extent.[15] One type of imperfection is the spatial in-homogeneity of the decoupler field, and it is particularly important when the $I$ spins are at large offsets where the composite pulse is unable to compensate for flip angle errors. It was also found that the decoupling performance was surprisingly sensitive to small errors in the 90° phase shift required to generate the composite $(\pi/2)_x(\pi)_y(\pi/2)_x$ pulses. An error of as little as 0.5° can cause residual splittings as large as 0.2 Hz at certain offsets; this is confirmed[21] by calculations of the scaling factor $\lambda$. In contrast, sequences that employ only (nominal) 180° phase shifts are much less sensitive to any small phase error.[21]

It is known[22,23] for an alternative composite inversion pulse that employed only 180° phase shifts and which might also be tolerant of small flip angle errors caused by miscalibration or spatial inhomogeneity of $B_2$. Levitt's concept of a reversed precession pulse paved the way. This hypothetical pulse would have the property that the $I$ spins precess in the opposite sense to their natural Larmor precession in a magnetic field. For a particular $I$-spin chemical shift, the effect of reversed precession can be achieved by reversal of the sense of the decoupler offset, but of course this is impractical for a spectrum of many lines. Levitt *et al.* showed that a similar effect can be achieved with a composite pulse over a certain range of offsets.[15]

It is well known[24] that a $\pi/2$ pulse is largely self-compensating for resonance off-set effects if judged by its capacity to take a vector from $+z$ to the $xy$ plane, without regard to the phase dispersion in the $xy$ plane. This could form the first half of a very efficient inversion pulse if we could only find a reversed precession $\pi/2$ pulse where the dephasing effect is reversed. Now consider the combination

$$(3\pi/2)_{-x}(3\pi/2)_x \approx 1 \qquad\qquad \textbf{7.31}$$

where 1 represents an exact cycle. The expression is exact at an offset $\Delta B = 0$ and also at $\Delta B = \pm\, 0.88\, B_2$ where the effective field is such that the flip angles become $2\pi$ radians. In the intermediate regions this combination is only approximately cyclic but we may anticipate that these deficiencies will be corrected by the expansion schemes described above. The expression can be rewritten as

$$(\pi)_{-x}(3\pi/2)_x \approx [(\pi/2)_{-x}]^{-1} \qquad\qquad \textbf{7.32}$$

where $[(\pi/2)_{-x}]^{-1}$ represents a reversed precession $(\pi/2)$ pulse about the $-x$ axis. A self-compensating inversion pulse may therefore be constructed by combining this with a normal $\pi/2$ pulse:

$$R = (\pi/2)_x\,[(\pi/2)_{-x}]^{-1} \approx (\pi/2)_x\,(\pi)_{-x}\,(3\pi/2)_x \qquad\qquad \textbf{7.33}$$

In a convenient shorthand notation where the nominal flip angles are written as multiples of $\pi/2$, this becomes

$$R = 1\,\bar{2}\,3 \qquad\qquad \textbf{7.34}$$

With this sequence, spin inversion is effective over a wide band, although there are appreciable dips in the profile near $\Delta B = \pm\,0.3\,B_2$. This composite pulse is known as WALTZ for obvious reasons.

The first stage of expansion proceeds exactly as before by setting up the sequence

$$R\bar{R}\bar{R}R = 1\,\bar{2}\,3 \quad \bar{1}\,2\,\bar{3} \quad \bar{1}\,2\,\bar{3} \quad \bar{1}\,2\,\bar{3} \qquad \textbf{7.35}$$

This sequence is WALTZ-4. We can now combine adjacent pulses with the same sense (but we must not subtract pulses of opposite sense because of the effect of the tilt angle) and rewrite the sequence as

$$R\bar{R}\bar{R}R = 1\,\bar{2}\,4\,\bar{2}\,3 \quad \bar{1}\,2\,\bar{4}\,2\,\bar{3} \qquad \textbf{7.36}$$

Use is now made of the permutation rule stated above. The WALTZ-4 sequence is almost cyclic, having a small overall rotation about an axis nearly parallel to the $z$ axis. If a $(\pi/2)_x$ pulse is moved from the beginning to the end of the sequence:

$$K\bar{K} = \bar{2}\,4\,\bar{2}\,3\,\bar{1} \quad 2\,\bar{4}\,2\,\bar{3}\,1 \qquad \textbf{7.37}$$

the new rotation axis is brought into the $xy$ plane. This is a very good approximation over a wide range of offsets because of the particular properties of a $\pi/2$ pulse. Compensation can be achieved by combining the sequence $K\bar{K}$ with its phase inverted counterpart $\bar{K}K$ to give WALTZ-8:

$$K\bar{K}\bar{K}K = \bar{2}\,4\,\bar{2}\,3\,\bar{1} \quad 2\,\bar{4}\,2\,\bar{3}\,1 \quad 2\,\bar{4}\,2\,\bar{3}\,1 \quad \bar{2}\,4\,\bar{2}\,3\,\bar{1} \qquad \textbf{7.38}$$

Each element $K$ is a better spin inversion sequence than $R$. Note that the compensation is effective at all offsets within the operating band even though the rotation axis of the $\pi/2$ pulse varies appreciably with offset.

Since the residual rotation of WALTZ-8 is again about an axis very nearly parallel to $+z$ the same procedure can be repeated, moving a $(\pi/2)_{-x}$ pulse from the end to the beginning of the sequence and combining this with its phase-inverted counterpart, giving WALTZ-16:

$$Q\bar{Q}\bar{Q}Q = \bar{3}\,4\,\bar{2}\,3\,\bar{1}\,2\,\bar{4}\,2\,\bar{3} \quad 3\,\bar{4}\,2\,\bar{3}\,1\,\bar{2}\,4\,\bar{2}\,3 \quad 3\,\bar{4}\,2\,\bar{3}\,1\,\bar{2}\,4\,\bar{2}\,3 \quad \bar{3}\,4\,\bar{2}\,3\,\bar{1}\,2\,\bar{4}\,2\,\bar{3} \quad \textbf{7.39}$$

As before, the symmetry of the sequence allows it to be written in the form $Q\bar{Q}\bar{Q}Q$ where $Q$ has an excellent phase inversion profile.

In practice WALTZ-16 has proved an efficient and robust broadband decoupling sequence and has been widely used. Figure 7.3 shows the first experimental results as a function of the normalized offset parameter $\Delta B/B_2$; the practical figure of merit in this case is $\Xi \approx 1.8$, although somewhat higher values ($\Xi \approx 2.0$) are found on more modern spectrometers. Calculations of the scaling factor $\lambda$ (Figure 7.4) vividly illustrate the improvement achieved in going from WALTZ-8 to WALTZ-16, confirming that the residual splittings (for $J_{IS} = 200$ Hz) are well below 0.2 Hz within the operating bandwidth. These simulations correspond very closely with the observed experimental behaviour. They predict a tolerance to a $\pm 5\%$ missetting of

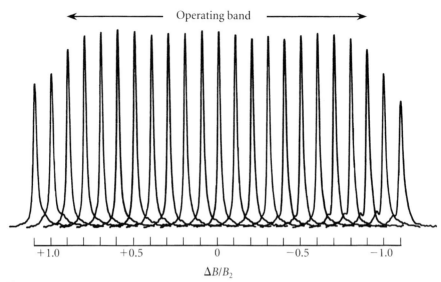

**Figure 7.3** Experimental performance of the WALTZ-16 decoupling sequence as a function of the normalized offset parameter $\Delta B/B_2$. This gives a figure of merit $\Xi \approx 1.8$.

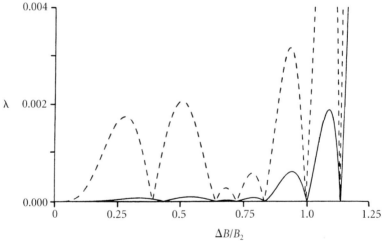

**Figure 7.4** Calculated scaling factors $\lambda$ for the WALTZ-8 (broken line) and WALTZ-16 (full line) decoupling sequences as a function of the normalized offset parameter $\Delta B/B_2$. A value of $\lambda = 0.001$ corresponds to a residual splitting of only 0.2 Hz if $J_{IS} = 200$ Hz.

all the flip angles without appreciable degradation in performance.[23] When account is taken of the spatial inhomogeneity of the $B_2$ field, the simulations give a better fit to experiment since the additional broadening hides the small residual splittings predicted by the theory for a perfectly uniform $B_2$ field. Calculations also confirm the insensitivity to phase errors between the two decoupler channels; a 170° phase difference gave results almost indistinguishable from those predicted for the ideal 180° phase shift.

Expansion of WALTZ-16 to WALTZ-32 proved disappointing, a result attributed to the deleterious effect of a slight amplitude imbalance between the $+x$ and $-x$ decoupler channels, which has the effect of turning the axis of residual rotation of WALTZ-16 away from the $z$ direction, an error which is not compensated by the expansion procedure. This serves as a warning that what actually matters is the experimental performance rather than any theoretical simulation; there are many insidious practical imperfections that could conceivably interfere with these decoupling supercycles. Many of these are examined in the comprehensive review by Shaka and Keeler.[21]

## 7.4.2 Cycling sidebands

So far we have tacitly assumed that the $S$-spin free induction decay is sampled stroboscopically in synchronism with the decoupler period. We have seen above that the timing point for sampling may be anywhere within the decoupler period provided that it is synchronized and records corresponding points in each successive period. In effect we are monitoring the slight deviation from exact cyclicity and following how this small-angle rotation builds up in time to give the apparent frequencies $F_1$ and $F_2$. Ideally we would sample only once during the entire period of a supercycle, or failing that, once per period of the basic cycle. However, for the practical case of most interest—$^{13}C$ spectroscopy with proton decoupling where the sampling rate may be as high as 40 kHz—it is not feasible to satisfy the stroboscopic condition even for the basic proton decoupler period $\tau$. Consequently, many $^{13}C$ data points are acquired within each decoupler period, revealing periodic signal deviations from the ideal smooth sinusoid at the frequency $F_1 - F_2$. After Fourier transformation, these low-level parasitic modulations generate cycling sidebands in the decoupled spectrum,[25] loosely analogous to the well-known spinning sidebands. Fortunately they are of very low amplitude, but they could eventually constitute a limitation on the sensitivity of the method. This is the reason why windowed decoupling sequences have not proved popular; in principle they can have a better figure of merit than the corresponding windowless sequences but they induce much stronger cycling sidebands because of the strong signal modulation caused by the decoupler switching.

The height of the cycling sidebands can be considerably reduced if the decoupling is rendered asynchronous with the $S$-spin acquisition[18] by building a hard-wired decoupler that runs quite independently of the spectrometer acquisition computer. An alternative scheme makes use of the fact that there are many equivalent forms of a decoupling sequence with the same general performance. Hence we may continually permute some element without altering the decoupling efficiency and yet radically change the disposition of the cycling sidebands.[25] Unfortunately this does not change the total integrated sideband intensity but merely splits the existing sidebands into many (apparently noisy) components of extremely low amplitude.

### 7.4.3 **Ultrabroadband decoupling**

Spectroscopists have developed an insatiable appetite for decoupling over wider and wider frequency bands. This is partly due to the inexorable progress to higher magnetic fields and partly to the necessity for removing splittings from heteronuclear species such as $^{13}$C and $^{15}$N in the proton spectra of biological molecules that have been deliberately enriched in these isotopes. The bandwidth requirement for decoupling $^{13}$C is about five times higher than for proton decoupling at the same applied magnetic field. Furthermore, for the same decoupler coil filling factor and the same quality factor ($Q$), the radiofrequency power required to achieve the same decoupler field $\gamma_c B_2/2\pi$ is increased by a factor $(\gamma_H/\gamma_C)^2 = 16$. Together these factors amount to a 400-fold increase in decoupler power. The situation is often exacerbated by the ionic nature of many aqueous solutions of biological samples where the dielectric loss aggravates the heating effect.

The WALTZ-16 decoupling sequence produces residual splittings of the $^{13}$C resonances that are less than 0.1 Hz over the operating bandwidth; with a line broadening of 0.25 Hz, the effect on peak heights is then small enough to be neglected. This implies that very high resolution conditions can be used when there is interest in (say) small $^{13}$C–$^{13}$C splittings or couplings to a heteronuclear species (other than protons). For more routine applications of $^{13}$C spectroscopy these requirements may be too stringent. A more realistic line broadening might be between 1 and 2 Hz, and this greatly relaxes the requirement for very small residual splittings. We might then place the emphasis on other desirable attributes, for example a higher figure of merit ($\Xi$) and an increased tolerance to the effects of spatial inhomogeneity of the decoupler field $B_2$.

Several such decoupling sequences have been proposed,[26–32] loosely based on the concepts outlined above. The improvements come about by concentrating on the effective bandwidth of the composite inversion pulses $R$ and $\bar{R}$ rather than by changing the expansion procedure, which usually remains essentially the same as the MLEV prescription. If we allow more realistic $S$-spin linewidths (1.5 Hz is generally accepted as a standard) and also relinquish the requirement that the nominal pulse flip angles be simple multiples of one another, this provides more freedom in the design process. One such broadband decoupling scheme is GARP[28] which was derived by numerical optimization, paying particular attention to a high figure of merit and good tolerance toward miscalibration or spatial inhomogeneity of the $B_2$ field. With $S$-spin linewidths of 1.5 Hz, GARP offers a figure of merit $\Xi \approx 4.8$, a significant improvement over WALTZ-16.

More recent numerical optimization programs using simulated annealing[29] have demonstrated a scheme (SUSAN-1) with an improved figure of merit $\Xi \approx 6.2$, albeit with a sequence almost double the length of GARP. (Recall that a decoupling cycle must be repeated fast compared with $J_{IS}$.)

### 7.4.4 **Adiabatic fast passage**

Probably the most promising approach to broadband decoupling is to employ adiabatic rapid passage[4] (§ 2.4.2) to invert the spins, since this is known to make very efficient use of the available radiofrequency power, and it has the unusual property of being quite insensitive to the amplitude (and spatial inhomogeneity) of $B_2$ once the adiabatic condition has been satisfied:

$$\left|\frac{d\theta}{dt}\right| << \omega_{\text{eff}} \qquad \textbf{7.40}$$

where $\omega_{\text{eff}}$ is the effective field, the resultant of $\Delta\omega = \gamma\Delta B$ and $\omega_2 = \gamma B_2$ (all expressed in frequency units). The angle $\theta$ is the inclination of the effective field with respect to the $+x$ axis of the rotating frame and it runs from $+\pi/2$ to $-\pi/2$ during the adiabatic sweep. It is usual to define an adiabaticity factor $Q$ given by

$$Q = \frac{\omega_{\text{eff}}}{\left|\dfrac{d\theta}{dt}\right|}. \qquad \textbf{7.41}$$

It has been suggested that $Q$ should always be greater than about five.[33] Simulations shown in § 2.4.2 indicate what happens to the spin inversion as $Q$ is reduced below this value. Since $\omega_{\text{eff}}$ increases with offset, the most critical point of the sweep is the exact resonance condition (where $Q = Q_0$). We shall see below that, even with $Q_0 = 1$, effective broadband decoupling can be achieved when suitable phase cycles are used.

Fujiwara et al.[31,32] have employed a stepped-frequency scheme that mimics an adiabatic passage, and have demonstrated an appreciable increase in effective decoupling bandwidth (Table 1). The frequency is swept in the form that approximates a tangent function. The resulting family of decoupling sequences achieve bandwidths similar to the inversion bandwidths of the constituent adiabatic pulses. For example, the sequence MPF-9, using a composite pulse made up of nine frequency-switched elements, has a figure of merit $\Xi \approx 12.3$.

Several workers have taken a lead from McCall and Hahn[34] and implemented adiabatic passage with a frequency sweep in the form of a hyperbolic tangent, with the amplitude modulated as a hyperbolic secant.[33,35] The advantage is that the theoretical frequency-domain profile for spin inversion is almost rectangular, consisting of a central flat region, sharp transition regions and negligible out-of-band inversion. Decoupling sequences based on this idea have been described by Starčuk et al.[36] and by Bendall.[37] The corresponding figures of merit are set out in Table 1.

The hyperbolic secant has acquired an almost mystical significance for adiabatic passage; it is certainly an elegant mathematical solution. But if we analyse the problem from a purely practical point of view, noting that a wide range of different chemical shifts must be affected approximately equally, then the hyperbolic secant seems less appropriate. On the contrary, it suggests that the adiabatic sweep should be made up of a very wide section of constant radiofrequency amplitude and constant sweep rate, so that the spin inversion parameters are roughly constant. This leads to the idea of stretched adiabatic pulses, where the term is used in the same sense as a stretch limousine or stretch jetliner—the nose and tail remain the same but a long centre section is added. The adiabaticity factor is the same for all offsets

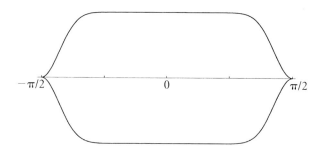

**Figure 7.5** The sausage-shaped amplitude profile used for WURST decoupling. It follows the profile defined by Equation 7.42, with θ swept from $-\pi/2$ to $+\pi/2$.

in the broad central region. The extremities of the sweep cover much narrower frequency ranges and serve only to carry the magnetization vector from the $+z$ axis or to the $-z$ axis as rapidly as possible, consistent with the adiabaticity condition. It is convenient to employ an amplitude function with a flat central region and a smooth rise or fall at the ends.[38] One promising envelope is the sausage-shaped function (Figure 7.5) defined by

$$A = \pm A_0 \{1 - |\sin\theta|^n\} \qquad\qquad \textbf{7.42}$$

where $n$ is a large even integer, for example 40. It has proved feasible to operate stretched adiabatic pulses with $Q_0$ as low as unity, provided that a suitable supercycle is used to compensate the imperfections in the spin inversion. In practice the $(0°, 150°, 60°, 150°, 0°)$ phase sequence of Tycko et al.[39] nested within the $(0°, 0°, 180°, 180°)$ phase sequence of Levitt[8] has proved very effective. Since the key features of this decoupling scheme are wideband, uniform rate and smooth truncation, it has been called WURST.[40,41] Sequences with constant radiofrequency level, without any smoothing of the extremities, give rise to an undesirable oscillatory component in the frequency-domain decoupling profile.[41,42]

Just as in line narrowing by spinning the sample (which causes spinning sidebands), magic cycles or supercycles that compensate for poor spin inversion inevitably cause cycling sidebands, and the harder they have to work, the more intense the sidebands. Fortunately the frequencies (and amplitudes) of the cycling sidebands can be predicted. Figures of merit $\Xi$ exceeding 50 are attainable with WURST decoupling, but under these conditions there are strong cycling sidebands. (of the order of 3% of the main decoupled signal) which eventually limit the sensitivity of the method.

For most experimental purposes we would choose to work with a slightly higher $Q$ value and therefore a lower figure of merit $\Xi$; the cycling sidebands then decrease quite rapidly to more acceptable levels. We are forced to the conclusion that there can be no single optimum decoupling regime; we must tailor the parameters to fit the practical case under investigation. When the decoupling bandwidth at low radiofrequency power is the overriding consideration, and the signal to noise is

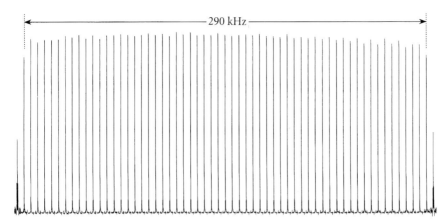

**Figure 7.6** Decoupling of protons from $^{13}$C in enriched methyl iodide ($J_{IS}$ = 150 Hz) with the WURST scheme. The 400 MHz proton resonance was recorded as the $^{13}$C decoupler was varied over a range of ±150 kHz in 5kHz steps. The effective decoupler field was $\gamma_C B_2(rms)/2\pi$ = 6.3 kHz, and the sweep duration was 1.5 ms. Imperfections were compensated with the 20-step supercycle described in the text. This gives a figure of merit $\Xi$ = 46.

poor, then we might be tempted to sail close to the wind, setting $Q$ very low and benefitting from the very high $\Xi$ value.

An experimental demonstration[43] of the WURST scheme (with $n$ = 240) is shown in Figure 7.6 for decoupling $^{13}$C in enriched methyl iodide ($J_{IS}$ = 150 Hz) while recording the proton spectrum at 400 MHz. For amplitude-modulated $B_2$ profiles it is useful to make a slight modification to the expression for the figure of merit so that the performance of decoupling schemes with differently shaped pulse envelopes can be compared at the same radiofrequency power. This is achieved by breaking down the pulse envelope into a histogram of narrow ordinates of different intensities and calculating the root-mean-square ordinate $B_2(rms)$. This is then used in place of $B_2$ in the expression for $\Xi$. The effective decoupler level in this experiment was $\gamma_C B_2(rms)/2\pi$ = 6.3 kHz, although the maximum level was slightly higher, $\gamma_C B_2(max)/2\pi$ = 6.5 kHz. The 20-step nested supercycle described above was used to compensate any residual imperfections in spin inversion. The duration of the stretched adiabatic sweep was 1.5 ms, implying that just over four adiabatic sweeps were completed in a period $1/(J_{IS})$. The effective decoupling bandwidth $\Delta F^\star$ = 290 kHz, which corresponds to a figure of merit $\Xi$ = 46. This would be quite adequate to cover the entire range of $^{13}$C shifts in any future NMR spectrometer that might operate at 1 GHz for protons. Note, however the relatively strong cycling sidebands, which are most prominent near the edges of the operating band. Even a slight increase in $B_2$ level and a consequent reduction in the sweep duration significantly attenuates these sidebands, at the expense of a small reduction in $\Xi$.

The figure of merit $\Xi$ was intended for decoupling schemes where the effective bandwidth was directly proportional to the radiofrequency level $B_2$. However, it can be shown[43,44] that adiabatic decoupling has an effective bandwidth that is approximately proportional to the square of the radiofrequency level:

$$\Delta F^\star = \xi \frac{[\gamma B_2(\text{max})]^2 T}{2\pi Q_0} \qquad \textbf{7.43}$$

where $B_2(\text{max})$ is the peak value of the (amplitude-modulated) decoupler field, $T$ is the duration of the adiabatic sweep, and $Q_0$ the adiabaticity factor at resonance. The factor $\xi = \Delta F^\star/\Delta F$ is the ratio of the effective decoupling bandwidth to the sweep range; it is usually a number just below unity. This dependence of $\Delta F^\star$ on the square of $B_2$ is perhaps the main reason why adiabatic schemes outperform composite-pulse methods such as WALTZ and GARP, where this dependence is only linear. Consequently, the figure of merit $\Xi$ for adiabatic decoupling is directly proportional to $B_2$, and comparisons of different adiabatic schemes[42,45] can be misleading unless we specify $B_2$. Alternatively, we may define a different figure of merit:

$$\Phi = \frac{\Delta F^\star J_{IS}}{[\gamma B_2(\text{rms})/2\pi]^2} \qquad \textbf{7.44}$$

where $B_2(\text{rms})$ is the effective decoupler level defined above. The numerical values of $\Phi$ are of course quite different from those of $\Xi$; they are normally close to unity.

### 7.4.5  Instrumental imperfections

It would be a mistake to judge the various decoupling sequences solely on their different figures of merit, because the shortcomings of the spectrometer also play an important role in determining performance. An interesting example of this point came to light when the experimental performance of the MLEV sequences was being verified. Not only did the decoupling efficiency appear to be extremely sensitive to the accuracy of the 90° phase shift between channels, but it appeared to reach an optimum for a phase shift that is not quite equal to 90°. It is now clear from the Waugh analysis[17] that this small phase error serves to compensate the small residual rotation of the basic sequence, but at the time this observation seemed inexplicable. It was this kind of difficulty, rather than considerations about the figure of merit, that prompted the search for a practical alternative to the MLEV sequences.

Composite pulse decoupling sequences are normally designed with specific pulse flip angles in mind, so it is not surprising that the performance degrades if there is a flip angle error. This may arise either through poor calibration or through spatial inhomogeneity in the decoupler radiofrequency field. Fortunately the expansion procedures used to improve the figure of merit also increase the tolerance to flip angle errors. However, some sequences are more sensitive to this effect than others. A ±5% calibration error does not seriously degrade the WALTZ-16 performance (interestingly, a positive error is much more readily tolerated than a negative error), whereas alternative sequences[26,27] are quite seriously affected. The GARP sequence readily compensates for a 5% misset, since this was one of the design criteria. A discrepancy between the flip angles in different channels is a quite different kettle of fish, for this destroys the symmetry so essential to the compensation. Fortunately modern digital circuitry is quite capable of ensuring a high level of amplitude balance (better than 0.2%) and no operator intervention is required. Adiabatic pulses have the marked advantage of being quite insensitive to $B_2$ level beyond a certain minimum required by the adiabatic condition.

Reference has already been made to the hypersensitivity of the MLEV sequences to small phase errors between channels that are nominally orthogonal. However if the orthogonality is accurate, the phase inversion used to obtain the other two (orthogonal) channels can be grossly misset without impairing performance significantly. This is clearly related to the high tolerance to errors in the 180° phase shift that is one of the strong points of the WALTZ family, a 10° phase error giving an essentially imperceptible change in performance.[23] A related problem is the phase glitch induced by any rapid switching of the radiofrequency voltage in a tuned circuit.[46] This is a severe problem in multipulse work in solids where the radiofrequency pulses are often less than a microsecond in duration, but is far less serious for broadband decoupling in liquids, as simulations have demonstrated.[21]

Sample spinning moves a typical group of spins into a slightly different decoupler field $B_2$ and therefore has the potential to interfere with the delicate balance required for compensation. Fortunately the decoupler modulation is at a frequency of the order of 100 Hz, whereas practical spinning rates are much slower (10–20 Hz), so the interaction is not usually very severe. We expect any degradation in performance to be most apparent at the edges of the effective decoupler bandwidth, a prediction that is borne out in practice and corroborated by computer simulation.[21]

### 7.4.6 **Band-selective decoupling**

In general, the broadband decoupling sequences described above were designed without any concern for what happens outside the effective bandwidth. The excitation in these outer reaches is poorly-defined but by no means negligible and we would anticipate that radiofrequency power is being wasted to no good purpose. Consequently the radiofrequency heating effect can be reduced by modifying the existing sequences so as to minimize the excitation outside the operating band.[47] An approximate treatment suggests that this can be achieved by analysing the Fourier spectrum of the composite inversion pulses and then reshaping them by eliminating the higher-order Fourier terms. These rounded-off composite pulses[48] have approximately the same decoupling efficiency but generate clean coupled spectra outside the decoupling bandwidth, and they offer a reduction of radiofrequency heating by as much as 30%. They also permit band-selective decoupling, where empty regions of the $I$-spin spectrum are not irradiated, thus conserving power even more effectively.[49]

### 7.4.7 **J scaling**

Broadband decoupling simplifies the $^{13}C$ spectrum and enhances the sensitivity, but only at the expense of the multiplicity information often needed for assignment purposes. This can be retrieved by editing methods such as DEPT[50] or SEMUT[51] but it is also possible to retain the spin multiplet information by using the proton decoupler to scale down the $^{13}$C–H splittings (§ 4.3.4). The scaling factor $\lambda$ is chosen to be large enough that the direct couplings are resolved so that (for example) a methyl group gives a 1:3:3:1 quartet, but small enough that the adjacent multiplets do not overlap, and any long-range splittings fall within the instrumental line width. It is desirable that the scaling factor be constant, independent of the proton chemical shift. This $J$ scaling effect can be achieved by a time-shared version of WALTZ decoupling.[52,53] The decoupler is switched on for a period $t_w$ and off for a period $t_p$. Assuming that the residual splittings observed during continuous WALTZ decoupling are negligible, the scaling factor is then given by the expression:

$$\lambda = \frac{t_p}{t_p + t_w} \qquad\qquad \textbf{7.45}$$

Only decoupling schemes that are essentially cyclic may be used for this $J$ scaling technique. Sampling can be asynchronous, using a hardwired decoupler. The delay $t_p$ must be kept short (hundreds of microseconds) implying that $t_w$ should not be too long either; this suggests the use of WALTZ-8 rather than WALTZ-16. The effective bandwidth matches that of the corresponding broadband decoupling sequence. Naturally the sensitivity is less than for decoupled $^{13}$C spectroscopy since the intensity from each $^{13}$C site is divided between several multiplet components. Above all the method is extremely simple, requiring only a minor modification of the broadband WALTZ decoupler. Problems arise if the scaled proton shifts lead to strong coupling effects, just as in continuous-wave decoupling.

### 7.4.8  Homonuclear Hartmann–Hahn transfer

In the development of the theory of broadband decoupling, the simplifying assumption was made that a single $I$ spin was involved, that is to say, any $I$–$I$ couplings could be neglected. This seemed reasonable at the time; the practical situation of interest involves quite small proton–proton couplings which scarcely perturb the decoupler offset on the frequency scale under consideration (several kHz). Then a few isolated instances came to light where WALTZ-16 decoupling was less efficient that expected, the $^{13}$C lines having a significant broadening or even distortion. It turns out that the explanation is rather subtle.[54,55] The decoupling sequence has the property that it can induce homonuclear Hartmann–Hahn transfer[1,56] between the coupled protons, a process exploited in total correlation spectroscopy[2] (TOCSY), sometimes called the HOHAHA effect.[3] A proton may, in effect, change its precession frequency during a composite inversion pulse, upsetting the delicate requirement for cyclicity. Fortunately the effect is only detected under very high resolution conditions and only at relatively low settings of $B_2$ where $J_{HH}\tau$ becomes appreciable (where $\tau$ is the decoupler period).

   Shaka et al.[55] analysed this effect in detail and showed that multiple-pulse schemes such as MLEV and WALTZ can produce an effective Hamiltonian containing a combination of bilinear operators (such as $2I_{1x}I_{2x}$) with unequal coefficients; the interaction is no longer scalar. In particular, the homonuclear zero-quantum coherence ($2I_{1y}I_{2x} - 2I_{1x}I_{2y}$) is invariant to 180° phase shifts and can interfere with the heteronuclear decoupling process. Shaka et al.[55] were able to design new composite inversion pulses that minimized these undesirable cross-terms and assemble them into the classic $R\bar{R}\bar{R}R$ pattern to give decoupling in the presence of scalar interactions (DIPSI) sequences. The effective decoupling bandwidths are smaller than for WALTZ-16 but the spurious broadening effects are appreciably reduced.

One very positive outcome of these investigations has been to focus attention on the fact that broadband decoupling sequences can also serve as broadband isotropic mixing schemes for TOCSY or HOHAHA experiments. The DIPSI sequence is widely employed in this manner. Normally, with continuous monochromatic spin-lock fields $B_1$ and $B_2$, the homonuclear Hartmann–Hahn effect is very sensitive to radiofrequency offset. The use of broadband decoupling sequences greatly extends the effective bandwidth for efficient coherence transfer.[57,58] Shaka et al.[59] have addressed this problem directly and designed a composite zero-quantum pulse specifically to increase the operating bandwidth for Hartmann-Hahn cross-polarization. They show that the resulting supercycle FLOPSY-8 (flip-flop spectroscopy) has significantly improved performance compared with the earlier broadband decoupling sequences.

## 7.5 Discussion

Ernst's introduction of noise decoupling[6] was crucial to the initial acceptance of $^{13}$C spectroscopy by chemists, for not only did it guarantee an important sensitivity improvement, but it also greatly simplified the spectra. It served its purpose well for more than a decade (the 1970s) and at that time it was hard to imagine that there might be better solutions for broadband decoupling. The beauty of the deterministic decoupling procedures pioneered by Levitt, is that not only did they improve decoupling efficiency significantly, but also offered promise of even greater improvements in the figure of merit, as now demonstrated by adiabatic pulse decoupling (Table 1). It is also rather reassuring that we no longer have to rely on random (or pseudorandom) phenomena to solve a physical problem.

## References

1. S. R. Hartmann and E. L. Hahn, *Phys. Rev.* **128**, 2042 (1962).
2. L. Braunschweiler and R. R. Ernst, *J. Magn. Reson.* **53**, 521 (1983).
3. A. Bax and D. G. Davis, *J. Magn. Reson.* **65**, 355 (1985).
4. A. Abragam, *The Principles of Nuclear Magnetism,* Clarendon Press, Oxford, 1961.
5. A. J. Shaka, T. Frenkiel and R. Freeman, *J. Magn. Reson.* **52**, 159 (1983).
6. R. R. Ernst, *J. Chem. Phys.* **45**, 3845 (1966).
7. M. H. Levitt and R. Freeman, *J. Magn. Reson.* **33**, 473 (1979).
8. M. H. Levitt and R. Freeman, *J. Magn. Reson.* **43**, 502 (1981).

9. M. H. Levitt, R. Freeman and T. Frenkiel, *J. Magn. Reson.* **47**, 328 (1982).
10. A. J. Shaka and R. Freeman, *J. Magn. Reson.* **55**, 487 (1983).
11. J. S. Waugh. L. M. Huber and U. Haeberlen, *Phys. Rev.* **20**, 180 (1968).
12. P. Mansfield, *J. Phys. C.* **4**, 1444 (1971).
13. M. H. Levitt, R. Freeman and T. Frenkiel, *J. Magn. Reson.* **50**, 157 (1982).
14. U. Haeberlen and J. S. Waugh, *Phys. Rev.* **175**, 453 (1968).
15. M. H. Levitt, R. Freeman and T. Frenkiel, *Adv. Magn. Reson.* **11**, 47 (1983).
16. J. S. Waugh, *J. Magn. Reson.* **49**, 517 (1982).
17. J. S. Waugh, *J. Magn. Reson.* **50**, 30 (1982).
18. R. Freeman, T. Frenkiel and M. H. Levitt, *J. Magn. Reson.* **50**, 345 (1982).
19. W. A. Anderson and R. Freeman, *J. Chem. Phys.* **37**, 85 (1962).
20. F. Bloch and A. Siegert, *Phys. Rev.* **57**, 522 (1940).
21. A. J. Shaka and J. Keeler, *Prog. NMR Spectrosc.* **19**, 47 (1986).
22. A. J. Shaka, J. Keeler, T. Frenkiel and R. Freeman, *J. Magn. Reson.* **52**, 335 (1983).
23. A. J. Shaka, J. Keeler and R. Freeman, *J. Magn. Reson.* **53**, 313 (1983).
24. R. Freeman and H. D. W. Hill, *J. Chem. Phys.* **54**, 3367 (1971).
25. A. J. Shaka, P. B. Barker, C. J. Bauer and R. Freeman, *J. Magn. Reson.* **67**, 396 (1986).
26. B. M. Fung, *J. Magn. Reson.* **60**, 424 (1984).
27. V. Sklenář and Z. Starčuk, *J. Magn. Reson.* **62**, 113 (1985).
28. A. J. Shaka, P. B. Barker and R. Freeman, *J. Magn. Reson.* **64**, 547 (1985).
29. N. Sunitha Bai, N. Hari and R. Ramachandran, *J. Magn. Reson. A* **106**, 241 (1994).
30. V. J. Basus, P. D. Ellis, H. D. W. Hill and J. S. Waugh, *J. Chem. Phys.* **35**, 19 (1979).
31. T . Fujiwara and K. Nagayama, *J. Magn. Reson.* **86**, 584 (1990).
32. T. Fujiwara, T. Anai, N. Kurihara and K. Nagayama, *J. Magn. Reson. A.* **104**, 103 (1993).
33. J. Baum. R. Tycko and A. Pines, *Phys. Rev. A* **32**, 3435 (1985).
34. S. L. McCall and E. L. Hahn, *Phys. Rev.* **183**, 457 (1969).
35. M. S. Silver, R. J. Joseph and D. I. Hoult, *Phys. Rev. A* **31**, 2753 (1985).
36. Z. Starčuk, Jr., K. Bartušek and Z. Starčuk, *J. Magn. Reson. A* **107**, 24 (1994).
37. M. R. Bendall, *J. Magn. Reson. A* **112**, 126 (1995).
38. J. M. Böhlen and G. Bodenhausen, *J. Magn. Reson. A* **102**, 293 (1993).
39. R. Tycko, A. Pines, and R. Gluckenheimer, *J. Chem. Phys.* **83**, 2775 (1985).
40. E. Kupče and R. Freeman, *J. Magn. Reson. A* **115**, 273 (1995).
41. E. Kupče and R. Freeman, *J. Magn. Reson. A* **117**, 246 (1995).
42. R. Fu and G. Bodenhausen, *J. Magn. Reson. A* **117**, 324 (1995).
43. E. Kupče and R. Freeman, *J. Magn. Reson. A* **118**, 299 (1996).
44. E. Kupče and R. Freeman, *Chem. Phys. Lett.* **250**, 523 (1996).
45. R. Fu and G. Bodenhausen, *Chem. Phys. Lett.* **245**, 415 (1995).
46. M. Mehring and J. S. Waugh, *Rev. Sci. Instr.* **43**, 649 (1972).
47. M. A. McCoy and L. Müller, *J. Am. Chem. Soc.* **114**, 2108 (1992).
48. E. Kupče, J. Boyd and I. D. Campbell, *J. Magn. Reson. A* **110**, 109 (1994).
49. E. Kupče and R. Freeman, *J. Magn. Reson. A* **102**, 364 (1993).
50. D. M. Doddrell, D. T. Pegg and R. M. Bendall, *J. Magn. Reson.* **49**, 323 (1982).
51. H. Bildsøe, S. Dønstrup, H. J. Jakobsen and O. W. Sørensen, *J. Magn. Reson.* **53**, 154 (1983).
52. G. A. Morris, G. L. Nayler, A. J. Shaka, J. Keeler and R. Freeman, *J. Magn. Reson.* **58**, 155 (1984).
53. A. J. Shaka, J. Keeler, R. Freeman, G. A. Morris and G. L. Nayler, *J. Magn. Reson.* **58**, 161 (1984).
54. P. B. Barker, A. J. Shaka and R. Freeman, *J. Magn. Reson.* **65**, 535 (1985).
55. A. J. Shaka, C. J. Lee and A. Pines, *J. Magn. Reson.* **77**, 274 (1988).

56. G. C. Chingas, A. N. Garroway, R. D. Bertrand and W. B. Moniz, *J. Chem. Phys.* **74**, 127 (1981).
57. N. Chandrakumar, *J. Magn. Reson.* **71**, 322 (1987).
58. S. Subramanian and A. Bax, *J. Magn. Reson.* **71**, 325 (1987).
59. M. Kadkhodaie, O. Rivas, M. Tan, A. Mohebbi and A. J. Shaka, *J. Magn. Reson.* **91**, 437 (1991).

# 8

# Two-dimensional spectroscopy

## 8.1 Introduction

Imagine for a moment that an extraterrestrial civilization is monitoring life on Earth through a satellite-based telescope and doing its best to make sense of the observations. Some of our sporting activities might well pose a considerable challenge to their understanding. One bewildering observation would be in athletics —the relay race. Several runners go around an approximately oval track but they inexplicably stop when they meet a second group of athletes who, after a short period of confusion, then set off at full speed. What is going on? The telescope in question is limited in resolution and misses some details of the changeover but it is clear that the net result is a high degree of correlation between the colours worn by the fastest runners in the first stage and those of the leading runners in the second.

## 8.2 Homonuclear correlation spectroscopy

Two-dimensional correlation spectroscopy operates along similar lines. Several different groups of spins are identified in the first stage of the experiment (the evolution period) on the basis of their precession frequencies. This is followed by a short fixed mixing period. In the second stage (called the detection period) they acquire different precession frequencies, but certain strict correlation rules apply, telling us which groups are related by a specific interaction. The mysterious mixing period involves 'passing the baton to members of the same team.' The interaction in question may be the scalar coupling (COSY), the internuclear Overhauser effect (NOESY), slow chemical exchange (EXSY) or indeed any suitable link dreamed up by the spectroscopist.

Correlation spectroscopy was first performed by Jeener[1] as the archetypal two-dimensional technique, but for understandable instrumental reasons his spectra were noisy and were never published. Soon afterwards an elegant analysis and a comprehensive exploration of the topic were presented by Aue *et al.*[2] The technique was further popularized by Nagayama *et al.*[3] and by Bax *et al.*[4] and has developed into one of the most widely used two-dimensional experiments for assigning high resolution NMR spectra. The COSY experiment was soon followed by the NOESY[5] method (§ 9.3.1) which provides direct information about the proximity of protons, and yields invaluable structural information, particularly when applied to biological macromolecules in solution. The wide popularity of these two experiments can be explained by the graphic way in which the correlation information is displayed—a contour map with diagonal peaks marking the conventional spectrum and cross-peaks indicating the correlations. This kind of topological diagram is ideal for indicating interactions within a given spectrum.

## 8.2.1 Pseudocorrelation spectroscopy

We may think of the COSY experiment as being derived from the earlier custom of studying scalar interactions by double resonance—one group of spins $I$ is irradiated while observing the perturbation of a coupled group $S$. Translated into modern pulse technology, this would entail the application of a selective (soft) excitation pulse (§ 5.4.1) to one of the $I$-spin lines, followed by excitation of the entire spectrum with a nonselective (hard) pulse. In the vocabulary of product operators (§ 3.5.1) this would be written for a two-spin system as

$$\frac{1}{2}(I_z + 2I_zS_z) \xrightarrow{\tilde{I}_x} \frac{1}{2}(-I_y - 2I_yS_z) \xrightarrow{\tilde{I}_x \quad \tilde{S}_x} \frac{1}{2}(-I_z + 2I_zS_y) \qquad \textbf{8.1}$$

The term $2I_zS_y$ leads to the observable correlation response—an antiphase absorption-mode doublet of lines of splitting $J_{IS}$ centred at the chemical shift of the $S$ spins. If we choose to apply the soft pulse at the frequency of the other $I$-spin line, a similar doublet is observed but inverted in phase:

$$\frac{1}{2}(I_z - 2I_zS_z) \xrightarrow{\tilde{I}_x} \frac{1}{2}(-I_y + 2I_yS_z) \xrightarrow{\tilde{I}_x \quad \tilde{S}_x} \frac{1}{2}(-I_z - 2I_zS_y) \qquad \textbf{8.2}$$

In order to minimize intervention by the operator, we might choose to scan the frequency of the soft pulse in very small increments across the spectrum to make sure to irradiate each transition separately. This is called pseudo-two-dimensional spectroscopy.[6] It is a direct exploration of one frequency dimension ($F_1$) through a large number of double resonance experiments, each with a slightly different frequency of the soft pulse. The region immediately adjacent to the $S$-spin chemical shift might then look like this:

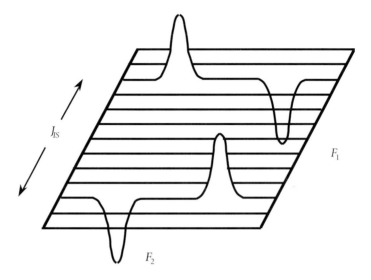

This is closely analogous to the antiphase square pattern of a typical cross-peak in a COSY spectrum (§ 2.6.1).

There is another group of four responses corresponding to signals that are both excited and detected at the $I$-spin chemical shift, involving no coherence transfer. They are dispersive in both frequency dimensions and all four are in the same sense. In practice these dispersion-mode responses can be suppressed without affecting the cross-peaks if the spin pinging technique[7] is used (§ 5.3.2). This is useful when it is necessary to observe cross-peaks close to the diagonal. Figure 8.1 shows a two-dimensional pseudo-COSY spectrum of $m$-bromonitrobenzene obtained by this procedure; all the cross-peaks are well-defined but the diagonal peaks have vanishing intensity. This spectrum illustrates the very fine detail that may be observed by the pseudo-COSY technique if the frequency increments are made small enough. Figure 8.2 shows expansions of the cross-peaks, obtained with soft pulses of duration 1.3 s, scanned through a range of 16 Hz with frequency increments of 0.2 Hz. There is clear evidence for the small (0.35 Hz) *para* coupling in Figure 8.2(b).

## 8.2.2 Classic COSY experiment

The true two-dimensional COSY spectrum (Figure 8.3) has essentially the same appearance as the pseudo-COSY spectrum, but it is obtained with far higher sensitivity. It employs two hard $\pi/2$ pulses separated by a variable evolution period ($t_1$),

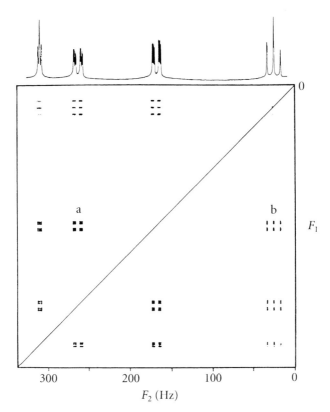

**Figure 8.1** Experimental pseudo-COSY spectrum of *m*-bromonitrobenzene obtained by stepping the frequency of a soft pulse in small increments across the $F_1$ frequency dimension. The spin-pinging sequence was used which has the effect of suppressing the dispersion mode signals on the principal diagonal.

during which all the spectral components precess freely at their characteristic frequencies.[1–4]

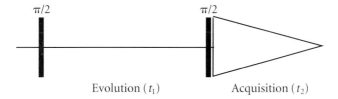

A vector picture of this experiment has been presented in § 2.6.1, emphasizing that the process can be understood in terms of population disturbances without recourse to the special terms 'coherence transfer' and 'mixing pulse'. The more formal product operator treatment is shown as a worked example in § 3.3.5. We may

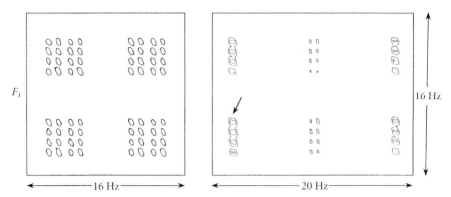

**Figure 8.2** Re-examination of the cross peaks (a) and (b) in the pseudo-COSY spectrum of metabromonitrobenzene shown in Figure 8.1. The $F_1$ frequency was scanned in increments of 0.2 Hz through a range of only 16 Hz. The soft pulse duration was 1.3 s. Note the perceptible splitting (indicated by the arrow) attributable to the small *para* coupling (0.35 Hz).

write the operator algebra in a shorthand form to show how *I*-spin magnetization is converted into antiphase *S*-spin magnetization in a two-spin system:

$$+I_z \xrightarrow{\tilde{I}_x} -I_y \xrightarrow{\tilde{I}_z} +I_x \xrightarrow{2\tilde{I}_z\tilde{S}_z} +2I_yS_z \xrightarrow{\tilde{I}_x} +2I_zS_z \xrightarrow{\tilde{S}_x} -2I_zS_y \qquad \textbf{8.3}$$

The $F_1$ dimension is explored indirectly by simultaneous excitation of all frequency components; these are later separated by Fourier transformation as a function of the evolution period $t_1$. The method enjoys a multiplex advantage in the $F_1$ dimension just as conventional Fourier transform spectroscopy benefits from a multiplex advantage in the $F_2$ dimension. Instead of plodding, step by tiny step, through the entire spectrum, we gather global information at each and every $t_1$ increment. Time-domain sampling is much more efficient than selective excitation in the frequency domain. See, however, Hadamard spectroscopy (§ 5.5.4).

If we make a few simple assumptions we can calculate an approximate expression for the sensitivity improvement. In pseudo-two-dimensional spectroscopy we record N traces, but only a very small number of these gather correlation information from a given *I*-spin transition. In fact it is the trace closest to the peak of the chosen *I*-spin line that determines the signal-to-noise ratio for the absorption mode responses. If we wish to make the most efficient use of the available time we must employ rather coarse frequency increments so that there is approximately one sample per linewidth. Finer digitization does not improve the signal-to-noise ratio significantly (it merely ensures that there is a sampling point closer to the peak). Thus, out of N traces, only about one trace carries a particular signal of interest. Consequently, N is determined by the ratio of the spectral width to the width of a typical line.

On the other hand, true two-dimensional spectroscopy gathers an appreciable signal at every $t_1$ increment. Its amplitude is slightly reduced by relaxation losses

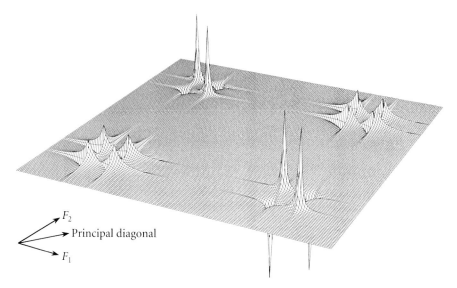

$F_2$

Principal diagonal

$F_1$

**Figure 8.3** Simulation of a COSY spectrum for a system of two coupled spins with $\delta_{IS} \approx 60$ Hz and $J_{IS} \approx 8$ Hz and a full Lorentzian line width of 1.5 Hz. The diagonal peaks form square patterns of double-dispersion mode responses sitting astride the principal diagonal ($F_1 = F_2$). The cross-peaks are square patterns of absorption peaks with phase alternation in both frequency dimensions.

during $t_1$, and because of the sinusoidal modulation, but this is normally only a small loss. In a given time, we accumulate $N$ times as many useful scans by two-dimensional spectroscopy, and since noise increases only as $\sqrt{N}$, the approximate improvement in sensitivity is given by $\sqrt{N}$. This enhancement factor could easily amount to between one and two orders of magnitude. That is why two-dimensional correlation spectroscopy is seldom performed by stepping the excitation frequency, except, perhaps, in those few cases where interest focuses on a very restricted region of the correlation spectrum.

Jeener[1] clearly anticipated this multiplex advantage of two-dimensional spectroscopy. This has been one of the mainstays of the popularity of the method—it is preferable to gather data that is changing with time (thus providing new information), rather than accumulate the same free induction decay from a selective experiment over and over again, and the sensitivity is virtually the same.

One modification of the classic COSY experiment has been widely adopted—double-quantum filtered correlation spectroscopy[8] (DQ-COSY). The pulse sequence may be expressed as

$$(\pi/2)_\phi - t_1 - (\pi/2)_{\phi+90°} - \Delta - (\pi/2)_\psi \text{ acquisition } (\pm) \qquad \textbf{8.4}$$

with the phase $\phi$ cycled $0°$, $90°$, $180°$ and $270°$ while the receiver phase is alternated. Quadrature detection in the $t_1$ dimension is implemented by setting $\psi$ to $0°$ and $90°$ in successive experiments. Double-quantum-filtered COSY is described in more detail in § 6.3.1, where a practical application is presented. Its advantage is that the diagonal peaks are mainly in absorption (when the cross-peaks are also phased for absorption) thus avoiding long dispersion tails and facilitating the detection of cross-peaks close to the diagonal. Double-quantum filtration also helps to discriminate against singlet responses from the solvent. Normally these improvements more than compensate for the 50% loss in intensity of the cross-peaks, compared with the ordinary COSY experiment.

### 8.2.3  COSY cross-peaks

The basic square array predicted for a COSY cross-peak becomes more complex in practical situations where there are several coupled spins. It is useful to define the two spins involved in the coherence transfer as active and all other spins as passive. A passive spin ($R$) creates two effective chemical shifts to replace the true shifts $\delta_I$ and $\delta_S$, giving two square patterns centred at the coordinates:

$$(\delta_I + \tfrac{1}{2}J_{IR}, \delta_S + \tfrac{1}{2}J_{SR}) \text{ and } (\delta_I - \tfrac{1}{2}J_{IR}, \delta_S - \tfrac{1}{2}J_{SR}) \qquad \textbf{8.5}$$

However, the second pulse of the COSY sequence also has an effect on the passive $R$ spins; if this pulse has a flip angle of $\pi/2$ it behaves mid-way between a 0 and a $\pi$ radian pulse, inverting 50% of the $R$ spins between evolution and detection. This generates two more square patterns at the coordinates:

$$(\delta_I + \tfrac{1}{2}J_{IR}, \delta_S - \tfrac{1}{2}J_{SR}) \text{ and } (\delta_I - \tfrac{1}{2}J_{IR}, \delta_S + \tfrac{1}{2}J_{SR}) \qquad \textbf{8.6}$$

Consequently the cross-peak is a rectangular array made up of four of the basic square patterns.

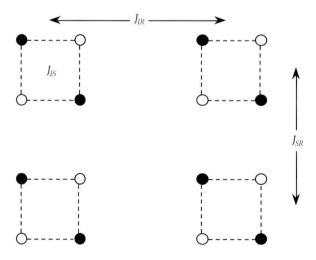

Each new passive spin that couples to spin $I$ doubles the number of component lines in the cross-peak. Similarly another passive coupling to spin $S$ splits the pattern again.

An interesting new effect[4] occurs if the flip angle β of the second pulse deviates from π/2 radians. If β << π/2 then the passive spin R remains largely unaffected by that pulse and acts merely as a spectator. Instead of a cross-peak made up of four square patterns, just two predominate, connected by a displacement vector V which is the resultant of $J_{IR}$ and $J_{SR}$, taking account of the relative signs of these two couplings. One square pattern corresponds to the R spin in an α state, while the other represents the β state. Thus if $J_{IR}$ and $J_{SR}$ have the same sign, the displacement vector V has a positive slope, whereas if they have opposite signs the slope is negative.

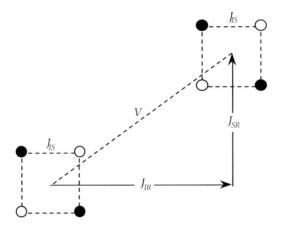

A flip angle of π/4 radians is commonly used for this determination; the experiment is then known as COSY-45. This scheme for determining the relative signs of two passive couplings[4,9] is closely analogous to the double resonance method that uses selective decoupling.

If, on the other hand, the flip angle β lies between π/2 and π, the majority of the R spins are inverted during the coherence transfer and the other two square patterns predominate. The signs of $J_{IR}$ and $J_{SR}$ are again obtained by inspection, but now a positive slope for the displacement vector implies opposite signs.

When it is necessary to examine the fine detail of a COSY cross-peak, it may be advantageous to employ band-selective (strictly, multiplet selective) excitation (§ 5.4.3) in order to use the available digitization to the best advantage. We may distinguish three different modes of band-selective COSY:

Strip-COSY

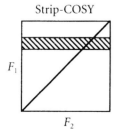

endo-COSY

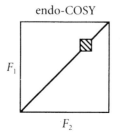

exo-COSY

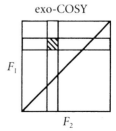

In the simplest regime, which we may call strip-COSY, only the excitation pulse is band-selective, the second pulse is hard, and we record the entire spectral width in the $F_2$ dimension. This exploits the fact that high resolution and fine digitization are readily achieved in this dimension without any significant time penalty. In the second mode (*endo*-COSY) the second pulse is also band-selective but is identical to the first pulse. This allows us to zoom in on a small square region straddling the principal diagonal. In the third mode (*exo*-COSY), the mixing stage consists of two independent band-selective pulses (§ 5.3.5) applied simultaneously.[10,11]

| | | |
|---|---|---|
| *I* spins: | soft $(\pi/2)$ — $t_1$ — soft $(\pi/2)$ | |
| *S* spins: | soft $(\pi/2)$ acquire | **8.7** |

This provides independent control of the excitation and acquisition bands so that any arbitrary rectangular zone of the full COSY spectrum may be examined with fine digitization in both frequency dimensions. This experiment,[12] sometimes called simply soft-COSY, may be used as the first step in the determination of accurate values of the spin–spin coupling constants. Alternatively, these splittings can be measured by line-selective excitation[7] that records a suitable section through the cross-peak in the $F_2$ dimension, where fine digitization is always available.

### 8.2.4 **Overlapping cross-peaks**

As molecular complexity increases, practical problems arise when two or more cross-peaks overlap through a near-degeneracy of the chemical shifts. Then it is difficult to disentangle the individual splittings, and the assignment may become ambiguous. One solution is to extend the experiment into a third frequency dimension (see 'multidimensional spectroscopy' below). Alternatively we can attack the overlap problem directly. In this case, it is important to have a well-resolved and finely digitized experimental COSY spectrum.

One experimental approach is through line-selective excitation. Suppose there are two overlapping cross-peaks (A and B) in a COSY spectrum. If we can find a suitable section that slices through lines belonging to the A spin without intersecting any lines belonging to B, then subsequent free evolution followed by a hard $(\pi/2)$ pulse spreads this coherence evenly throughout the A cross-peak but not through the B cross-peak, allowing a pure form of the former to be recorded.[13] A practical application is described in (§ 5.4.4).

A related method relies on data processing of the experimental COSY spectrum. One important symmetry property of a single, isolated cross-peak is that it can be completely reconstructed by multiplication of a trace taken parallel to the $F_1$ axis and another trace taken parallel to the $F_2$ axis.[14] Any two such traces suffice; they should pass through the appropriate responses but it is not necessary to intersect at the exact peak of the resonances. Suppose we record a section at $F_1 = a$, and another (orthogonal) section at $F_2 = b$, intersecting at the point $(a,b)$.

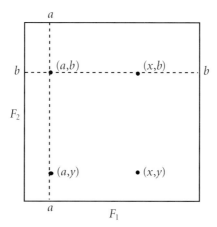

The intensity at any general point $(x,y)$ is given by the product $I(a,y) I(x,b)$ normalized by the intensity $I(a,b)$. It is preferable to choose $a$ and $b$ so that the traces have a good signal-to-noise ratio, particularly for the pixel $I(a,b)$ so as not to scale the reconstructed cross-peak improperly.

Suppose now that there are two partially overlapped COSY cross-peaks, A and B. If we can discover suitable orthogonal traces that pass only through A resonances, avoiding B resonances, then by multiplication we may reconstruct the A cross-peak with high fidelity, and by subtraction from the experimental data, leave only the B cross-peak.

In practical situations where the interpenetration is not too severe, part of the A cross-peak may be in the clear, with no significant overlap by B lines. If as much as one quadrant of the A cross-peak can be isolated in this manner, then the other three quadrants can be reconstructed according to the inherent antisymmetry of a COSY cross-peak. This again allows the B cross-peak to be isolated by subtraction. A related technique relies on obtaining information about the centre of the A cross-peak (from projected two-dimensional $J$ spectra) so that a symmetrization routine based on the two $C_2$ axes selects intensity from A resonances but rejects intensity from B resonances (§ 6.5.3).

## 8.2.5 Relayed correlation spectroscopy

It quite often happens that chemical shift degeneracies give rise to ambiguities in the classic COSY experiment. For example, we may not be able to distinguish a three-spin $IRS$ system (if $J_{IS}$ is vanishingly small) from the superposition of an $IR$ system and a $PS$ system where the chemical shifts of $R$ and $P$ overlap; there would be the same number and disposition of cross-peaks in both cases. A variation of correlation spectroscopy called RELAY[15,16] resolves this kind of difficulty by introducing a second stage of coherence transfer that takes place during a fixed mixing period $\tau_m$.

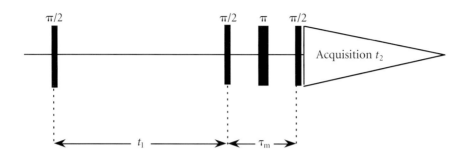

We consider a homonuclear *IRS* spin system with nonzero couplings $J_{IR}$ and $J_{RS}$ but with a vanishing long-range coupling $J_{IS}$. The method works best if $J_{IR}$ and $J_{RS}$ are comparable in magnitude, and the mixing period is set near the condition

$$\tau_m \approx 1/(2J_{IR}) \approx 1/(2J_{RS}) \qquad \textbf{8.8}$$

For the purposes of illustration we assume $J_{IR}$ is slightly larger than $J_{RS}$. We focus attention on the *R* spin multiplet ($R_y$) which appears as antiphase magnetization ($2I_zR_y$) at the beginning of the mixing period, since it has just been created by coherence transfer from the *I* spins. In the mixing period $\tau_m$ this evolves into inphase magnetization (with respect to *I*) but antiphase magnetization with respect to *S*:

$$2I_zR_y \xrightarrow{\ 2\tilde{I}_z\tilde{R}_z\ } -R_x \xrightarrow{\ 2\tilde{R}_z\tilde{S}_z\ } -2R_yS_z \qquad \textbf{8.9}$$

$$R_y \qquad\qquad 2I_zR_y \qquad\qquad 2R_yS_z$$

The same result is of course obtained if the evolution operators are applied in the reverse order:

$$2I_zR_y \xrightarrow{\ 2\tilde{R}_z\tilde{S}_z\ } -4I_zR_xS_z \xrightarrow{\ 2\tilde{I}_z\tilde{R}_z\ } -2R_yS_z \qquad \textbf{8.10}$$

The antiphase term $-2R_yS_z$ is converted by the last $(\pi/2)_x$ pulse into antiphase *S*-spin magnetization which becomes a cross-peak at the *S*-spin chemical shift. In practice it is preferable to include a refocusing pulse at the mid-point of the mixing period to refocus chemical shift effects, so the complete pulse sequence is:

$$(\pi/2)_x - t_1 - (\pi/2)_x - \tau_m/2 - (\pi)_x - \tau_m/2 - (\pi/2)_x\ \text{acquire} \qquad \textbf{8.11}$$

For those who prefer a geometrical picture, the experiment can be described in terms of the evolution of *R*-spin vectors during the mixing period $t_m$. Chemical shift effects are neglected on the grounds that they are refocused by the hard $\pi$ pulse. The four components of $R$-spin magnetization may be labelled $\beta\beta$, $\beta\alpha$, $\alpha\beta$ and $\alpha\alpha$, according to the spin states of $I$ and $S$, respectively.

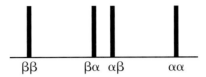

In a reference frame rotating at the $R$-spin chemical shift, $\alpha\alpha$ and $\beta\beta$ are the fast vectors and $\alpha\beta$ and $\beta\alpha$ the slow ones. They evolve during the mixing period as follows:

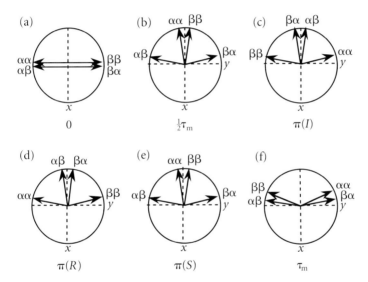

where the effect of the hard $\pi_x$ pulse has been broken down into a cascade of three selective $\pi_x$ pulses applied to $I$, $R$ and $S$ in turn. Note that the overall effect of all three inversion pulses is to return the vectors to their original positions. The mixing period simply converts an arrangement that is antiphase with respect to $I$ into one that is antiphase with respect to $S$. The subsequent $(\pi/2)_x$ pulse can be regarded as a cascade of soft $(\pi/2)_x$ pulses; the first creates a population disturbance on the energy levels shared with the $S$ spin and the second converts this into an antiphase $S$-spin signal.

The same manipulation can be achieved without a mixing period if a selective $\pi$ pulse is applied at the $R$-spin chemical shift to invert the two central components of the $R$-spin multiplet, leaving the outer components unaffected:

$$2I_zR_y \xrightarrow{\text{soft } \pi \text{ pulse}} -2R_yS_z \qquad \textbf{8.12}$$

This can be used in a repetitive manner to transfer coherence along a chain of coupled spins.[17]

## 8.2.6 Total correlation spectroscopy

A general extension of the idea of relayed coherence transfer is provided by the total correlation spectroscopy (TOCSY) technique.[18] It can be argued that some chemical applications are best served if coherence is distributed indiscriminately throughout the entire network of coupled spins, not merely to the directly coupled neighbours as in the COSY technique. The mechanism for this general transfer is quite distinct from the COSY scheme. It relies on a mixing Hamiltonian $\bar{\mathcal{H}}_m$ that is dominated by the full isotropic terms:

$$\bar{\mathcal{H}}_m = \sum_{i<j} 2\pi J_{ij} \mathbf{I}_i \cdot \mathbf{S}_j)$$   **8.13**

instead of the more familiar term $2\pi J_{ij} I_{iz} S_{jz}$. Several different repetitive pulse sequences have been proposed[18] for $\bar{\mathcal{H}}_m$. It was soon realized[19–21] that the mechanism for coherence transfer is the homonuclear Hartmann-Hahn effect[22–26] and it has therefore also come to be known as the HOHAHA technique.

Under this type of isotropic mixing Hamiltonian, the spin system evolves in collective spin modes, such as

$$\Sigma = \tfrac{1}{2}(I_y + S_y) \quad \text{and} \quad \Delta = \tfrac{1}{2}(I_y - S_y)$$   **8.14**

instead of the more familiar single spin modes. There is an oscillatory exchange of $I$ and $S$ coherence of the type discussed in (§ 5.5.6) for soft pulse excitation. In an extended spin network, where the target spin $S$ is coupled to more distant spins, part of the coherence moves on instead of returning to the source spin $I$. Eventually coherence propagates throughout the entire spin system. The effect is reminiscent of that first reported by James Thurber's aunt, who was careful to fit electric light bulbs in all available sockets 'to prevent electricity leaking all over the house.'

In total correlation spectroscopy the simplest mixing scheme employs a continuous spin-lock field $B_1$ applied for a period comparable with the inverse of the relevant spin–spin coupling constants. In practice an unmodulated $B_1$ field is very sensitive to off-resonance effects that interfere with the Hartmann–Hahn matching condition. Consequently, to cover a wide spectral width, it is usually modulated according to one of the broadband decoupling schemes,[27,28] or one of the pulse sequences specifically designed for the purpose, DIPSI[29] or FLOPSY.[30] These isotropic mixing techniques are discussed in more detail in § 7.4.8.

Total correlation spectroscopy is now widely used to complement the COSY technique. The mechanism of coherence transfer differs from that of the COSY experiment, and TOCSY has the important practical advantage that all the components of a cross-peak are in-phase, and there is no mutual cancellation when the coupling constant is poorly resolved. Hence the popularity of total correlation spectroscopy in studies of biological macromolecules, where the natural line widths are the limiting factor in resolution. Conversely, indiscriminate diffusion of a given amount of coherence throughout the coupling network necessarily dilutes the signal transferred to a given site, so the sensitivity tends to be lower than that of the directed multistage Hartmann–Hahn transfer[31] using soft pulses (§ 5.5.7).

### 8.2.7  Overhauser effect and chemical exchange

One of the most effective uses of two-dimensional spectroscopy has been to study the nuclear Overhauser effect, particularly in molecules of biochemical importance. A typical macromolecule may generate hundreds or even thousands of NOESY cross-peaks, providing a surfeit of information about interproton distances, some of which give valuable structural clues about the three-dimensional structure in solution. Often the information is used only qualitatively to establish distance constraints, which may be used in concert with molecular dynamics calculations. Since this subject is treated in detail in § 9.3, we simply outline the experiment at this point.

The radiofrequency pulse sequence for chemical exchange spectroscopy[5] (EXSY) and nuclear Overhauser spectroscopy (NOESY) may be written as

$$(\pi/2)_x - t_1 - (\pi/2)_x - \tau_m - (\pi/2)_x \text{ acquire} \qquad \textbf{8.15}$$

As before, the evolution period $t_1$ serves only to label the magnetizations from various sites according to their chemical shifts. Mixing takes place during the fixed interval $\tau_m$, during which the magnetization remains aligned along the $z$ axis. The mechanism is either cross-relaxation (NOESY) or a physical translation of a spin from one molecule to another or between two distinct conformations of the same molecule (EXSY). By the time the receiver is switched to detection, the magnetization component in question has acquired a different precession frequency and therefore generates a cross-peak in the two-dimensional spectrum. Normally we measure the rate of growth of the cross-peak as a function of $\tau_m$ in a series of two-dimensional spectra. We see that this is the two-dimensional analogue of the Forsén–Hoffman experiment for studying slow chemical exchange.[32]

## 8.3  Heteronuclear correlation spectroscopy

Heteronuclear correlation is based on the principles outlined above, but with some important variations. One difference is that in heteronuclear correlation experiments there is scope for decoupling the spin–spin interactions in both frequency dimensions in order to concentrate intensity into singlets at the chemical shift frequencies. The low natural abundance of heteronuclei such as $^{13}$C and $^{15}$N presents a new dilemma—should the magnetization transfer proceed from the high-γ spins to the low-γ spins or *vice versa*? For example, H → $^{13}$C transfer exploits the higher Boltzmann population differences of protons, but $^{13}$C → H transfer benefits from the higher detection sensitivity of protons.

### 8.3.1  Round-trip coherence transfer

This dilemma is usually resolved by arranging a double transfer of the type $I \rightarrow S \rightarrow I$, which takes advantage of the large polarization of the source spins and their high

detection efficiency. We may calculate the relative efficiencies for polarization transfer experiments on the basis of the Boltzmann populations of the source spins (proportional to $\gamma_{source}$) and the efficiency with which the spectrometer detects the
observed spins (proportional to the square of $\gamma_{observed}$). The receiver noise is
assumed to increase in proportion to the square root of the radiofrequency. This
gives an approximate expression for the relative signal-to-noise ratio:

$$S/N \propto \gamma_{source} \, (\gamma_{observed})^{\frac{3}{2}} \tag{8.16}$$

If we take the case where the $I$ spins are protons then we can tabulate the sensitivi-
ties relative to direct observation of the $S$ spins:

| Type of transfer | Relative sensitivity | $S = {}^{13}C$ | $S = {}^{15}N$ |
|---|---|---|---|
| $S$ (direct) | $(\gamma_S)^{\frac{5}{2}}$ | 1 | 1 |
| $I \rightarrow S$ | $\gamma_I \, (\gamma_S)^{\frac{3}{2}}$ | 4 | 10 |
| $S \rightarrow I$ | $\gamma_S \, (\gamma_I)^{\frac{3}{2}}$ | 8 | 30 |
| $I \rightarrow S \rightarrow I$ | $(\gamma_I)^{\frac{5}{2}}$ | 32 | 300 |

Incomplete spin-lattice relaxation of the source spin between scans may be included
as a term $[1 - \exp(-T/T_1)]$ where $T$ is the interval between scans.

This is, of course, only part of the story, because sensitivity is influenced by multiplet splitting, losses due to spin–spin relaxation, and artefacts due to incomplete
suppression of strong alien signals. Nevertheless it is clear that round-trip polarization transfer has a lot to offer, particularly for low-$\gamma$ nuclei such as ${}^{15}N$.

## 8.3.2  Preparation period

A common feature of natural-abundance heteronuclear correlation experiments is
the need to suppress the intense proton signals from ${}^{12}C$ or ${}^{14}N$ sites. Presaturation
is not an option under these circumstances since that would also destroy the proton
polarization required for the double transfer. We need to distinguish the interesting
protons attached to ${}^{13}C$ or ${}^{15}N$ from their much more abundant, but boring cousins.
One useful trick is to prepare the system with a bilinear rotation decoupling (BIRD)
module[33] that inverts the spin populations of protons attached to ${}^{12}C$ while leaving
the protons on ${}^{13}C$ unaffected. This method was analysed by the vector model in
§ 4.2.8, where it was called the BIRD-MG module:

$I$ spins:          $(\pi/2)_x$ — $\tau$ — $(\pi)_y$ — $\tau$ — $(\pi/2)_x$
$S$ spins:                                    $(\pi)$                                   (8.17)

which we distinguish from the BIRD-CP module that has the reverse effect. The $\tau$
interval is set at the condition $\tau = 1/(2J_{IS})$.

For the BIRD-MG sequence there is no overall change at the ${}^{13}C$ sites:

$$+I_z \xrightarrow{\tilde{I}_x} -I_y \xrightarrow{(\pi/2)\,2\tilde{I}_z\tilde{S}_z} +2I_xS_z \xrightarrow{\pi\,\tilde{I}_y} -2I_xS_z \xrightarrow{\pi\,\tilde{S}_x}$$

$$+2I_xS_z \xrightarrow{(\pi/2)\,2\tilde{I}_z\tilde{S}_z} +I_y \xrightarrow{\tilde{I}_x} +I_z \tag{8.18}$$

whereas for the $^{12}C$ sites there is a population inversion of the protons:

$$+I_z \xrightarrow{\tilde{I}_x} -I_y \xrightarrow{\pi \tilde{I}_y} -I_y \xrightarrow{\pi \tilde{S}_x} -I_y \xrightarrow{\tilde{I}_x} -I_z \qquad \textbf{8.19}$$

The trick is to allow a delay of $T_1\ln2$ after spin inversion so that the protons on $^{12}C$ just reach the null condition on their exponential recovery curve. The main part of the correlation experiment is then initiated in the normal manner without interference from the abundant $^{12}C$ protons. The TANGO module[34] can be used in a similar manner. It excites protons directly attached to $^{13}C$ by rotating them through $\pi$ radians about an axis tilted at 45° with respect to the $x$ and $z$ axes, but behaves as an identity operator for protons coupled only through long-range interactions (§ 4.2.8).

### 8.3.3 Heteronuclear single-quantum correlation

One type of round-trip polarization transfer was pioneered by Bodenhausen and Ruben,[35] and is sometimes called the 'Overbodenhausen' experiment or heteronuclear single-quantum correlation (HSQC). Figure 8.4 shows a simple implementation of this experiment, based on two INEPT[36] polarization transfer sequences. (The INEPT technique is described according to the vector picture in § 4.2.6.) Here we adopt the alternative product operator approach (§ 3). It simplifies the algebra considerably if we first recognize that chemical shift effects are refocused in the $\Delta$ interval, so they can be safely omitted, leaving only the divergence due to spin–spin coupling. The fixed $\Delta$ interval is set equal to $1/(2J_{IS})$ and gives rise to antiphase

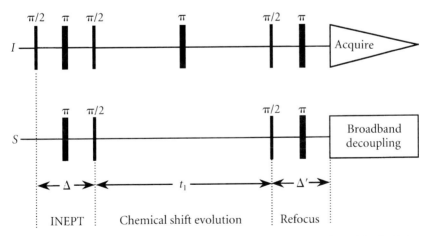

**Figure 8.4** Pulse sequence for round-trip transfer of polarization ($I \rightarrow S \rightarrow I$) in the heteronuclear single-quantum correlation (HSQC) experiment. The INEPT sequence with $\Delta = 1/(J_{IS})$ transfers $I$-spin polarization to the $S$ spins, which then precess freely during the evolution period. After polarization transfer back to the $I$ spins, a period $\Delta'$ is allowed for refocusing, so that broadband $S$-spin decoupling can be used without cancelling the $I$-spin signal.

magnetization components. The first polarization transfer stage creates antiphase $S$-spin magnetization:

$$+I_z \xrightarrow{\tilde{I}_x} -I_y \xrightarrow{(\pi/4)\,2\tilde{I}_z\tilde{S}_z} -I_y\cos(\pi/4) + 2I_xS_z\sin(\pi/4) \xrightarrow{\pi\,\tilde{I}_x}$$

$$+I_y\cos(\pi/4) + 2I_xS_z\sin(\pi/4) \xrightarrow{\pi\,\tilde{S}_x} I_y\cos(\pi/4) - 2I_xS_z\sin(\pi/4)$$

$$\xrightarrow{(\pi/4)\,2\tilde{I}_z\tilde{S}_z} -2I_xS_z \xrightarrow{\tilde{I}_y} +2I_zS_z \xrightarrow{\tilde{S}_x} -2I_zS_y \qquad \textbf{8.20}$$

This antiphase $S$-spin magnetization $(-2I_zS_y)$ precesses freely during the evolution period, thus monitoring the $S$-spin chemical shift. We are not really interested in $J_{IS}$ at this stage, so a considerable simplification is achieved by decoupling—allowing the divergence due to $J_{IS}$ to be refocused by the $\pi$ pulse applied to the $I$ spins at the mid-point of the evolution period.[37] Because of the identity

$$\exp(+i\pi J_{IS}2I_zS_zt_1)\,\exp(+i\pi I_x)\,\exp(+i\pi J_{IS}2I_zS_zt_1)$$
$$= \exp(+i\pi I_x)\,\exp(-i\pi J_{IS}2I_zS_zt_1)\,\exp(+i\pi J_{IS}2I_zS_zt_1) = \exp(+i\pi I_x) \qquad \textbf{8.21}$$

we may ignore the heteronuclear spin–spin splitting during the evolution period and write

$$-2I_zS_y \xrightarrow{\phi\,\tilde{S}_z} -2I_zS_y\cos\phi + 2I_zS_x\sin\phi$$

$$\xrightarrow{\pi\,\tilde{I}_x} 2I_zS_y\cos\phi - 2I_zS_x\sin\phi$$

$$\xrightarrow{\phi\,\tilde{S}_z} 2I_zS_y\cos^2\phi - 2I_zS_x\cos\phi\sin\phi - 2I_zS_x\sin\phi\cos\phi - 2I_zS_y\sin^2\phi$$
$$= +2I_zS_y\cos2\phi - 2I_zS_x\sin2\phi \qquad \textbf{8.22}$$

where $2\phi = 2\pi\delta_S t_1$. In this manner we monitor only the chemical shift evolution. The vector model may also be used to demonstrate this effect (§ 2.4.5).

At the end of the evolution period, simultaneous $\pi/2$ pulses on the $I$ and $S$ spins induce a transfer of antiphase magnetization from the $S$ spins back to the $I$ spins:

$$+2I_zS_y\cos2\phi - 2I_zS_x\sin2\phi \xrightarrow{\tilde{S}_x} +2I_zS_z\cos2\phi - 2I_zS_x\sin2\phi \xrightarrow{\tilde{I}_x}$$
$$-2I_yS_z\cos2\phi + 2I_yS_x\sin2\phi \qquad \textbf{8.23}$$

The last term leads to unobservable coherences, leaving only the antiphase $I$-spin magnetization $-2I_yS_z$, modulated by the chemical shift of the $S$ spins.

If we are content to observe the coupled $I$-spin spectrum, acquisition may then begin immediately. More commonly we would like to record the broadband decoupled spectrum, so the antiphase magnetization must be allowed to evolve until it is in-phase again, otherwise it will vanish when the heteronuclear decoupler is switched on. Again the chemical shift effects are refocused (to avoid introducing frequency-dependent phase shifts) but the spin–spin coupling causes a convergence onto the $x$ axis:

$$-2I_yS_z\cos2\phi \xrightarrow{\;(\pi/4)\,2\tilde{I}_z\tilde{S}_z\;} -[2I_yS_z\cos(\pi/4) - I_x\sin(\pi/4)]\cos2\phi$$

$$\xrightarrow{\;\pi\,\tilde{I}_x\;\;\pi\,\tilde{S}_x\;} -[2I_yS_z\cos(\pi/4) - I_x\sin(\pi/4)]\cos2\phi$$

$$\xrightarrow{\;(\pi/4)\,2\tilde{I}_z\tilde{S}_z\;} +2I_x\cos(\pi/4)\sin(\pi/4)\cos2\phi = +I_x\cos2\phi \qquad \textbf{8.24}$$

For the usual case of H$\rightarrow$ $^{13}$C $\rightarrow$ H correlation spectroscopy, this second fixed interval $\Delta'$ must have a compromise setting, since there may be one, two, or three protons directly attached to the $^{13}$C site in question. In general this will cause a slight perturbation of the relative proton intensities. Because the decoupler must cover a wide range of $^{13}$C chemical shifts, an ultrabroadband scheme is advisable (§ 7.4.3).

This analysis has ignored the proton–proton coupling $J_{HH}$ on the grounds that it is very small compared with $^1J_{CH}$. If the active coupling for polarization transfer is comparable with $J_{HH}$ then the process becomes less efficient owing to the excitation of proton multiple-quantum coherence at time $\Delta$. Then the methods outlined below are preferable.

## 8.3.4 **Heteronuclear multiple-quantum correlation**

The executive stage in the INEPT sequence is the simultaneous pair of $(\pi/2)$ pulses applied to the $I$ and $S$ spins in order to generate an antiphase $S$-spin signal. If we represent these two pulses as a pulse cascade[38] and assume that the ordering is

$$-2I_xS_z \xrightarrow{\;\tilde{I}_y\;} +2I_zS_z \xrightarrow{\;\tilde{S}_x\;} -2I_zS_y \qquad \textbf{8.25}$$

then the intermediate is longitudinal two-spin order $(2I_zS_z)$, and this is rightly regarded as a polarization transfer experiment. However, if we assume the reverse ordering

$$-2I_xS_z \xrightarrow{\;\tilde{S}_x\;} +2I_xS_y \xrightarrow{\;\tilde{I}_y\;} -2I_zS_y \qquad \textbf{8.26}$$

then, although the same result is achieved, the mechanism of the coherence transfer is formally quite different, since the intermediate state consists of a mixture of heteronuclear zero- and double-quantum coherence $(2I_xS_y)$. The difference is only evident in practice if time is allowed between the two pulses for evolution of the heteronuclear multiple-quantum coherence.

In a remarkably perceptive early paper, Müller[39] showed how this scheme could be exploited to achieve the sensitivity benefits of round-trip coherence transfer by allowing the heteronuclear two-spin coherence to evolve during $t_1$. The undesirable contribution to the multiple-quantum frequencies from the proton chemical shift was removed by a proton $\pi$ pulse at the mid-point of the evolution period, which has the effect of interchanging zero- and double-quantum terms. This experiment in fact antedates the Overbodenhausen experiment.[35] It is now normally known as heteronuclear multiple-quantum correlation (HMQC)[40] and is often used as an alternative to heteronuclear single-quantum correlation (HSQC).[41,42]

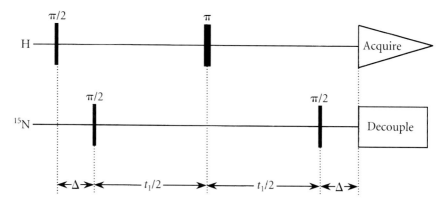

**Figure 8.5** Heteronuclear multiple-quantum correlation (HMQC) of protons and $^{15}$N. Antiphase proton magnetization at the end of the $\Delta$ interval is converted into heteronuclear multiple-quantum coherence during the evolution period. This precesses at $(\nu_N + \nu_H)$ for a time $t_1/2$ and then at $(\nu_N - \nu_H)$ for the second period $t_1/2$, before being converted into antiphase proton magnetization by the $\pi/2$ pulse. The refocusing interval $\Delta$ brings these components into phase so that broadband decoupling can be applied.

There is now a whole family of HMQC experiments with slightly different pulse sequences and slightly different properties. They are widely used for studying large biochemical molecules, often with enrichment of $^{13}$C or $^{15}$N, or both. Since such samples are commonly studied in (nondeuterated) aqueous solution, presaturation cannot be used to suppress the intense $^{12}$C–H or $^{14}$N–H signals. This task is handled by the multiple-quantum filtration that is implicit in the HMQC technique. We describe here one typical example (based on indirect detection of $^{15}$N) to illustrate the general principles involved (Figure 8.5).

The first stage is the creation of antiphase proton magnetization by allowing transverse magnetization to diverge under the $^{15}$N–H coupling for a time $\Delta = 1/(2J_{NH})$. This is converted into a mixture of heteronuclear zero- and double-quantum coherence by a $\pi/2$ pulse applied to $^{15}$N. (Since the magnetogyric ratio of $^{15}$N is negative, the zero-quantum coherence actually has the higher frequency in this situation.) During the ensuing evolution period ($t_1$), the $^{15}$N chemical shift information would normally be encoded into the multiple-quantum frequencies $(\nu_N + \nu_H)$ and $(\nu_N - \nu_H)$. It is convenient to eliminate the proton chemical shift contribution in this dimension by inserting a proton $\pi$ pulse at the mid-point of the evolution period, which interchanges zero- and double-quantum coherences.[39] At the end of the evolution period, a second $\pi/2$ pulse applied to $^{15}$N converts the multiple-quantum coherence into antiphase proton magnetization, and the round-trip transfer is complete. However, in most cases we would prefer to detect proton spectra without $^{15}$N-H splittings, so a further delay $\Delta$ is introduced to allow realignment of the antiphase proton vectors and then broadband $^{15}$N decoupling is switched on, just before signal acquisition. Selection of signals derived from heteronuclear double-quantum coherence is achieved by cycling the phase of the first $^{15}$N pulse $(+x, +y, -x, -y)$

and the receiver reference phase in the opposite sense ($+x$, $-y$, $-x$, $+y$). This is the basic cycle; in practice phase cycling is usually more complex, designed to correct several possible instrumental shortcomings.

Initially it was considered that the HMQC technique was to be preferred over the HSQC method for $^{15}$N–H correlation experiments in proteins. It certainly contains fewer pulses and therefore might be expected to be less sensitive to pulse imperfections. However the transverse relaxation rate, which determines the linewidth in the $F_1$ dimension, is faster for multiple-quantum coherence than for single-quantum coherence.[41,42] Furthermore, the HMQC version exhibits a proton–proton splitting (often unresolved) in the $F_1$ dimension that is absent in the HSQC scheme. Both these factors ensure that the HSQC technique has narrower lines in the $F_1$ dimension, and consequently higher sensitivity.

An ever-present danger in all heteronuclear correlation experiments arises from strong coupling in the proton spectrum. These effects are easily overlooked because it is the interaction between satellite signals that is of importance, not the protons attached to $^{12}$C or $^{14}$N (see Figure 1.5). The phenomenon is most simply described as virtual coupling[43] (or a virtual nuclear Overhauser effect[44]) and gives rise to spurious correlation peaks that could suggest an unfortunate misassignment.

### 8.3.5 **Long-range correlation**

The $^{13}$C–H and $^{15}$N–H heteronuclear correlation experiments described above exploit the single-bond couplings, which usually have magnitudes in the range 125–170 Hz, and 86–95 Hz, respectively. An interesting variant of this type of experiment is to limit the transfer to the much smaller long-range couplings. This multiple-bond correlation experiment (HMBC)[45] can provide useful additional information to complement conventional direct correlation spectroscopy. In particular, long-range $^{13}$C–H correlation is very important for the sequential assignment of aminoacids since it connects protons separated by nonprotonated carbon atoms, for example carbonyl groups in peptides.

In order to develop the required antiphase magnetization ($2I_xS_z$), long-range correlation spectroscopy must employ much longer $\Delta$ periods, rendering it more susceptible to relaxation losses. One implementation of HMBC is to remove the second refocusing period from the HMQC sequence, together with the broadband S-spin decoupling (Figure 8.6). There is then less signal loss through transverse relaxation. The splitting due to proton–proton coupling in the $F_1$ dimension is normally not resolved since the longest evolution time $t_1$(max) is kept quite short; this allows phasing for pure absorption-mode in this frequency dimension. Unfortunately, since there is appreciable precession due to proton shifts and proton–proton coupling in the $\Delta$ interval, the spectra cannot be phased for pure absorption in the $F_2$ dimension, so the absolute-value mode is used.

Note that magnetization vectors associated with direct connectivities ($^1J_{CH}$) diverge through many rotations during the relatively long interval $\Delta$, so the corresponding peaks may be present or absent, depending on the residual angle of

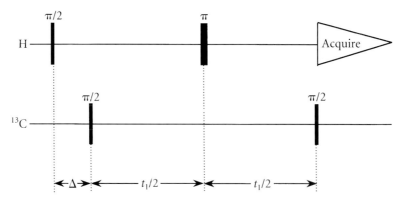

**Figure 8.6** Basic sequence for long-range C–H correlation spectroscopy (HMBC). The $\Delta$ interval is sufficiently long that magnetization vectors diverging due to long-range couplings reach an antiphase condition. During the evolution period $t_1$, the proton contribution to the multiple-quantum frequencies is eliminated by the $\pi$ pulse which interconverts $(\nu_C + \nu_H)$ and $(\nu_C - \nu_H)$. The coupled spectrum is recorded in the absolute-value mode in the $F_2$ dimension

divergence. There are several possible refinements of the basic sequence that can be used to minimize the undesirable effects of direct coupling. The initial proton excitation pulse or the first $^{13}$C mixing pulse may be replaced by the TANGO module[34] (§ 4.2.8) which acts only as the identity operation for directly bound spins. Alternatively the proton $\pi$ pulse may be replaced by the BIRD-MG module[33] (§ 4.2.8) which inverts only those spins connected through a long-range coupling. A further possibility is to introduce a low-pass $J$ filter[45] (§ 6.3.5).

### 8.3.6  Constant-time experiments

The principal drawback of the long-range correlation experiment is signal loss through relaxation during the relatively long interval $\Delta$, which is of the order of 40–100 ms. Some mitigation can be achieved by nesting the evolution period into the fixed interval $\Delta_1$, the so-called constant time version.[4,46,47] Round-trip constant-time, long-range correlation experiments can be quite complex, so for simplicity we illustrate the case of a one-step H $\rightarrow$ $^{13}$C transfer (Figure 8.7). Proton shift evolution is achieved by moving the two $\pi$ pulses (together) within the fixed interval $\Delta_1$. When the proton $\pi$ pulse is at the mid-point of the $\Delta_1$ interval, it refocuses the chemical shift exactly, but when it reaches the end of that interval, full chemical shift precession occurs. The choice of $\Delta_1$ limits the maximum value of $t_1$ and hence the resolution in the $F_1$ dimension. The pair of $\pi$ pulses has no effect on the C–H couplings. Although C–H and H–H coupling both influence the amplitude of magnetization transferred at the end of the $\Delta_1$ interval, the fact that $\Delta_1$ is constant ensures that there is no modulation of this signal and therefore no corresponding splittings in the $F_1$ dimension. The refocusing interval $\Delta_2$ permits broadband proton decoupling, and the BIRD-MG module (§ 4.2.8) suppresses the undesirable effects of one-bond coupling, as described above. An interesting extension of this experi-

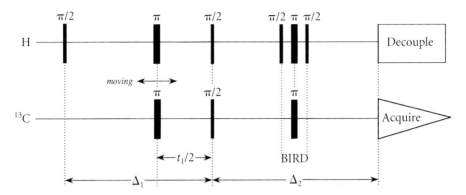

**Figure 8.7** Long-range correlation via the constant-time method. Proton chemical shift evolution is achieved by moving the pair of π pulses from the mid-point of the $\Delta_1$ interval to the end. The BIRD-MG module suppresses the undesirable effects of one-bond coupling. If the first $^{13}C$ π pulse is eliminated, the long-range C–H splittings appear in the $F_1$ dimension.

ment permits the measurement of the coupling constant $^nJ_{CH}$, simply by eliminating the π pulse on the $^{13}C$ spins. Then the moving proton π pulse partially refocuses the C–H coupling at the end of the $\Delta_1$ interval, and the resulting modulation gives rise to a long-range C–H splitting in the $F_1$ dimension.[47]

### 8.3.7 Editing according to multiplicity

It can be a useful aid to assignment if $^{13}C$ sites can be identified according to the number of directly attached protons. One way to accomplish this is to edit the heteronuclear correlation spectra according to multiplicity (§ 3.4.6). We noted above (Equations 8.25 and 8.26) that the relative timing of the two 'executive' π/2 pulses of the INEPT experiment can be critical—if there is a delay between the π/2 pulse applied to the S spins ($^{13}C$) and that applied to the I spins (protons) then heteronuclear multiple-quantum coherence is created in this interval.

$$2I_xS_z \xrightarrow{\bar{S}_x} -2I_xS_y \qquad 8.27$$

This can be put to good use for editing if we set the interval (τ) after the $\bar{S}_x$ pulse equal to $1/(2J_{IS})$. Suppose there are n protons directly attached to that $^{13}C$ site. One of these is involved in the zero- and double-quantum coherence term $-2I_xS_y$. Spin–spin coupling to the remaining $n-1$ protons creates antiphase (and doubly antiphase) multiple-quantum coherence according to one of the three schemes:

CH group: $$-2I_xS_y \longrightarrow -2I_xS_y \qquad 8.28$$

CH$_2$ group: $$-2I_xS_y \xrightarrow{2\bar{I}'_zS_z} +4I'_zI_xS_x \qquad 8.29$$

CH$_3$ group: $$-2I_xS_y \xrightarrow{2\bar{I}'_zS_z} +4I'_zI_xS_x \xrightarrow{2\bar{I}''_zS_z} +8I'_zI''_zI_xS_y \qquad 8.30$$

A simple $I \to S$ transfer is set up with the sequence

$^1$H $\qquad (\pi/2) - t_1 - \qquad - \tau - (\beta) - \tau -$ decouple
$^{13}$C $\qquad\qquad\qquad (\pi/2) \qquad\qquad\qquad$ acquire $\qquad$ **8.31**

then by analogy with the DEPT experiment (analyzed in detail in § 3.4.6) the relative intensities of the $^{13}$C signals are given by

CH group: $\qquad\qquad\qquad\qquad \sin\beta$ $\qquad\qquad$ **8.32**

CH$_2$ group: $\qquad\qquad\qquad\qquad \sin 2\beta$ $\qquad\qquad$ **8.33**

CH$_3$ group: $\qquad\qquad \frac{3}{4}(\sin\beta + \sin 3\beta)$ $\qquad\qquad$ **8.34**

Then the usual linear combinations of spectra recorded with different settings of $\beta$ give edited versions of the correlation experiment.

## 8.4 Forbidden transitions

The introduction of an evolution period ($t_1$) raises the exciting possibility of monitoring forbidden transitions that are not directly observable in the spectrometer—the multiple-quantum coherences (§ 1.3.2). They are detected indirectly by reconversion into single-quantum coherence before signal acquisition. As the experiment is repeated at increasing settings of the evolution period, the multiple-quantum coherence evolves through a progressively larger angle and the detected signal is encoded with this information. After two-dimensional Fourier transformation, the multiple-quantum frequencies appear in the $F_1$ dimension. This facility triggered a widespread renewal of interest in multiple-quantum phenomena (§ 1.3).

### 8.4.1 Carbon–carbon connectivity

The structure of an organic molecule is determined principally by the connectivity of the carbon framework, give or take an occasional nitrogen or oxygen atom. This is the basis of a two-dimensional correlation experiment known as INADE-QUATE[48,49] in which directly bound carbon–carbon links are identified (§ 6.1.4). The values of $^1J_{CC}$ normally fall within the range 30–60 Hz. Each $^{13}$C–$^{13}$C link is monitored independently, since the 1% natural abundance of $^{13}$C ensures that groups of more than two $^{13}$C spins in any given molecule are very rare indeed. Unfortunately this also means that only one molecule in 8100 gives rise to the signals of interest, so the intrinsic sensitivity is poor. A low level of nonspecific enrichment in $^{13}$C would be an advantage in such cases; increasing the abundance uniformly by a factor $x$ increases the sensitivity by a factor $x^2$.

Double-quantum coherence provides two invaluable features to aid the detection and assignment of these carbon–carbon linkages. First of all, it is only the coupled $^{13}$C–$^{13}$C pairs that can sustain double-quantum coherence; the much more abundant isolated $^{13}$C spins cannot. This allows the application of a very effective double-quantum filter to extract the desired weak signal from $^{13}$C–$^{13}$C pairs, using either a phase cycle or matched field gradients (§ 6.2.2). Secondly, in the two-dimensional version of the experiment, each $^{13}C-^{13}C$ subspectrum appears at a frequency in the $F_1$ dimension determined by its characteristic double-quantum precession frequency $(\delta_I + \delta_S)$ Hz. In the $F_2$ dimension there is a four-line pattern (an AX or occasionally an AB spectrum) centred at the mean chemical shift $(\delta_I + \delta_S)/2$ Hz. In the two-dimensional spectrum, these four-line subspectra must therefore straddle the skew diagonal of slope 2.

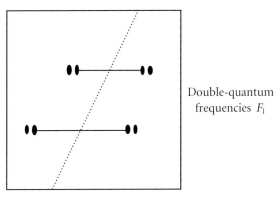

Double-quantum
frequencies $F_1$

Conventional spectrum $F_2$

This, together with the inherent symmetry of the subspectra, is a powerful aid to assignment.

The INADEQUATE technique allows us to identify the four main topological features of the carbon framework—the straight chain, the branched chain, the closed ring and the quaternary carbon site, simply from the interrelations between the $^{13}C-^{13}C$ subspectra. These correspondences are so direct that there seems little danger of misinterpretation of the type of connectivity (Figure 8.8).

The pulse sequence for the two-dimensional version[49] of this INADEQUATE experiment[48] may be written:

$$(\pi/2)_x - \tau - (\pi)_y - \tau - (\pi/2)_x - t_1 - (\pi/2)_\phi, \text{acquire } (\psi) \qquad \textbf{8.35}$$

Broadband proton decoupling is applied throughout. The fixed interval $\tau$ is set at $1/(4J_{CC})$ to optimize the generation of double-quantum coherence. Long-range $^{13}C-^{13}C$ couplings are too small to generate any appreciable amounts of double-quantum coherence. The basic phase cycle (discussed in § 6.1.4) is designed to follow the phase of the signal derived from double-quantum coherence while rejecting the unwanted signal from isolated $^{13}C$ spins:

| $\phi$ | $\psi$ |
|---|---|
| $+x$ | $+x$ |
| $-y$ | $+y$ |
| $-x$ | $-x$ |
| $+y$ | $-y$ |

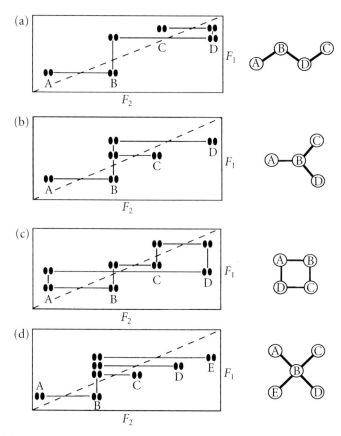

**Figure 8.8** Topological features that may be deduced from the INADEQUATE spectrum: (a) a straight chain, (b) a branched chain, (c) a closed ring, (d) a quaternary carbon site.

In practice more elaborate phase cycles are usually employed to compensate instrumental imperfections.

Note that some structural problems can be solved by establishing a single feature, say a chain branching point. One early application of the INADEQUATE experiment was the unambiguous proof that a photodimer formed from a $\Delta^{4,6}$-diene-3-keto steroid exclusively adopted a head-to-tail rather than head-to-head configuration.[50]

The INADEQUATE technique may also be employed with other nuclear species, for example protons, although the simple two-spin nature of the subspectra is normally lost under these circumstances. Double-quantum filtered proton correlation spectroscopy[8] (DQ-COSY) has found widespread acceptance since it facilitates the detection of cross-peaks close to the diagonal, and helps suppress singlet solvent resonances, notably water (§ 8.2.2).

## 8.4.2 Multiple-quantum spectroscopy

Although double-quantum spectroscopy seems to have stolen the show, there are nevertheless interesting effects to be seen with other kinds of forbidden transitions, for example zero- or triple-quantum effects. Here we must allow for the possibility that not all the quantum jumps need be in the same sense (§ 1.3.2). Thus in a coupled three-spin system we can excite the triple-quantum coherence and three-spin single-quantum coherence:

$$\alpha\alpha\alpha \longleftrightarrow \beta\beta\beta$$
$$\alpha\alpha\beta \longleftrightarrow \beta\beta\alpha$$
$$\beta\alpha\alpha \longleftrightarrow \alpha\beta\beta$$
$$\alpha\beta\alpha \longleftrightarrow \beta\alpha\beta \qquad \textbf{8.36}$$

The last three are sometimes called combination lines by analogy with infrared spectroscopy (§ 1.3.5). In general these are described as '$n$-spin $p$-quantum coherences', with $n \geq p$. They exhibit a $p$-fold sensitivity to static field inhomogeneity, applied field gradients and radiofrequency phase shifts. The complexity of $p$-quantum spectra increases as $p$ approaches $n/2$ for $n$ even, or $(n\pm1)/2$ for $n$ odd. If $p=n$ there is only a single $p$-quantum transition.

For many years multiple-quantum spectra were regarded as something of a spectroscopic curiosity. Two-dimensional methods have rendered them much more accessible, revealing some interesting properties. For example, the three-spin single-quantum coherence spectrum of a coupled three-spin ($ISR$) system consists of a singlet response for each group, which could be useful for the assignment of crowded spectra. An example is presented in Figure 8.9 of the proton spectrum of 2,3-dibromopropanoic acid in acetone-d$_6$.[51] For strong proton signals of this type, it is perfectly practicable to excite three-spin coherence by means of a line-selective $\pi/2$ pulse followed by a hard $\pi/2$ pulse:

$$I_z + 2I_zS_z + 2I_zR_z + 4I_zS_zR_z \xrightarrow{\text{soft } \tilde{I}_x} -I_y - 2I_yS_z - 2I_yR_z - 4I_yS_zR_z$$

$$\xrightarrow{\tilde{I}_y \tilde{S}_y \tilde{R}_y} -I_y - 2I_yS_x - 2I_yR_x - 4I_yS_xR_x \qquad \textbf{8.37}$$

We are particularly interested in the last term. There follows an evolution period $t_1$, during which precession is caused only by the chemical shifts. The coupling constants are not involved, as we saw from the expressions for the energy levels set out in § 1.3.5. If we write $\phi_I = 2\pi\nu_It_1$, $\phi_S = 2\pi\nu_St_1$, $\phi_R = 2\pi\nu_Rt_1$, then

$$-4I_yS_xR_x \xrightarrow{\tilde{I}_z \tilde{S}_z \tilde{R}_z} +4I_xS_yR_y \sin\phi_I \sin\phi_S \sin\phi_R \qquad \textbf{8.38}$$

A final hard $\pi/2$ pulse converts this into a doubly antiphase $I$-spin signal:

$$4I_xS_yR_y \sin\phi_I \sin\phi_S \sin\phi_R \xrightarrow{\tilde{I}_x \tilde{S}_x \tilde{R}_x} 4I_xS_zR_z \sin\phi_I \sin\phi_S \sin\phi_R \qquad \textbf{8.39}$$

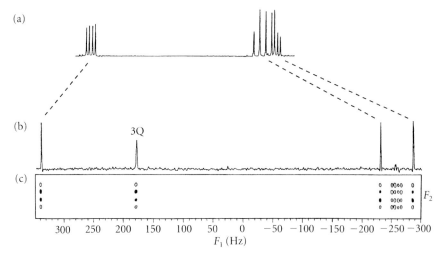

**Figure 8.9** Three-spin coherence spectra from protons in a coupled *ISR* spin system: (a) conventional spectrum; (b) projection of the two-dimensional spectrum shown in (c). The three-spin single-quantum coherences fall at offsets $(v_R - v_I + v_S)$, $(v_R + v_I - v_S)$ and $(-v_R + v_I + v_S)$. Note that these separations are twice the chemical shift differences. To illustrate this point, spectra (b) and (c) are shown on a reversed frequency scale. There is also a triple-quantum response (3Q) and a 16-line multiplet near $-260$ Hz corresponding to a COSY diagonal peak.

Through standard identities this can be expressed as

$$I_x S_z R_z \{\sin(\phi_R - \phi_I + \phi_S) + \sin(\phi_R + \phi_I - \phi_S) + \sin(-\phi_R + \phi_I + \phi_S) + \sin(-\phi_R - \phi_I - \phi_S)\}$$

$$\textbf{8.40}$$

Three of these responses fall at the three-spin single-quantum frequencies in $F_1$:

$$+v_R - v_I + v_S$$
$$+v_R + v_I - v_S$$
$$-v_R + v_I + v_S \qquad \textbf{8.41}$$

while the fourth falls at the triple-quantum frequency, $+v_R + v_I + v_S$. In the $F_2$ dimension they are all identical doubly antiphase patterns $(+ - - +)$. We only record a thin strip from the two-dimensional spectrum (Figure 8.9). Because of the sign alternation, conventional projections vanish, but we can use a $(+ - - +)$ mask to obtain the desired projection onto the $F_1$ axis. This gives three singlet responses for the three-spin single-quantum coherences. If we examine the expressions for the separations between these responses, we see that they are $2|v_I - v_S|$, $2|v_R - v_S|$ and $2|v_I - v_R|$, which are just twice the chemical shift differences in the conventional spectrum (§ 1.3.5). The two-dimensional spectrum also contains a 4 by 4 spin multiplet corresponding to the COSY diagonal peak, but it exhibits no sign alternation and thus disappears in this type of masked projection.

Particular orders of multiple-quantum coherence may be separated from other orders by appropriate multiple-quantum filters[52] based on a phase cycle (§ 6.3.1) or a sequence of matched field gradients (§ 6.2.2). Homonuclear zero-quantum coherence is a special case since it cannot be easily distinguished from $z$ magnetization. It also possesses the interesting property (since $p=0$) of being unaffected by the spatial inhomogeneity of the applied magnetic field.

## 8.5 J spectroscopy

The scope of two-dimensional spectroscopy is very wide; the only fundamental requirements are that different conditions prevail during the evolution and detection periods, or that some polarization or coherence transfer occurs. If we can arrange for different NMR parameters to be active in these two intervals, then a separation is possible, for example chemical shifts from spin–spin couplings. Since a spin echo experiment can be set up to refocus chemical shifts while retaining the divergence due to spin–spin coupling,[53] it can be incorporated into the evolution period of a two-dimensional experiment to suppress the chemical shift effects in the $F_1$ dimension.[54]

### 8.5.1 Heteronuclear J spectroscopy

The simplest case arises in heteronuclear systems, for example where it is desirable to separate $^{13}C$ chemical shifts from C–H couplings. Echo modulation[55] requires that both spin species experience $\pi$ pulses (§ 4.2.3):

$^1H$                                              $\pi$                    decouple

$^{13}C$              $(\pi/2)_x - t_1/2 - (\pi)_y - t_1/2 -$ acquire $(t_2)$                    **8.42**

The CH multiplet structure appears in the $F_1$ dimension and the $^{13}C$ chemical shifts in the $F_2$ dimension. The number of $t_1$ increments need only be sufficient to cover a limited frequency range, the maximum width of a C–H spin multiplet (about 500 Hz).

If the intention is merely to determine the number of directly attached protons, fewer $t_1$ increments may be used, thus shortening the experiment. Nishida $et\ al.$[56] have shown that as few as five increments are enough to distinguish the doublets, triplets and quartets that arise from CH, $CH_2$ and $CH_3$ sites. This makes the method competitive with other schemes for multiplicity determination.

If the proton–proton coupling is first-order, two-dimensional heteronuclear $J$ spectra are exactly symmetrical about the axis $F_1 = 0$. Consequently, pure absorption-mode spectra can be obtained by calculating the real (cosine) Fourier transform with respect to the evolution time $t_1$, essentially superimposing the $J$ spectrum on a reflected version of itself so that dispersion-mode signals cancel. When there is strong proton–proton coupling (in the $^{13}C$ satellite spectrum) this symmetry is broken[57] and pure absorption can only be achieved by performing two experiments, one with reversed precession.

Another scheme[58] generates the echo modulation by having the proton decoupler active during one-half of the evolution period. By combining the results of two such gated decoupler experiments with the decoupler alternately on/off and off/on, the dispersion-mode contributions are eliminated. A third strategy[59] achieves the effect of reversed precession by applying a $\pi$ pulse immediately before acquisition on alternate scans of the pulse sequence illustrated in Equation 8.42.

### 8.5.2 **Homonuclear J spectroscopy**

The principal difference here is our inability to employ broadband decoupling during signal acquisition, which means that the $F_2$ dimension contains both chemical shift and spin coupling information. The pulse sequence is

$$(\pi/2)_x - \tfrac{1}{2}t_1 - (\pi)_y - \tfrac{1}{2}t_1 - \text{acquire}\,(t_2) \qquad\qquad \textbf{8.43}$$

Since both chemical shift and spin–spin coupling are active during acquisition, the individual spin multiplet patterns are aligned along 45° diagonals with their centres on the axis $F_1 = 0$. It is as if a Venetian blind had been opened at 45°. This is called a two-dimensional $J$ spectrum. It allows us to read off the chemical shifts (even in crowded spectra) and examine each spin multiplet separately.

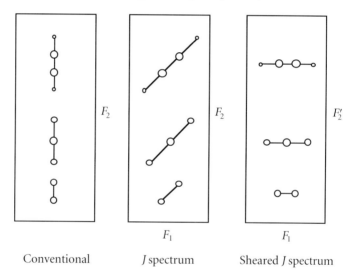

Conventional        $J$ spectrum        Sheared $J$ spectrum

    In an alternative mode of presentation, the Venetian blind can be opened completely by shearing the data matrix through 45°. All the spin multiplets then run horizontally, leaving only chemical shifts in the new $F_2$ dimension ($F_2'$). Figure 8.10 shows the sheared two-dimensional $J$-spectrum (absolute-value mode) of a tricyclodecanone derivative containing eleven coupled protons and a hydroxy group.[60] The individual spin multiplets are well separated and the chemical shift frequencies can be obtained directly from the $F_2'$ ordinates. This spectrum also illustrates a complication that arises in strongly coupled spin systems (protons D and E in this example). It shows up as a spurious group of lines at the mean chemical shift

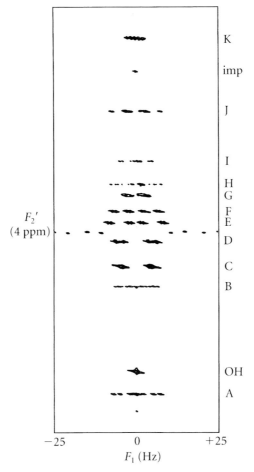

**Figure 8.10** Two-dimensional proton *J* spectrum (400 MHz) of a tricyclodecanone derivative. Normally the spin multiplets run along 45° diagonals but the data matrix has been sheared (clockwise) through 45° so that the multiplets are parallel to the $F_1$ axis. Strong coupling between protons D and E give rise to a spurious set of eight lines at the mean chemical shift of D and E. There is a weak signal from an impurity, indicated by 'imp'.

$\frac{1}{2}(\delta_D + \delta_E)$ in $F_2'$, with splittings in the $F_1$ dimension that are a function of both chemical shift and spin–spin coupling parameters.

### 8.5.3 Decoupled proton spectra

One of the key attributes of $^{13}C$ spectroscopy is the simplicity of the spectra; all C–H splittings are decoupled, leaving only chemical shift information. An equivalent simplification can be achieved in proton spectroscopy by projecting the homonuclear two-dimensional *J* spectrum in a direction parallel to the spin multiplet structure, leaving only the chemical shifts.[54] (The proton–proton couplings are still accessible separately by examination of the appropriate cross-sections.)

Unfortunately this intriguing scheme is fraught with practical difficulties stemming from the peculiar two-dimensional line shape that results from Fourier transformation of phase-modulated signals. The response is a superposition of equal amounts of two-dimensional absorption and two-dimensional dispersion, the phase twist line shape[61] discussed in more detail in § 8.7.1. The difficulty arises when we try to project the phase twist line[62] by computing integrals along 45° diagonals. This type of projection simply vanishes, the negative dispersion lobes exactly cancelling the positive absorption lobe. As a corollary, a 45° cross-section shows a positive central response symmetrically flanked by negative excursions; this has undesirable consequences for displaying the spin multiplet structure.

Many different strategies have been attempted to circumvent this problem in order to display decoupled proton spectra. Projection of the absolute-value mode gives lines with very prominent tails and undesirable interference effects in regions of overlap. A scheme known as skyline projection records the maximum ordinate along the line of projection, rather than the integral; it circumvents the cancellation problem but it has the disadvantage that it cannot reproduce the correct intensities.[63] Pseudoecho weighting[64] and the constant-time spin echo method[4,65] give signal intensities that are affected by the magnitudes of the coupling constants and consequently seriously distort the relative intensities within the projected spectrum. Dispersion-mode contributions can be eliminated by a peak-finding routine[66] based on a scheme used in radioastronomy, but this is unsatisfactory in crowded or noisy spectra.

Better success has been achieved by modifying the spin echo sequence to purge certain signal components just before signal acquisition, separating overlapping two-dimensional multiplets by means of a $C_4$ symmetry filter.[60,67] A related scheme[68] superimposes a reflected $J$ spectrum upon itself by suppressing frequency discrimination in the $F_1$ dimension, following this with a similar filtration stage based on $C_4$ symmetry. These last two methods offer absorption-mode proton spectra without spin–spin splittings but at the expense of a considerable amount of post-acquisition data processing. A rather more rapid method[69] performs a two-dimensional line-shape transformation in the frequency domain which greatly attenuates the dispersion-mode tails and permits the 45° projection. However, it does distort relative intensities, so it is advisable to extract more reliable values of the intensities from the absolute-value spectrum. This technique has been used to derive a decoupled proton spectrum of chrysanthelline A, which exhibits 25 distinct chemical shifts in a region of less than 1 ppm (Figure 8.11). A well-defined singlet is obtained for each of the 25 chemically distinct sites, but it is clear that the intensities still lack the desired uniformity. It is doubtful whether these variations in intensity can be attributed to variations in the relaxation times of the different sites. Clearly there is scope for further improvement in this technique.

Decoupled proton spectra are principally useful as an aid to assignment of crowded spectra. Once a given proton spectrum has been reduced to the basic

(c)

(b)

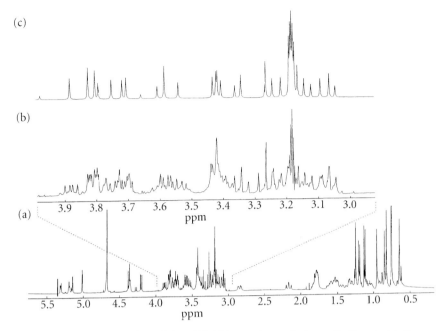

(a)

**Figure 8.11** A crowded region of the 400 MHz spectrum of chrysanthelline A: (a) the conventional proton spectrum; (b) an expansion of the crowded region; (c) the decoupled proton spectrum. There are 25 chemically distinct sites in this region, together with a quintet from the residual protons of the methanol-d$_4$ solvent. Spectrum (c) has been derived by projection of the two-dimensional J spectrum after a line-shape transformation.

chemical shifts, it is in a convenient form for further manipulations, for example the determination of spin–lattice relaxation times by inversion recovery.[60] Decoupled proton spectra have also been used to monitor the slow interconversion of α-D-glucose into the equilibrium mixture of α and β forms.[68] Perhaps one of the most promising applications is to use the chemical shift information in other forms of two-dimensional spectroscopy, for example correlation spectra. These are now so ubiquitous that automated data reduction is becoming an essential aid—an alternative to a painstaking assignment by hand. This is particularly challenging in congested spectra where several cross-peaks may overlap. Although there are well-known pattern recognition programs for locating COSY or NOESY cross-peaks, they are obliged to search the entire two-dimensional data matrix, and then search in further dimensions to find the appropriate coupling constants. Prior knowledge of the proton shifts allows us to establish a two-dimensional chemical shift grid (§ 6.5.3). All genuine cross-peaks must be centred at an intersection of this chemical shift grid, although, of course, not all intersections are occupied. This greatly reduces the dimensionality of the search algorithm and allows the two-dimensional spectrum to be replaced by a simple table of shift correlations, together with estimates of the confidence level (the relative integrated intensity of the cross-peak in question).[70]

## 8.6 **Multidimensional spectroscopy**

Two-dimensional experiments may be concatenated to generate three- or four-dimensional spectra[71,72] using further stages of Fourier transformation, for example:

$$S(t_1, t_2, t_3) \rightarrow f(F_1, F_2, F_3) \qquad \textbf{8.44}$$

The goal is to aid the resolution of overcrowded spectra by spreading the information into new evolution dimensions. The basic building blocks are usually the classic COSY, NOESY or TOCSY two-dimensional experiments, together with heteronuclear coherence transfer schemes. Although the three-dimensional schemes necessarily demand more instrument time and more extensive data storage, there is only a rather small penalty in sensitivity compared with the corresponding two-dimensional experiment performed in the same total time.

However, we must be wary of embarking on a three-dimensional experiment when the constituent two-dimensional experiments would yield the same information. For one thing, there are obvious practical problems associated with the display when the dimensionality is higher than two. Three-dimensional perspective diagrams have a certain aesthetic appeal but there is no reliable way to deduce the frequency coordinates from such a display. Consequently the information from a three-dimensional experiment is almost always displayed in the form of two-dimensional contour plots representing slices through the data matrix $f(F_1, F_2, F_3)$. At the time of writing, computer data storage is still a limitation if high definition is required in all the frequency dimensions; $10^9$ words of storage is soon reached in three-dimensional spectroscopy. A more serious problem is that the introduction of a second (or third) evolution dimension prolongs the data gathering phase of the experiment, soon reaching the practical limit of a few days of instrument time (see § 5.5.5). To alleviate these problems, band-selective excitation schemes (§ 5.4.3) may be employed to reduce the effective spectral widths, and phase cycles should be cut to the bare minimum or replaced by pulsed-field-gradient methods.[73]

### 8.6.1 **Typical three-dimensional experiments**

In general, a three-dimensional experiment is set up by combining two pulse sequences used in two-dimensional spectroscopy, by deleting the first acquisition period and the second preparation period:

preparation → 1st evolution → mixing → 2nd evolution → mixing → acquisition
$\qquad\qquad\qquad t_1 \qquad\qquad\qquad\qquad\qquad t_2 \qquad\qquad\qquad\qquad t_3$

By convention, $t_2$ now becomes an evolution period and $t_3$ is used for acquisition. The corresponding frequency axes are $F_1$, $F_2$ and $F_3$. The overall duration of the experiment is determined principally by the product of the number of increments in $t_1$ and the number in $t_2$. A certain parsimony is therefore desirable.

We take the homonuclear COSY-COSY experiment as an example, where coherence is transferred through the scalar coupling in two successive stages. Consider the simplest general case of three coupled spins $I$, $S$ and $R$. There are basically five types of response. Body diagonal peaks are generated by coherence that remains on the source spin throughout, with no coherence transfer. If the spectrum is represented by a cube, these responses are centred on the principal body diagonal ($F_1 = F_2 = F_3$).

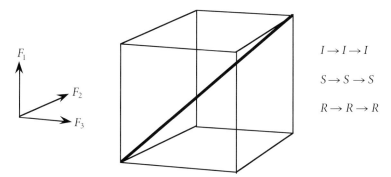

$$I \rightarrow I \rightarrow I$$

$$S \rightarrow S \rightarrow S$$

$$R \rightarrow R \rightarrow R$$

Cross-diagonal peaks are of three kinds. In the first, coherence is transferred by the first mixing pulse but remains unaffected by the second mixing pulse. These responses lie on the first mixing plane ($F_2 = F_3$):

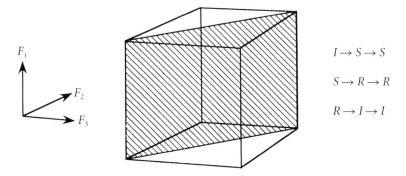

$$I \rightarrow S \rightarrow S$$

$$S \rightarrow R \rightarrow R$$

$$R \rightarrow I \rightarrow I$$

For the second type of cross-diagonal peak, coherence is transferred only by the second mixing pulse, but not the first. These responses lie on the second mixing plane ($F_1 = F_2$):

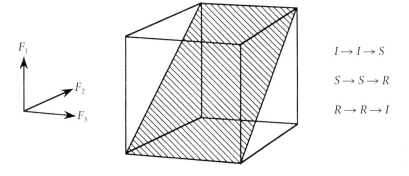

$$I \rightarrow I \rightarrow S$$

$$S \rightarrow S \rightarrow R$$

$$R \rightarrow R \rightarrow I$$

The third type involves back-transfer of coherence to the same spin, and the resulting responses lie on the back-transfer plane ($F_1 = F_3$):

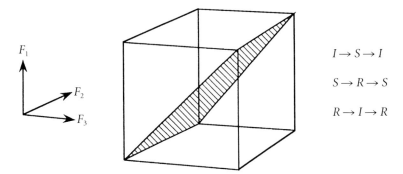

$$I \rightarrow S \rightarrow I$$

$$S \rightarrow R \rightarrow S$$

$$R \rightarrow I \rightarrow R$$

In COSY-COSY experiments, these peaks are of little interest, but they can be useful if two different experiments are combined, for example in COSY-NOESY.

Finally, the responses of interest involve two successive coherence transfers involving all three spins:

$$I \rightarrow S \rightarrow R$$
$$S \rightarrow R \rightarrow I$$
$$R \rightarrow I \rightarrow S$$

These true cross-peaks lie on none of the planes illustrated above. They give two pieces of connectivity information, and indicate that $I$, $R$, and $S$ form part of a coupled group (which may of course be part of a still larger group).

As mentioned above, three-dimensional spectra are normally displayed in the form of contour diagrams on planes perpendicular to one of the frequency axes, just as if a line-selective experiment had been performed in one of the three frequency dimensions. For example, we might take a plane section perpendicular to the $F_1$ axis at the height of the $I$-spin resonance:

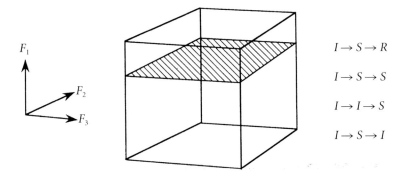

$$I \rightarrow S \rightarrow R$$

$$I \rightarrow S \rightarrow S$$

$$I \rightarrow I \rightarrow S$$

$$I \rightarrow S \rightarrow I$$

The resulting two-dimensional contour diagram shows all those responses that originate from $I$-spin coherence and are involved in coherence transfer to $S$ or $R$, or back to $I$. Similarly, a suitable plane perpendicular to the $F_2$ axis would monitor all coherences that pass through a particular intermediate spin, for example the $S$ spin in the transfer $I \rightarrow S \rightarrow R$. Finally, a plane perpendicular to the $F_3$ axis would monitor all coherences terminating on a particular spin. All three types of section have their uses.

### 8.6.2 Isotope labelling

Perhaps the most promising application of multidimensional spectroscopy is to exploit heteronuclei such as $^{13}C$ and $^{15}N$ to resolve ambiguities in the two-dimensional proton spectra of proteins, using samples biosynthetically enriched to a high level in one or both of these nuclei.[74,75] In effect, this records a two-dimensional proton spectrum (for example NOESY) for each $^{13}C$ or $^{15}N$ site. For example, polarization transfer by heteronuclear multiple-quantum coherence (HMQC) or heteronuclear single-quantum coherence (HSQC) might be employed in a round-trip transfer which begins and ends with proton magnetization. Separate three-dimensional experiments for $^{13}C$ and $^{15}N$ labelling are required if all proton resonances are to be observed. Alternatively a four-dimensional experiment can be set up to exploit the presence of both heteronuclei,[76] although restrictions on the experimental duration tend to limit the resolution when four frequency dimensions are employed. However, since the protons in these doubly labelled proteins are bound either to $^{13}C$ or $^{15}N$ but not to both, proton coherences can be generated independently from both isotopomers.[77] Polarization transfer (PT) is effected simultaneously to both $^{13}C$ and $^{15}N$, which act as independent labels before the polarization is returned to the protons.[78]

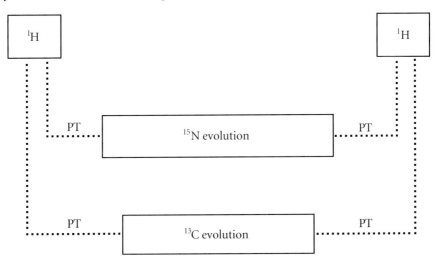

By creating proton–carbon two-spin order ($2I_zS_z$) for a variable period, it is possible to increment the $^{13}C$ and $^{15}N$ evolution periods independently, giving two frequency axes of different widths—a four-dimensional experiment that behaves like a three-dimensional one. Pulsed field gradients (§ 6.2.2) are used to shorten the total experimental duration. This type of simplification may turn out to be essential for the study of really large biological macromolecules.

## 8.7 Two-dimensional lineshapes

Lineshape is an important factor in all types of high-resolution spectroscopy since it can determine whether the fine structure is poorly resolved, distorted, or hidden altogether. As we move into higher dimensions, the question becomes more complex. We limit the present discussion to two-dimensional NMR.

In the general case, if the time-domain NMR signal $S(t_1, t_2)$ has followed a single coherence transfer path, it may be written as

$$S(t_1, t_2) = M \exp(-i\Omega_1 t_1) \exp(-\lambda_1 t_1) \exp(-i\Omega_2 t_2) \exp(-\lambda_2 t_2) \qquad \textbf{8.45}$$

If, for simplicity, we can assume the ideal condition that instrumental broadening can be neglected, then the decay terms are pure exponentials with $\lambda_1 = \lambda_2 = 1/T_2$, and the corresponding Fourier transforms are Lorentzian. Equation 8.45 describes a single response of amplitude $M$ at frequency coordinates $(\Omega_1, \Omega_2)$. After a complex Fourier transformation we obtain

$$f(\omega_1, \omega_2) = M \left( \frac{1}{i\Delta\Omega_1 + \lambda_1} \right) \left( \frac{1}{i\Delta\Omega_2 + \lambda_2} \right) \qquad \textbf{8.46}$$

where $\Delta\Omega_1$ and $\Delta\Omega_2$ are the frequency offsets from the centre of the line. It is convenient to separate the absorptive and dispersive terms:

$$f(\omega_1, \omega_2) = M(A_1 - iD_1)(A_2 - iD_2) = M[(A_1 A_2 - D_1 D_2) - i(A_1 D_2 + A_2 D_1)] \qquad \textbf{8.47}$$

where the following absorptive ($A$) and dispersive ($D$) line profiles have been defined:

$$A_1 = \frac{\lambda_1}{(\Delta\Omega_1)^2 + \lambda_1^2}$$

$$A_2 = \frac{\lambda_2}{(\Delta\Omega_2)^2 + \lambda_2^2}$$

$$D_1 = \frac{\Delta\Omega_1}{(\Delta\Omega_1)^2 + \lambda_1^2}$$

$$D_2 = \frac{\Delta\Omega_2}{(\Delta\Omega_2)^2 + \lambda_2^2} \qquad \textbf{8.48}$$

Phasing the two-dimensional spectrum requires the adjustment of the instrumental phase parameters in both frequency dimensions to select the real part of Equation 8.47:

$$\text{Real}[f(\omega_1, \omega_2)] = M(A_1 A_2 - D_1 D_2) \qquad \textbf{8.49}$$

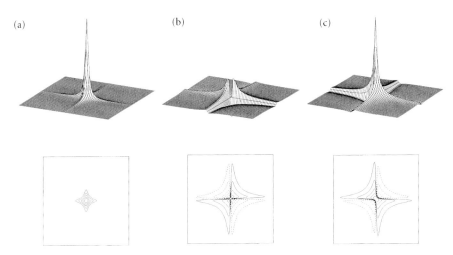

**Figure 8.12** Two-dimensional line shapes assuming equal (Lorentzian) line widths in both frequency dimensions: (a) double absorption $M(A_1A_2)$; (b) double dispersion $M(D_1D_2)$. (c) phase twist $M[(A_1A_2 - D_1D_2)]$.

We see that this is the superposition of a two-dimensional absorption line $M(A_1A_2)$ on a two-dimensional dispersion line $M(D_1D_2)$. These are shown in Figures 8.12(a) and (b). The superposition (Figure 8.12(c)) is known as a phase twist line shape.[61] If we think of it as a solid object and take successive slices through it parallel to one of the frequency axes, these one-dimensional profiles are initially almost pure dispersion but they acquire increasing amounts of absorption until, at the line centre, the profile is pure absorption (Figure 8.13). If we continue to take slices at increasing offsets, the profile become increasingly dispersive in the opposite sense. It is as if the receiver phase were being progressively rotated as we move through the line; hence the name phase twist.

As mentioned above in the section on decoupled proton spectra (§ 8.5.3), the phase twist line shape has the unusual property that its 45° integral projection vanishes. The negative lobes exactly cancel the positive central peak. In practical cases, truncation of the skirts of the line introduces some artifacts into the 45° projection, but in principle the projection is identically zero. This implies that the two-dimensional dispersion line shape (Figure 8.12(b)) has a 45° projection that is the same as the 45° projection of the (inverted) two-dimensional absorption-mode profile, a somewhat unexpected property.

There are serious distortions if two phase-twisted lines come into close proximity, and the strong dispersive tails inhibit resolution. Consequently valiant efforts have been made to achieve pure absorptive line shapes in two-dimensional spectroscopy.

## 8.7.1 **Pure-phase spectra**

The standard practice in Fourier transform NMR is to operate with the transmitter frequency near the centre of the spectral width. In this manner we use the available

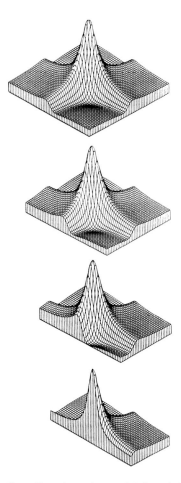

**Figure 8.13** Successive sections through a phase-twist line. At large offsets (top) the slice profile is essentially dispersive but it acquires increasing amounts of absorption until at the exact centre (bottom) the profile is purely absorptive.

transmitter power to its full advantage and avoid detecting spectral regions where there are no signals but only thermal noise. It is therefore essential to discriminate the sign of the NMR frequencies, that is to say, the sense of the nuclear precession in the frame rotating at the transmitter frequency. Otherwise, signals from the right-hand side of the spectrum would be folded onto the left-hand side and *vice versa*, and the only way to avoid confusion would be to move the transmitter to one extreme edge of the spectrum.

Sign discrimination is easily achieved in the $F_2$ dimension by the usual quadrature detection—the simultaneous acquisition of two signals 90° out of phase (§ 6.1.1). Whereas a single detector channel can only record an amplitude modulated signal $S_c(t_2) = M\cos\Omega_2 t_2$, two channels in quadrature record both

$S_c(t_2) = M\cos\Omega_2 t_2$ and $S_s(t_2) = M\sin\Omega_2 t_2$, and hence establish the sense of rotation of the magnetization vector. However, because there is no actual signal detection during the evolution period $t_1$, sign discrimination in the $F_1$ dimension is rather more difficult.[79]

The problem really arises because, in the majority of two-dimensional experiments, the signal that evolves during $t_1$ is amplitude-modulated rather than phase modulated. It may be represented by a pair of equal counter-rotating vectors in the $xy$ plane.

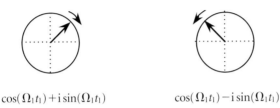

$$\cos(\Omega_1 t_1) + i\sin(\Omega_1 t_1) \qquad\qquad \cos(\Omega_1 t_1) - i\sin(\Omega_1 t_1)$$

It is quite feasible to select one or other of these vectors, the left- or right-handed precession, by a suitable phase cycle. Let us suppose, for the sake of argument, that the quadrature detector has been set up to detect right-handed precession during $t_2$. Then left-handed precession during $t_1$ gives rise to the so-called echo or $n$-type response, whereas right-handed precession during $t_1$ generates the anti-echo or $p$-type response.[80,81]

Unfortunately, the resulting phase-modulated signals generate the undesirable phase-twist lineshape (discussed above), which is incompatible with really high resolution spectroscopy. The chirality of the precession translates into a related chirality of the two-dimensional response, which exists in a left- or right-handed form when we really need the equivalent of a racemate. So we are forced to the conclusion that both phase-modulated components must be acquired separately and then recombined in such a way that the two phase twists are opposed and thus cancel. We can see from Figure 8.12(b) that a double-dispersion response vanishes if it is superimposed on its mirror image. This kind of cancellation was first achieved[58] in the special case of heteronuclear $J$ spectroscopy where two modes of operation of the pulse sequence produce the two senses of phase modulation during $t_1$, allowing the two final spectra to be combined to give pure absorption. A more general solution was suggested independently by two different groups—States et al.[82] in the context of NOESY, and Marion and Wüthrich[83] in the context of COSY. Later, Keeler and Neuhaus[79] demonstrated that although these two approaches appear to be quite dissimilar, this is mainly a consequence of the type of instrument used, and both methods are based on the same principle.

The key is to record two versions of the two-dimensional data set, one with cosine modulation during $t_1$, the other with sine modulation:

$$S_c(t_1, t_2) = \cos(\Omega_1 t_1)\exp(i\Omega_2 t_2) \qquad\qquad \textbf{8.50}$$

$$S_s(t_1, t_2) = i\sin(\Omega_1 t_1)\exp(i\Omega_2 t_2) \qquad\qquad \textbf{8.51}$$

For simplicity of notation we neglect the relaxation terms. These two data sets must be stored separately. Suppose, for the moment, that we subject both to double Fourier transformation and obtain the corresponding set of two two-dimensional spectra. We take the real (cosine) transform with respect to $t_1$ and the complex transform with respect to $t_2$:

$$\mathrm{Re}[f_c'(\omega_1, \omega_2)] = \tfrac{1}{2}[A_1^+ A_2 + A_1^- A_2] \qquad \textbf{8.52}$$

This represents two equal two-dimensional absorption signals, $\tfrac{1}{2}[A_1^+ A_2]$ at a positive frequency in $F_1$, and $\tfrac{1}{2}[A_1^- A_2]$ at an equal negative frequency. The second data set, Equation 8.51, is treated to give the imaginary part:

$$\mathrm{Im}[f_s'(\omega_1, \omega_2)] = -\tfrac{1}{2}[A_1^+ A_2 - A_1^- A_2] \qquad \textbf{8.53}$$

which also represents two absorption responses, but one of them is inverted. In the difference spectrum, the line at a negative $F_1$ frequency vanishes, leaving only a single response $A_1^+ A_2$. We have achieved the desired sign discrimination and a pure-phase spectrum. In practice the combination of the two data sets is carried out prior to the second Fourier transformation stage.

At first sight the scheme proposed by Marion and Wüthrich[83] appears to be quite different. The effect of quadrature detection in the $t_1$ domain is obtained by an extension of the Redfield trick[84]—the phase of the coherence evolving during $t_1$ is advanced by 90° for each $t_1$ increment. This time-proportional phase incrementation, (TPPI) scheme[85,86] is implemented by shifting the phase of all the pulses prior to the start of the evolution period. For single-quantum coherence these radio-frequency phase increments are 90°, whereas for $p$-quantum coherence they are $90°/p$. The effect is to displace the effective receiver frequency away from the transmitter frequency by half the spectral width. We may think of this as imposing an additional continuous rotation of the receiver reference frame, although it is in fact a set of discontinuous 90° phase jumps. A signal that is amplitude-modulated in $t_1$ is then subjected to real Fourier transformation and gives an absorption mode response but, since all coherences now have frequencies of the same sign, no spurious folded signals appear in the spectrum.

The similarity to the scheme of States et al.[82] emerges if we write the phase-incremented sequence as:

$$a \, b \, \bar{a} \, \bar{b} \, a \, b \, \bar{a} \, \bar{b} \, a \, b \, \bar{a} \, \bar{b} \ldots \qquad \textbf{8.54}$$

where $a$ and $b$ represent orthogonal channels and an overbar denotes phase inversion. The signals in the $a$ and $b$ channels constitute cosine-modulated and sine-modulated data sets, respectively (while the phase alternation within each set merely reverses the ordering of the frequencies in the final spectrum). We can imagine a hypothetical situation in which the $a$ and $b$ data sets are kept separate and transformed into two distinct two-dimensional spectra. The spectrum from the cosine-modulated data would contain two positive absorption signals, whereas that from the sine-modulated data would have one positive and one negative signal. Subtrac-

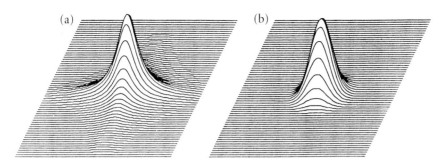

**Figure 8.14** (a) A two-dimensional Lorentzian line. (b) A two-dimensional Gaussian line, which has rotational symmetry about the vertical axis.

tion (as before) generates a spectrum with sign discrimination and pure phase. The scheme is therefore based on the same principle as that of States *et al.*, although the practical implementation is quite different.

It is now standard practice to use one or other of these methods to obtain pure absorption spectra from signals that are amplitude-modulated as a function of $t_1$. Certain echo-modulation experiments, and techniques that employ read pulses with flip angles different from $\pi/2$, generate phase-modulated signals that cannot be handled by these data processing schemes.

Even when care has been taken to acquire pure-phase spectra, there remains another slight infelicity. A purely absorptive two-dimensional Lorentzian line is not quite ideal for our purposes since, even when $\lambda_1 = \lambda_2$, the lineshape is not a solid of revolution. The contour lines are not circular but have a characteristic four-pointed star shape,[55] as can be appreciated from Figure 8.12(a). A much more convenient shape is the two-dimensional Gaussian response (Figure 8.14(b)) usually obtained by a Lorentzian to Gaussian data transformation.[87]

## 8.8 Discussion

The advent of two-dimensional spectroscopy has signalled a renaissance in NMR methodology comparable with that engendered by the Fourier transform revolution. Not only does it offer new methods of correlation, assignment and structure determination, but it has revived interest in formerly esoteric studies such as multiple-quantum coherence, and has put nuclear Overhauser and chemical exchange measurements on a much sounder footing. We should also remember that several useful techniques owe their inspiration to two-dimensional spectroscopy, for example INEPT (§ 4.2.6) and the modern broadband decoupling schemes (§ 7.4). Finally, the recent remarkable successes in protein structure determination are principally due to a great deal of hard work on the applications of multidimensional spectroscopy.[88]

# References

1. J. Jeener, in *NMR and More. In Honour of Anatole Abragam* Eds: M. Goldman and M. Porneuf, Les Editions de Physique, Les Ulis, France, 1994.
2. W. P. Aue, E. Bartholdi and R. R. Ernst, *J. Chem. Phys.* **64**, 2229 (1976).
3. K. Nagayama, K. Wüthrich and R. R. Ernst, *Biochem. Biophys. Res. Commun.* **90**, 305 (1979).
4. A. Bax and R. Freeman, *J. Magn. Reson.* **44**, 542 (1981).
5. J. Jeener, B. H. Meier, P. Bachmann and R. R. Ernst, *J. Chem. Phys.* **71**, 4546 (1979).
6. S. Davies, J. Friedrich and R. Freeman, *J. Magn. Reson.* **75**, 540 (1987).
7. P. Xu, X. L. Wu and R. Freeman, *J. Magn. Reson.* **84**, 198 (1989).
8. U. Piantini, O. W. Sørensen and R. R. Ernst, *J. Am. Chem. Soc.* **104**, 6800 (1982).
9. A. Bax and R. Freeman, *J. Magn. Reson.* **45**, 177 (1981).
10. Ē. Kupče and R. Freeman, *J. Magn. Reson. A* **112**, 134 (1995).
11. Ē. Kupče and R. Freeman, *J. Magn. Reson. A.* **112**, 261 (1995).
12. R. Brüschweiler, J. C. Madsen, C. Griesinger, O. W. Sørensen and R. R. Ernst, *J. Magn. Reson.* **73**, 380 (1987).
13. P. Xu, X. L. Wu and R. Freeman, *J. Magn. Reson.* **89**, 198 (1990).
14. L. McIntyre and R. Freeman, *J. Magn. Reson.* **89**, 632 (1990).
15. G. W. Eich, G. Bodenhausen and R. R. Ernst, *J. Am. Chem. Soc.* **104**, 3731 (1982).
16. P. H. Bolton and G. Bodenhausen, *Chem. Phys. Lett.* **89**, 139 (1982).
17. J. Friedrich, S. Davies and R. Freeman, *J. Magn. Reson.* **80**, 168 (1988).
18. L. Braunschweiler and R. R. Ernst, *J. Magn. Reson.* **53**, 521 (1983).
19. A. Bax and D. G. Davis, *J. Magn. Reson.* **63**, 207 (1985).
20. A. Bax, D. G. Davis and S. K. Sarkar, *J. Magn. Reson.* **63**, 230 (1985).
21. A. Bax and D. G. Davis, *J. Magn. Reson.* **65**, 355 (1985).
22. S. R. Hartmann and E. L. Hahn, *Phys. Rev.* **128**, 2042 (1962).
23. D. P. Weitekamp, J. R. Garbow and A. Pines, *J. Chem. Phys.* **77**, 2870 (1982).
24. P. Caravatti, L. Braunschweiler and R. R. Ernst, *Chem. Phys. Lett.* **100**, 305 (1983).
25. L. Müller and R. R. Ernst, *Mol. Phys.* **38**, 963 (1979).
26. G. C. Chingas, A. N. Garroway, R. D. Bertrand and W. B. Moniz, *J. Chem. Phys.* **74**, 127 (1981).
27. M. H. Levitt, R. Freeman and T. Frenkiel, *J. Magn. Reson.* **47**, 328 (1982).
28. A. J. Shaka, J. Keeler and R. Freeman, *J. Magn. Reson.* **53**, 313 (1983).
29. A. J. Shaka, C. J. Lee and A. Pines, *J. Magn. Reson.* **77**, 274 (1988).
30. M. Kadkhodaie, O. Rivas, M. Tan, A. Mohebbi and A. J. Shaka, *J. Magn. Reson.* **91**, 437 (1991).
31. Ē. Kupče and R. Freeman, *J. Am. Chem. Soc.* **114**, 10671 (1992).
32. S. Forsén and R. A. Hoffman, *J. Chem. Phys.* **39**, 2892 (1963).
33. J. R. Garbow, D. P. Weitekamp and A. Pines, *Chem. Phys. Lett.* **93**, 514 (1982).
34. S. Wimperis and R. Freeman, *J. Magn. Reson.* **58**, 348 (1984).
35. G. Bodenhausen and D. J. Ruben, *Chem. Phys. Lett.* **69**, 185 (1980).
36. G. A. Morris and R. Freeman, *J. Am. Chem. Soc.* **101**, 760 (1979).
37. G. Bodenhausen and R. Freeman, *J. Magn. Reson.* **28**, 471 (1977).
38. G. Bodenhausen and R. Freeman, *J. Magn. Reson.* **36**, 221 (1979).
39. L. Müller, *J. Am. Chem. Soc.* **101**, 4481 (1979).
40. A. Bax, R. H. Griffey and B. L. Hawkins, *J. Magn. Reson.* **55**, 301 (1983).

41. A. Bax, M. Ikura, L. E. Kay, D. E. Torchia and R. Tschudin, *J. Magn. Reson.* **86**, 304 (1990).
42. T. J. Norwood, J. Boyd, J. E. Heritage, N. Soffe, and I. D. Campbell, *J. Magn. Reson.* **87**, 488 (1990).
43. G. A. Morris and K. I. Smith, *J. Magn. Reson.* **65**, 506 (1985).
44. K. E. Kövér and G. Batta, *J. Magn. Reson.* **74**, 397 (1987).
45. A. Bax and M. F. Summers, *J. Am. Chem. Soc.* **108**, 2093 (1986).
46. H. Kessler, C. Griesinger, J. Zarbock and H. R. Loosli, *J. Magn. Reson.* **57**, 331 (1984).
47. C. Bauer, R. Freeman and S. Wimperis, *J. Magn. Reson.* **58**, 526 (1984).
48. A. Bax, R. Freeman and T. A. Frenkiel, *J. Am. Chem. Soc.* **103**, 2102 (1981).
49. A. Bax, R. Freeman, T. A. Frenkiel and M. H. Levitt, *J. Magn. Reson.* **43**, 478 (1981).
50. R. Freeman, T. Frenkiel and M. B. Rubin, *J. Am. Chem. Soc.* **104**, 5545 (1982).
51. X. L. Wu, P. Xu and R. Freeman, *J. Magn. Reson.* **88**, 417 (1990).
52. A. J. Shaka and R. Freeman, *J. Magn. Reson.* **51**, 169 (1983).
53. E. L. Hahn and D. E. Maxwell, *Phys. Rev.* **88**, 1070 (1952).
54. W. P. Aue, J. Karhan and R. R. Ernst, *J. Chem. Phys.* **64**, 4226 (1976).
55. R. Freeman and H. D. W. Hill, *J. Chem. Phys.* **54**, 301 (1971).
56. T. Nishida, C. Enzell and J. Keeler, *J. Chem. Soc. Chem. Comm.* 1489 (1985).
57. G. Bodenhausen, R. Freeman, G. A Morris and D. L. Turner, *J. Magn. Reson.* **28**, 17 (1977).
58. R. Freeman, S. P. Kempsell and M. H. Levitt, *J. Magn. Reson.* **34**, 663 (1979).
59. P. Bachmann, W. P. Aue, L. Müller and R. R. Ernst, *J. Magn. Reson.* **28**, 29 (1977).
60. P. Xu, X. L. Wu and R. Freeman, *J. Magn. Reson.* **95**, 132 (1991).
61. G. Bodenhausen, R. Freeman, R. Niedermeyer and D. L. Turner, *J. Magn. Reson.* **26**, 133 (1977).
62. K. Nagayama, P. Bachmann, K. Wüthrich and R. R. Ernst, *J. Magn. Reson.* **31**, 133 (1978).
63. B. Blümich and D. Ziessow, *J. Magn. Reson.* **49**, 151 (1982).
64. A. Bax, R. Freeman and G. A. Morris, *J. Magn. Reson.* **43**, 333 (1981).
65. A. Bax, A. F. Mehlkopf and J. Smidt, *J. Magn. Reson.* **35**, 167 (1979).
66. A. J. Shaka, J. Keeler and R. Freeman, *J. Magn. Reson.* **56**, 294 (1984).
67. P. Xu, X. L. Wu and R. Freeman, *J. Am. Chem. Soc.* **113**, 3596 (1991).
68. M. Woodley and R. Freeman, *J. Magn. Reson. A* **109**, 103 (1994).
69. M. Woodley and R. Freeman, *J. Magn. Reson. A* **111**, 225 (1994).
70. M. Woodley and R. Freeman, *J. Am. Chem. Soc.* **117**, 6150 (1995).
71. G. W. Vuister and R. Boelens, *J. Magn. Reson.* **73**, 328 (1987).
72. C. Griesinger, O. W. Sørensen and R. R. Ernst, *J. Magn. Reson.* **73**, 574 (1987).
73. R. E. Hurd, *J. Magn. Reson.* **87**, 422 (1990).
74. D. Marion, L. E. Kay, S. W. Sparks, D. A. Torchia and A. Bax, *J. Am. Chem. Soc.* **111**, 1515 (1989).
75. E. R. P. Zuiderweg and S. W. Fesik, *Biochemistry*, **27**, 3568 (1989).
76. L. E. Kay, G. M. Clore, A. Bax and A. M. Gronenborn, *Science*, **249**, 411 (1990).
77. O. W. Sørensen, *J. Magn. Reson.* **89**, 210 (1990).
78. R. Boelens, M. Burgering, R. H. Fogh and R. Kaptein, *J. Biomol. NMR* **4**, 201 (1994).
79. J. Keeler and D. Neuhaus, *J. Magn. Reson.* **63**, 454 (1985).
80. K. Nagayama, A. Kumar, K. Wüthrich and R. R. Ernst, *J. Magn. Reson.* **40**, 321 (1980).
81. T. H. Mareci and R. Freeman, *J. Magn. Reson.* **48**, 158 (1982).
82. D. J. States, R. A. Haberkorn and D. J. Ruben, *J. Magn. Reson.* **48**, 286 (1982).
83. D. Marion and K. Wüthrich, *Biochem. Biophys. Res. Commun.* **113**, 967 (1983).

84. A. G. Redfield and S. D. Kunz, *J. Magn. Reson.* **19**, 250 (1975).
85. G. Drobny, A. Pines, S. Sinton, D. Weitekamp and D. Wemmer, *Symp. Faraday Soc.* **13**, 49 (1979).
86. G. Bodenhausen, R. L. Vold and R. R. Vold, *J. Magn. Reson.* **37**, 93 (1980).
87. R. R. Ernst, *Adv. Magn. Reson.* **2**, 1 (1966).
88. K. Wüthrich, *NMR of Proteins and Nucleic Acids*, Wiley, New York, 1986.

# 9

# Nuclear Overhauser effect

## 9.1 Introduction

At the scientific meeting where Overhauser[1] first proposed that NMR signals might be enhanced by a very large factor simply by saturating the conduction electrons in a metal, many of the physicists present could not believe it was possible. It was felt that this somehow violated the second law of thermodynamics by creating nuclear spin order some $10^3$ times higher than it had any right to be. Overhauser's suggestion was soon proved correct by experiments[2] on lithium, and the concept of dynamic nuclear polarization was born.[3] Later it became clear that an analogous cross-relaxation experiment could be performed between two different species of nuclear spins.

For many years this nuclear Overhauser effect was regarded simply as a means to enhance sensitivity. A spin species $I$ with high magnetogyric ratio $\gamma_I$, and therefore favourable Boltzmann populations, transfers this benefit to a second species $S$ of low magnetogyric ratio $\gamma_S$ through a rearrangement of populations when the $I$ spins are saturated. This presupposes that the relaxation of the $S$ spins is determined predominantly by dipole–dipole interaction with the $I$ spins. The original example, where the $I$ spin was an unpaired electron, promised and indeed delivered[4] very high enhancement factors, but as a general method for enhancing the sensitivity of high resolution NMR it turned out to be unsatisfactory. Not only were there practical difficulties associated with the provision of simultaneous microwave and radiofrequency fields, but also the enhancements were very dependent on the nature of the sample. However, it soon became standard procedure[5] to exploit the much more modest enhancement of $^{13}C$ signals (roughly threefold) through saturation of the proton resonances, since there was normally a broadband proton decoupler already in place for simplification of the spectra.

### 9.1.1 Molecular structure studies

It was some time before the benefits of proton–proton Overhauser experiments were appreciated, since there is little point in seeking sensitivity enhancement by this route. Yet with hindsight we can see that here was the magic tape-measure for which we had all been waiting. The small proton–proton enhancement can be put to excellent use as a structural tool, using the fact that the magnitude of the effect falls off as the inverse sixth power of the internuclear distance, since it is based on the dipole–dipole interaction which has an inverse cube dependence on distance.

Of course it was easy to think of theoretical objections to the applicability of the Overhauser method. If the molecule is not rigid then different protons have different reorientational correlation times and, *in principle*, the various intramolecular nuclear Overhauser enhancements are not comparable. The reorientation of the internuclear vector may not be isotropic; rotation may be easier about certain axes than about others. Only the dipole–dipole relaxation mechanism gives rise to the Overhauser effect, so there are unknown leakage terms that compete with the effect being measured. The relaxation of any given proton is determined not merely by the dipolar interaction with its immediate neighbour but also by interactions with several other near neighbours. The extreme narrowing approximation (which assumes that molecular reorientation is fast in comparison with the Larmor frequency) is not always valid, particularly for large molecules in intense magnetic fields. The enhancement does not necessarily remain on the target spin, but may be passed on to other adjacent spins in a process called spin diffusion. Dipolar interactions between equivalent spins (e.g. within a methyl group) may also complicate the situation.

It is largely due to the faith and persistence of a few pioneers[6–9] that these apparently reasonable objections were dispelled, and structure determinations on molecules of biological importance in aqueous solution were put on a sound footing. The key is the fact that these proton systems are overdetermined; there are normally more nuclear Overhauser effects ripe for measurement than there are unknown internuclear distances of interest. Furthermore, the inverse sixth power dependence of the enhancement on distance effectively rules out long-range interactions altogether. To take an extreme example, the detection of a single cross-peak in two-dimensional nuclear Overhauser spectroscopy (NOESY) can prove once and for all that a protein chain loops back upon itself, and what is more, it identifies the two aminoacid residues involved. The recording of a NOESY spectrum establishes distance constraints which, when taken together with calculations of molecular dynamics, normally suggest a good candidate for the molecular structure in solution. This may or may not coincide with the structure determined by X-ray studies of a single crystal.

### 9.1.2 **Molecular reorientation**

For effective spin–lattice relaxation of a nuclear spin, its immediate environment must generate fluctuating magnetic fields at the Larmor frequency. These cause NMR transitions, but unlike an applied radiofrequency field, which induces absorption and emission with equal probabilities, these local radiofrequency fields favour downward transitions which bring the spin system back to thermal equilibrium with its surroundings (conventionally known as the lattice). This is because, in most circumstances, the lattice is cold compared with the spin system, and exchange of energy is more likely in the direction spin → lattice rather than the reverse. How does this coupling of the spins to the outside world come about? Which motions of the lattice are important? We consider the various possibilities in turn.

The relative translational motion of two molecules is not normally a significant factor in spin–lattice relaxation since the distances are too large. Exceptions occur when we are dealing with the very strong magnetic moment of an unpaired electron spin in a paramagnetic impurity or in dissolved oxygen gas. Vibrational motion is also quite ineffective because the fluctuations are at very high frequencies and the residual component at the Larmor frequency is extremely weak. This leaves only the rotation of the molecule. Since the spins always remain aligned along the magnetic field as the molecule rotates, the dipole–dipole field fluctuates according to the relative inclination of the two spins, and this is a powerful relaxation mechanism.

Molecular reorientation modulates the dipolar field generated by the $I$ spin at the $S$-spin site and *vice versa*. The rate of reorientation (tumbling) is important; it is defined in terms of a rotational correlation function which is the ensemble average

$$g(\tau) = \overline{R(t)\,R(t+\tau)} \qquad \textbf{9.1}$$

where $R(t)$ represents the molecular orientation at time $t$, and $R(t+\tau)$ the new orientation at time $\tau$ later. The correlation function $g(\tau)$ is unity at $\tau=0$ and is generally assumed to decay exponentially as a function of $\tau/\tau_c$, where the time constant $\tau_c$ is called the rotational correlation time, the time taken for a typical molecule to tumble through an angle of one radian. The Fourier transform of $g(\tau)$ is called the spectral density function:

$$J(\omega) = \int_{-\infty}^{+\infty} g(\tau)\exp(-i\omega\tau)\,d\tau \qquad \textbf{9.2}$$

If the correlation function $g(\tau)$ is exponential then $J(\omega)$ is a Lorentzian curve:

$$J(\omega) = \frac{2\tau_c}{1 + \omega^2\tau_c^2} \qquad \textbf{9.3}$$

The spectral density function is usually plotted on a logarithmic frequency scale, emphasizing that it is relatively flat over a considerable range of reorientational frequencies and then falls off near the frequency $1/\tau_c$. Fluctuating dipolar fields have negligible intensities at frequencies significantly beyond this cut-off frequency and cannot therefore induce the NMR transitions necessary for relaxation.

For example, a large molecule in a viscous medium reorients in a sluggish fashion and the dipolar interactions have little effect at the relatively high Larmor frequencies (Figure 9.1(a)). In contrast, the spectral density function for a small molecule extends to high frequencies, but has a lower amplitude since the area under the curve is constant (Figure 9.1(c)). Relaxation is most effective when the spectral density function cuts off just at the Larmor frequency (Figure 9.1(b)). There is a minimum in the spin–lattice relaxation time at this condition. The details of the nuclear Overhauser effect depend rather critically on the time-scale of the reorientation.

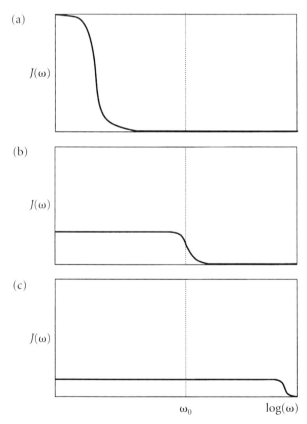

**Figure 9.1** Spectral density functions $J(\omega)$ plotted against frequency (logarithmic scale) for the cases that the molecular reorientation is (a) slow (b) intermediate and (c) fast in comparison with the Larmor frequency $\omega_0$. Since the area under these curves is constant, intermediate reorientation rates give the largest component at the Larmor frequency (b).

### 9.1.3 Two-spin system

The principles governing the Overhauser effect are most easily explained in terms of a two-spin system. The appropriate energy level diagram for an $IS$ spin system with the relaxation transition probabilities[10-12] for dipole–dipole interaction may be expressed as

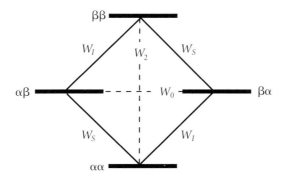

These transition probabilities reflect the efficiencies of the various processes of spin–lattice relaxation. They are derived from the corresponding spectral density functions $J(\omega)$, some fundamental physical constants and the relevant internuclear distance.

The zero-quantum transition probability $W_0$ represents a concerted flip of the two spins in opposite senses, a flip-flop transition ($\alpha\beta \longleftrightarrow \beta\alpha$, ) and is given by the expression

$$W_0 = \frac{1}{10}\left(\frac{\mu_0}{4\pi}\right)^2 \gamma_I^2 \, \gamma_S^2 \, h^2 \, r^{-6} \; \frac{\tau_c}{1 + (\omega_I - \omega_S)^2 \tau_c^2} \qquad \textbf{9.4}$$

where $\mu_0/4\pi$ is the conversion factor into SI units, $h$ is Planck's constant divided by $2\pi$, and $r$ is the internuclear distance. Then there are the more familiar cases of single-quantum transition probabilities, where the $I$ (or $S$) spin flips independently:

$$W_I = \frac{3}{20}\left(\frac{\mu_0}{4\pi}\right)^2 \gamma_I^2 \, \gamma_S^2 \, h^2 \, r^{-6} \; \frac{\tau_c}{1 + (\omega_I)^2 \tau_c^2} \qquad \textbf{9.5}$$

$$W_S = \frac{3}{20}\left(\frac{\mu_0}{4\pi}\right)^2 \gamma_I^2 \, \gamma_S^2 \, h^2 \, r^{-6} \; \frac{\tau_c}{1 + (\omega_S)^2 \tau_c^2} \qquad \textbf{9.6}$$

Finally there is the case of a concerted flip of both spins in the same sense, $\alpha\alpha \longleftrightarrow \beta\beta$, the double-quantum transition probability:

$$W_2 = \frac{3}{5}\left(\frac{\mu_0}{4\pi}\right)^2 \gamma_I^2 \, \gamma_S^2 \, h^2 \, r^{-6} \; \frac{\tau_c}{1 + (\omega_I + \omega_S)^2 \tau_c^2} \qquad \textbf{9.7}$$

For the homonuclear case under consideration, $\omega_I$ and $\omega_S$ are essentially the same and we shall simply write them as $\omega_0$, the Larmor frequency. It is readily seen that these expressions depend on the rate at which the molecule reorients ($\tau_c^{-1}$) compared with $\omega_0$ and with the sum and difference frequencies ($\omega_I \pm \omega_S$).

We shall see below that it is the two *cross-relaxation* transition probabilities $W_2$ and $W_0$ that give rise to the Overhauser effect, the former being dominant for fast reorientation (relatively small molecules) and the latter for slow tumbling (large molecules).

### 9.1.4 Population dynamics

When the spin populations are disturbed from Boltzmann equilibrium the system reacts to restore equilibrium. The rate equations are analogous to those that govern the currents in a complex electrical network—the deviation of the populations from equilibrium behaves like a voltage, the flux of spins behaves like a current, and the relaxation transition probability like a conductance. The equations are therefore analogous to Kirchhoff's laws for an electrical circuit.

It is of course quite possible to perturb the two $I$-spin transitions differentially in what is known as a selective population transfer (§ 1.2.2), but in the experiment under consideration these asymmetrical perturbations are avoided; both $I$-spin transitions are assumed to be saturated (or inverted) with the same efficiency, and both $S$ transitions retain equal intensities throughout. We may therefore label the deviations of the spin populations from equilibrium in terms of just two variables $a$ and $b$, where

$$\frac{a}{b} = \frac{S_z - S_z^0}{I_z - I_z^0}$$

9.8

This gives the population deviations:

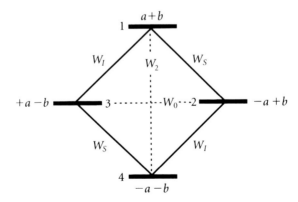

The flow of spins out of level 1 may then be written as the sum of the products of population differences and transition probabilities:

$$-\frac{dn_1}{dt} = +2aW_S + 2bW_I + 2(a+b)W_2$$

9.9

Similarly the flow of spins out of level 2 is the sum

$$-\frac{dn_2}{dt} = -2aW_S + 2bW_I - 2(a-b)W_0 \qquad \textbf{9.10}$$

Consequently,

$$-\frac{d(n_1-n_2)}{dt} = 2a\,(2W_S + W_2 + W_0) + 2b\,(W_2 - W_0) \qquad \textbf{9.11}$$

Written in a slightly different form, this is the famous Solomon equation:[10,11]

$$\frac{dS_z}{dt} = -\rho\,(S_z - S_z^0) - \sigma\,(I_z - I_z^0) \qquad \textbf{9.12}$$

were $\rho = (2W_S + W_2 + W_0)$ and $\sigma = (W_2 - W_0)$. We shall see that the nuclear Overhauser enhancement depends on this important cross-relaxation parameter $\sigma$.

### 9.1.5 Steady-state Overhauser enhancement

Most of the early experiments were performed by saturating one site (the $I$ spins) while monitoring the intensity at a neighbour site (the $S$ spins). For this steady-state experiment $I_z = 0$ and the maximum enhancement of the $S$-spin signal (no other relaxation paths) can be calculated by setting the derivative $dS_z/dt$ to zero:

$$\frac{S_z - S_z^0}{I_z^0} = \frac{\sigma}{\rho} \qquad \textbf{9.13}$$

Since $I_z^0/S_z^0 = \gamma_I/\gamma_S$, the maximum nuclear Overhauser enhancement $\eta$ is given by

$$\eta = \frac{S_z - S_z^0}{S_z^0} = \frac{\sigma}{\rho}\frac{\gamma_I}{\gamma_S} \qquad \textbf{9.14}$$

We see immediately that the enhancement vanishes if there is no cross-relaxation ($\sigma = 0$) and that the sign of the enhancement depends on the relative signs of the gyromagnetic ratios and on the sign of $\sigma = W_2 - W_0$. (Remember that $^{15}N$ and $^{29}Si$ have negative gyromagnetic ratios.)

In the case of a homonuclear system, $\gamma_I = \gamma_S$, and the enhancement expression simplifies to

$$\eta = \frac{S_z - S_z^0}{S_z^0} = \frac{\sigma}{\rho} \qquad \textbf{9.15}$$

Consequently the observed enhancement depends on the cross-relaxation rate $\sigma$, and we shall see below that this depends on the rate of molecular tumbling. We consider first of all the case of fast motion.

## 9.2 Fast molecular tumbling

The limit of fast molecular motion is sometimes called the extreme narrowing condition,[12] and can be expressed as:

$$(\omega_I - \omega_S)\tau_c \ll 1$$

$$\omega_I\tau_c \ll 1$$

$$\omega_S\tau_c \ll 1$$

$$(\omega_I + \omega_S)\tau_c \ll 1 \qquad \textbf{9.16}$$

In this regime the spectral density function is virtually the same at all four relevant frequencies and Equations 9.4–9.7 are considerably simplified, giving

$$W_2 : W_S : W_I : W_0 :: 12 : 3 : 3 : 2 \qquad \textbf{9.17}$$

We can appreciate that the double-quantum transition probability $W_2$ then tends to be the dominant influence in determining the population distribution, that is to say, whenever an $I$ spin relaxes it carries an $S$ spin with it in the same sense.

The extreme narrowing condition is satisfied for relatively small molecules in nonviscous media, the realm of the organic chemist. For such molecules the maximum homonuclear nuclear Overhauser enhancement is 50%. Most observed proton–proton Overhauser enhancements are much smaller than this, because there are competing relaxation mechanisms and dipolar interactions with protons outside the two-spin system under consideration. Many structural ambiguities (for example *cis–trans* isomerism) have been resolved by nuclear Overhauser experiments using either selective saturation or spin inversion. All that is required is to show that the cross-relaxation between one pair of spins ($I$ and $S$) is significantly greater than that between another pair of spins ($R$ and $S$). The majority of experiments aimed at molecular structure determination involve the proton–proton Overhauser effect.

The most common example of a heteronuclear Overhauser effect is the roughly threefold $(1 + \eta)$ enhancement of $^{13}C$ signals when the protons are saturated. This has become an entirely routine procedure and the enhancement is normally taken for granted. It is seldom employed as a structural tool, except perhaps to note the relatively low enhancements of quaternary $^{13}C$ sites because the nearest protons are relatively far away, and also of methyl groups because of the fast internal rotation. For $^{15}N$ spectroscopy, the maximum enhancement is $-5$. In situations like this, where $\eta$ is negative, there is the danger that competing relaxation mechanisms, while not actually dominating, may bring the system close to the condition $1 + \eta = 0$, where the observed signal is vanishingly small.

### 9.2.1 NOE difference spectroscopy

Steady-state nuclear Overhauser enhancements in homonuclear systems often amount to only a few percent of the intensity of the parent signal, so they are often presented as difference spectra by subtracting a data set obtained with the irradiation field extinguished or placed far from any resonance.[13] Although this does not improve the sensitivity of the method, it makes these small enhancements much easier to recognize, particularly in crowded spectra, and emphasizes the fact that the threshold for detecting small nuclear Overhauser effects is normally determined by the level of spectral artifacts rather than by the thermal noise. For example, a slight shift in resonance frequency or phase between the two scans gives rise to a difference-mode artifact made up of a close pair of positive and negative excursions rather like a dispersion-mode response. It is therefore important to avoid drifts in field/frequency regulation, or Bloch–Siegert shifts (§ 1.4.3) that may differ between one scan and the next. An appreciable improvement can be achieved by avoiding difference spectroscopy altogether, using pulsed field gradients to purge unwanted signal components and to select the $S$-spin signals generated by cross-relaxation.[14] Artifacts resulting from spectrometer instabilities are then greatly reduced and much smaller nuclear Overhauser effects can be detected.

## 9.2.2 Leakage

Molecules of interest are seldom as simple as a two-spin system, so we must consider other paths for dipolar relaxation and other (nondipolar) relaxation mechanisms. For the present purposes it is convenient to lump all these external relaxation interactions together, describing them by the blanket term leakage.[15,16] Later we shall single out certain external dipolar interactions and treat them separately. We may write

$$\eta = \frac{S_z - S_z^0}{S_z^0} = \frac{\sigma}{\rho + \rho^\star} \qquad \textbf{9.18}$$

where the leakage term $\rho^\star$ represents the rate of $S$-spin relaxation due to all mechanisms other than the dipole–dipole interaction with the $I$ spin.

The difficulty posed by leakage may be circumvented by comparing relative nuclear Overhauser enhancements of the $S$ spin in two separate experiments involving cross-relaxation with different spins $I$ and $R$, since the leakage term $\rho^\star$ is then common to both measurements. The extension of Equation 9.12 for three interacting spins $ISR$ gives:

$$\frac{dS_z}{dt} = -[\rho_{IS} + \rho_{SR} + \rho^\star] (S_z - S_z^0) - \sigma_{IS} (I_z - I_z^0) - \sigma_{SR} (R_z - R_z^0) \qquad \textbf{9.19}$$

This treatment of multispin interactions in terms of a pairwise summation of two-spin interactions neglects the fact that the motions of the three spins are in fact correlated. This complication is very difficult to evaluate in the general case, and it is commonly ignored.

The steady-state enhancement of the $S$-spin signal is obtained by setting the derivative $(dS_z/dt)$ to zero, and using the condition for $I$-spin saturation $(I_z = 0)$ with, for a homonuclear system, $S_z^0 = I_z^0 = R_z^0$:

$$\eta_S\{I\} = \frac{S_z - S_z^0}{S_z^0} = \frac{[\sigma_{IS} - \sigma_{SR}(R_z - R_z^0)/R_z^0]}{\rho_{IS} + \rho_{SR} + \rho^\star} \qquad \textbf{9.20}$$

where the notation $\{I\}$ indicates that the $I$ spin is saturated. Since Equation 9.20 contains the expression for the $R$-spin Overhauser enhancement upon saturation of the $I$ spins, it can be rewritten as

$$\eta_S\{I\} = \frac{[\sigma_{IS} - \eta_R\{I\}\sigma_{SR}]}{\rho_{IS} + \rho_{SR} + \rho^*} \qquad 9.21$$

If the molecular geometry is such that we can neglect the cross-relaxation between $I$ and $R$, then $\eta_R\{I\} = 0$, and the result is greatly simplified:

$$\eta_S\{I\} = \frac{\sigma_{IS}}{\rho_{IS} + \rho_{SR} + \rho^*} \qquad 9.22$$

and similarly, for saturation of the $R$ spins,

$$\eta_S\{R\} = \frac{\sigma_{SR}}{\rho_{IS} + \rho_{SR} + \rho^*} \qquad 9.23$$

Thus the ratio of the two enhancements gives the inverse ratio of the two internuclear distances raised to the sixth power:

$$\frac{\eta_S\{I\}}{\eta_S\{R\}} = \frac{\sigma_{IS}}{\sigma_{SR}} = \frac{r_{SR}^6}{r_{IS}^6} \qquad 9.24$$

Not only has the leakage effect been eliminated, but also the rotational correlation time of the molecule. In principle, the accuracy of the distance determination is high, since a 6% experimental error in the ratio of the nuclear Overhauser enhancements translates into only a 1% error in the relative internuclear distances. In many practical applications the structural question is often resolved in a cruder fashion by merely observing which enhancement is the larger.

### 9.2.3 Three-spin effects

In general it is not permissible to make the simplification (assumed above) that in the $ISR$ spin system the dipole–dipole interaction between $I$ and $R$ is negligible. There is then a new phenomenon called the three-spin effect which describes an additional (negative) contribution to the $S$-spin enhancement due to the deviation of the $R$-spin population in the positive sense:

$$\eta_R\{I\} = \frac{(R_z - R_z^0)}{R_z^0} \qquad 9.25$$

We must reconsider Equation 9.21 without making the simplifying assumption that $\eta_R\{I\} = 0$.

We can use this expression to draw some important practical conclusions. The first is that the nuclear Overhauser enhancement observed at the $S$ spins on saturation of the $I$ spins is *not* simply a function of the internuclear distance $r_{IS}$. In general, the presence of further spins that have cross-relaxation with the $S$ spins reduces the observed enhancement of the $S$-spin signal by increasing the denominator of Equation 9.21. Furthermore, saturation of the $I$ spins influences the $R$-spin populations, and this reacts back on the $S$-spin signal through cross relaxation between $R$ and $S$.

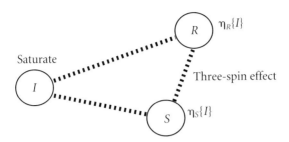

The corollary is that it is possible (with certain geometries) for the enhancement $\eta_S\{I\}$ to be negative (although necessarily small) if the three-spin effect outweighs the direct effect. The observed enhancement depends in general on the relative positions of all the interacting nuclei, and not just on the separation of $I$ and $S$. Fortunately in many practical situations one dipole–dipole interaction is dominant, and then we can obtain a good approximation to the relevant internuclear distance.

### 9.2.4 Spin diffusion

If the nuclear Overhauser experiment is performed on a linear three-spin system $(I - S - R)$, that is to say, one where the indirect $I \rightarrow R$ interaction may be neglected, then (in the fast motion regime) the positive enhancement on the $S$ spins generates a negative enhancement on the $R$ spins. This allows us to distinguish between the direct and indirect effects. For proton systems, since the direct $I \rightarrow S$ enhancement is quite small, the indirect interaction is smaller still. We shall see below that, in the limit of slow molecular reorientation, the direct Overhauser enhancement is negative. Consequently, the indirect effect is also negative (as if we had saturated the $S$ spins) and we can no longer use the sense of the enhancement to distinguish direct and indirect effects.

Real molecules are not usually three-spin systems. In the general case the population disturbance transmitted to the $S$ spins (the direct effect) may spread out through the molecule by several further cross-relaxation steps. Since the more distant spins are affected later than the adjacent spins, this has come to be known as spin diffusion, a rather ambiguous term.

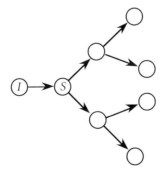

Spin diffusion becomes a much more important consideration in the slow tumbling regime (see below). Not only does it reduce the size of the direct effect, but it could well lead to erroneous conclusions about proximity. The general solution to this problem is to restrict these investigations to transient experiments, where the measurement is complete before significant spin diffusion can occur.

### 9.2.5 Transient nuclear Overhauser experiments

Leakage and spin diffusion effects can be minimized by using the truncated driven nuclear Overhauser experiment[17] where the rate of buildup of the enhancement is measured at times short enough that the initial rate approximation is valid. If the Solomon equation is written for saturation of the $I$ spins ($I_z = 0$) and the initial condition $S_z = S_z^0$, then the initial rate of buildup of the $S$-spin signal is given by

$$\left\{\frac{dS_z}{dt}\right\}_0 = \sigma\,(I_z^0) \qquad\qquad \textbf{9.26}$$

This gives the cross-relaxation rate $\sigma$ directly, and hence the internuclear distance (provided that the correlation time is known). Leakage effects have no time to compete effectively, provided that the time evolution is short compared with the intrinsic relaxation time of the $S$ spins.

Leakage is usually caused by interactions with rather short correlation times, for example, relaxation by dissolved (paramagnetic) oxygen molecules. For this reason, the leakage problem becomes less severe for slowly tumbling molecules, because the dipolar cross-relaxation parameter $\sigma$ is large in the slow motion regime and more easily dominates the external relaxation effects. However, the leakage problem is replaced by the increasing influence of spin diffusion, where the enhancement is passed on along a chain of interacting spins.

The most severe population disturbance is caused by a spin inversion pulse; this can be the basis of a transient nuclear Overhauser experiment if the $\pi$ pulse is frequency selective. The sequence is therefore:

$$\text{soft } (\pi) - \tau - \text{hard } (\pi/2) \text{ acquire} \qquad\qquad \textbf{9.27}$$

where the soft pulse is tuned to the $I$-spin resonance and does not directly influence the $S$ spins. During the $\tau$ interval, cross-relaxation causes an enhancement of the $S$-spin magnetization, while the intrinsic spin–lattice relaxation acts in the opposite sense to bring all populations back eventually to Boltzmann equilibrium. The behaviour is therefore analogous to the Forsén–Hoffman experiment[18] for measuring slow chemical exchange except that the sign of the enhancement is reversed. (Remember we are dealing here with fast molecular reorientation.) If the initial rate of buildup of the $S$-spin signal is measured using $\tau$ delays short enough that the external relaxation effects can be neglected, then

$$\left\{\frac{dS_z}{dt}\right\}_0 = 2\sigma(I_z^0) \qquad\qquad \textbf{9.28}$$

Note that this is just twice the initial rate of a truncated driven nuclear Overhauser experiment, since the initial perturbation is twice as large.

This behaviour is quite similar to the two-dimensional version of the nuclear Overhauser experiment (§ 8.2.7) which employs the sequence

$$(\pi/2)_x - t_1 - (\pi/2)_x - \tau_m - (\pi/2)_x \text{ acquire} \qquad \textbf{9.29}$$

We might regard the first section, $(\pi/2)_x - t_1 - (\pi/2)_x$, as an inversion pulse with an efficiency that is modulated as a cosine function, since if $2\pi\delta t_1 = 2n\pi$ radians (where n is an integer) then it is an effective $\pi$ pulse, whereas if $2\pi\delta t_1 = (2n+1)\pi$ it has no net effect. The mean enhancement is reduced by a factor two, and we have the initial rate of buildup of cross-peak intensity given by

$$\left\{\frac{dS_z}{dt}\right\}_0 = -\sigma(I_z^0) \qquad \textbf{9.30}$$

where $I_z^0$ now represents the intensity of the diagonal peak at $\tau_m = 0$. The negative sign indicates that in the fast motion regime the crosspeaks are negative if the diagonal peaks are recorded as positive. There is no contradiction here, merely the convention that in the one-dimensional selective inversion experiment the $I$-spin resonance is negative-going while the $S$-spin enhancement is positive.

For a homonuclear two-spin system the rate equations are greatly simplified if we make the assumption of equal external relaxation rates $(\rho^\star)$ for both $I$ and $S$. In the fast motion regime the $IS$ dipole–dipole interaction contributes to the leakage rate, which ensures that the cross-peaks are always relatively weak. We may write expressions for the time development of the diagonal and cross-peaks as

$$P_{diag} = +\tfrac{1}{2} M_0 \{1 + \exp(-2\sigma\,\tau_m)\} \exp[-(\rho + \rho^\star - \sigma)\tau_m] \qquad \textbf{9.31}$$

$$P_{cross} = -\tfrac{1}{2} M_0 \{1 - \exp(-2\sigma\,\tau_m)\} \exp[-(\rho + \rho^\star - \sigma)\tau_m] \qquad \textbf{9.32}$$

These are general expressions, valid for both slow and fast molecular reorientation, except that $P_{cross}$ becomes positive for slow molecular motion (when $\sigma$ is negative).[19] They predict that $I \rightarrow S$ and $S \rightarrow I$ cross-peaks should have equal intensities, giving NOESY spectra a mirror symmetry about the principal diagonal. This is *not* generally the case for steady-state nuclear Overhauser experiments since the leakage terms for the $I$ and $S$ spins may be different.

## 9.3  Slow molecular tumbling

There has been a remarkable upsurge of interest in the nuclear Overhauser effect in large biochemical molecules in the last decade, bolstered by the fact that few other physical techniques offer much insight into the structure adopted by a molecule in the liquid phase. These studies have become more feasible through the steady improvement in magnetic field intensity and NMR sensitivity, reinforced by the increased resolving power of multidimensional spectroscopy (§ 8.2.7). Under these conditions (high magnetic field, high molecular weight) the reorientation of the molecule tends to be slow in comparison with the Larmor frequencies:

$$\omega_I \tau_c \gg 1$$
$$\omega_S \tau_c \gg 1$$
$$(\omega_I + \omega_S)\tau_c \gg 1$$

but
$$|\omega_I - \omega_S|\tau_c \ll 1 \qquad\qquad \textbf{9.33}$$

When these expressions are substituted into Equations 9.4–9.7 we see that only the zero-quantum transition probability $W_0$ is appreciable, whereas $W_I$, $W_S$ and $W_2$ can be neglected. The cross-relaxation rate ($\sigma = W_2 - W_0$) is then negative, as is the steady-state nuclear Overhauser enhancement, which reaches $\eta = -1$ in the limit.

### 9.3.1 Two-dimensional nuclear Overhauser spectroscopy

In the slow tumbling regime, most studies are conducted by two-dimensional nuclear Overhauser spectroscopy (NOESY) which offers several advantages over one-dimensional selective saturation or population inversion. In particular, the two-dimensional method provides all the enhancements in a single experiment.

The NOESY radiofrequency pulse sequence is that shown in Equation 9.29. The evolution period $t_1$ serves to label the magnetization of a given site $I$ according to its characteristic precession frequency. This magnetization is then returned to the $\pm z$ axis for a mixing period $\tau_m$ to allow cross-relaxation to occur with a neighbouring site $S$. The mechanism is relaxation *via* the energy-conserving zero-quantum transition probability (represented by $W_0$) which causes an exchange of magnetization very similar to the effect that occurs with slow chemical exchange.[18] If an $I$ spin relaxes, it forces an $S$ spin to relax at the same time since the only appreciable transition probability corresponds to $\alpha\beta \longleftrightarrow \beta\alpha$.

The final pulse reads the enhancement of the $S$-spin signal. Phase cycling can be used to filter out undesirable components attributable to scalar coupling, except for those passed through zero-quantum coherences. However, since the latter oscillate as a function of $\tau_m$, they can be largely suppressed by employing several different values of $\tau_m$. In the slow-motion regime, the cross-peaks are positive (they have the same sense as the diagonal peaks). Note that in the corresponding steady-state nuclear Overhauser experiment, the $I$ spins are saturated or inverted and the $S$-spin signal is negative-going.

For a two-spin ($IS$) system in the limit of slow molecular motion the zero-quantum transition probability dominates all the rest and we may set

$$W_I = W_S = W_2 = 0 \qquad\qquad \textbf{9.34}$$

Consequently $\sigma = -W_0$. Cross-relaxation then competes only with the external relaxation processes caused by interactions with spins other than $I$ and $S$, and represented by $\rho^*$. Setting $\gamma_I = \gamma_S = \gamma$, we write the cross relaxation rate as

$$\sigma = -\frac{1}{10}\left\{\frac{\mu_0}{4\pi}\right\}^2 \gamma^4 \hbar^2 r^{-6} \tau_c \qquad\qquad \textbf{9.35}$$

Then, since $\sigma = -\rho$, Equations 9.31 and 9.32 for the transient intensities of diagonal and cross-peaks may be rewritten as

$$P_{\text{diag}} = +\tfrac{1}{2} M_0 \{1 + \exp(-2\sigma\,\tau_m)\} \exp(-\rho^* \tau_m) \qquad \textbf{9.36}$$

$$P_{\text{cross}} = +\tfrac{1}{2} M_0 \{1 - \exp(-2\sigma\,\tau_m)\} \exp(-\rho^* \tau_m) \qquad \textbf{9.37}$$

Note the change of sign of Equation 9.37 and the disappearance of $\rho$ and $\sigma$ from the last term. The diagonal peak loses intensity monotonically as a function of the mixing period $\tau_m$, while the cross-peak first grows and then decays again as the external relaxation term takes over (Figure 9.2). Both signals remain positive throughout.

Since we are interested in the internuclear distance $r_{IS}$, the mixing period $\tau_m$ must be short enough that the enhancement is not propagated further through the network (negligible spin diffusion). The measurements are therefore made in the initial rate regime where

$$\sigma\,\tau_m \ll 1 \quad \text{and} \quad \rho^* \tau_m \ll 1 \qquad \textbf{9.38}$$

The exponentials may then be expanded and higher-order terms dropped. This gives a simplified expression for the time development of the cross-peak intensity:

$$P_{\text{cross}} = +M_0 \sigma\,\tau_m\,(1 - \rho^* \tau_m) \qquad \textbf{9.39}$$

The last term can also be dropped if external relaxation can be neglected. The extraction of the internuclear distance involves repetition of the two-dimensional experiment at a few different settings of the mixing period in order to extract the initial rate of change. It is usual to assume that the reorientation is isotropic (although this may not be true for floppy protein sidechains).

Experimentally there is the practical complication that it is quite difficult to

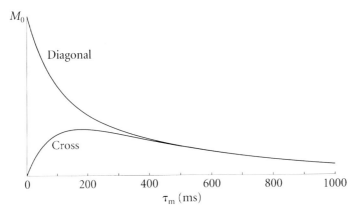

**Figure 9.2** Intensities of diagonal peaks and cross-peaks in two-dimensional nuclear Overhauser spectroscopy (NOESY) in the limit of slow molecular reorientation, plotted as a function of the mixing period $\tau_m$ in milliseconds, with $\sigma = 5\ \text{s}^{-1}$ and $\rho^* = 2\ \text{s}^{-1}$.

evaluate the volume of a two-dimensional cross-peak since the integral in the two frequency dimensions is rather sensitive to the choice of the truncation limits. It is particularly difficult where there is overlap between adjacent peaks. One practical alternative is simply to count the contour lines around the peak in question, essentially a measure of the peak height.

Where time and sensitivity considerations make it difficult or impossible to examine the build-up rate, one is forced to fall back on the measurement of a single NOESY spectrum at a mixing time short enough that only the direct cross-relaxation terms are appreciable and spin diffusion can be neglected. Normally the values of rotational correlation times are not available and so the internuclear distances are obtained by reference to a known distance, for example the proton–proton distance in a nonequivalent methylene group (1.8 Å). The unknown distance is then obtained by application of Equation 9.24, making the assumption that the molecule is rigid and that the motion is isotropic. A useful check on errors and misassignments is to compare the experimental NOESY spectrum with a simulation based on the postulated structure, obtained from molecular dynamics calculations.

The NOESY spectra of large biological molecules have proved remarkably useful for structural and conformational studies in solution.[9] The nuclear Overhauser enhancement data may be degraded by some unquantifiable perturbations (variation of correlation times, anisotropic reorientation, spin diffusion) but the results can be expressed in terms of rather lax distance constraints that are nevertheless very useful in determining the structure.[9] The lower limit of an internuclear distance may be assumed to be the sum of the relevant van der Waals radii, while an upper bound can be estimated from the nuclear Overhauser effect. Quite high latitudes can be accepted for the upper limits; sometimes the enhancements are simply classified as strong, medium or weak, and associated with upper limits of the internuclear distance of (for example) 2.5 Å, 3.5 Å, and 5.0 Å, respectively.[20]

Spin diffusion and relaxation losses reduce the magnitude of the observed nuclear Overhauser effect, whether measured as the intensity of a cross-peak or as the build-up rate in a transient experiment. The effect is most marked for the smaller nuclear Overhauser enhancements (the longer distances) which are usually the most interesting interactions, since they help define the folded structure of proteins. Allowances can be made for these spin diffusion effects, and the nuclear Overhauser buildup curves can, if necessary, be analysed in detail using the relaxation matrix approach.[21]

## 9.4  Intermediate molecular tumbling

We have seen that the homonuclear steady-state nuclear Overhauser enhancement passes from $\eta = +0.5$ for the extreme narrowing condition $\omega_0\tau_c \ll 1$ to $\eta = -1.0$ for the slow tumbling regime $\omega_0\tau_c \gg 1$. There is therefore a range of reorientation rates near to the zero-crossing point where the enhancement is very small and difficult to measure. From Equations 9.4–9.7, and neglecting leakage, we can derive an expression for the frequency-dependence of the steady-state enhancement factor:

$$\eta_S\{I\} = \frac{\sigma}{\rho} = \frac{W_2 - W_0}{2W_S + W_2 + W_0} = \frac{\left(\dfrac{6}{1+4\omega_0{}^2\tau_c{}^2} - 1\right)}{\left(1 + \dfrac{3}{1+\omega_0^2\tau_c^2} + \dfrac{6}{1+4\omega_0^2\tau_c^2}\right)} \qquad \textbf{9.40}$$

where we have used the fact that $(\omega_I - \omega_S)\,\tau_c \ll 1$. This expression may be simplified to give:

$$\eta_S\{I\} = \frac{\sigma}{\rho} = \frac{5 + \omega_0^2\tau_c^2 - 4\omega_0^4\tau_c^4}{10 + 23\omega_0^2\tau_c^2 + 4\omega_0^4\tau_c^4} \qquad \textbf{9.41}$$

This is a sigmoid curve (Figure 9.3) passing from $+0.5$ to $-1.0$ with a zero crossing at the critical value

$$\omega_0\tau_c = 1.12 \qquad \textbf{9.42}$$

In practice a measurement in this region is best avoided, because the observed enhancement may be very small. It may help to change the sample temperature or the solvent in order the change the correlation time $\tau_c$, or perform the experiment in a spectrometer at a different magnetic field (to change $\omega_0$). However we may also modify the Overhauser experiment itself and thus circumvent the difficulty altogether.

### 9.4.1 Rotating-frame nuclear Overhauser spectroscopy

Molecules that reorient at intermediate rates may be studied by rotating-frame cross-relaxation, called CAMELSPIN[22] or, in its two-dimensional version,

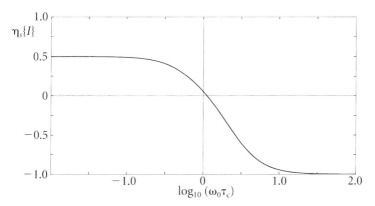

**Figure 9.3** The nuclear Overhauser enhancement $\eta$ (neglecting leakage) plotted as a function of $\omega_0\tau_c$ from 0.01 to 100 (logarithmic scale). The maximum enhancement for rapid molecular reorientation is $+0.5$, whereas the maximum enhancement for slow reorientation is $-1$. There is a zero crossing at $\omega_0\tau_c = 1.12$ where the Overhauser effect vanishes.

ROESY.[23] This technique offers the advantage that the enhancement factor is always positive irrespective of the molecular reorientation rate. It is readily distinguished from chemical exchange which gives a negative enhancement.

This technique employs a spin-locking field $B_1$ so the nuclei behave as if they were relaxing in the very weak applied field $B_1$, instead of in the much more intense spectrometer field $B_0$. In effect we are (almost) always working in the extreme narrowing regime:

$$\omega_1 \tau_c = \gamma B_1 \tau_c \ll 1 \qquad \textbf{9.43}$$

The only possible exception would be a molecule so large and sluggish that its rate of reorientation is below about $10^4$–$10^5$ radians per second. For fast molecular tumbling with respect to $\omega_0$, the predicted NOESY and ROESY enhancements are equal, but at slower rates of reorientation, the calculated rotating frame enhancement increases on a sigmoid curve, almost doubling its value in the limit.

Whereas it is quite safe to use NOESY for small molecules, and reasonably safe to apply it to very large molecules with strong negative enhancements, medium-sized molecules exhibit small or even vanishing nuclear Overhauser effects by this method. In fact, if there are variations in rotational correlation time within the molecule, both positive and negative NOESY enhancements may be observed in the same spectrum. ROESY possesses the important advantage that cross-peaks cannot be lost through this effect, whatever applied magnetic field available, since the ROESY enhancement curve never approaches zero.

The complications from the three-spin effect are less serious in ROESY than in NOESY, because with the former, indirect transfer through a third spin competes with the direct effect and diminishes the overall enhancement. This translates into an overestimate of the internuclear distances by the ROESY method, particularly for long distances. In contrast, in the slow tumbling regime, where the NOESY enhancements are negative, the three spin effect increases the magnitude of the enhancement and therefore underestimates the internuclear distances. This is dangerous since it may introduce large errors into the structure determination.[24]

## 9.5 **Discussion**

Although the Overhauser effect has been known since the early days of NMR spectroscopy, it is only in recent years that the seminal importance of cross-relaxation measurements for molecular structural studies has been appreciated. We had to wait for the introduction of two-dimensional spectroscopy (§ 8.2.7) and high-field spectrometers, before protein studies could reap the benefits of this powerful method. The emphasis has moved from structural tests on small molecules to the detailed study of biological macromolecules, a very active field at the present time. Nuclear Overhauser spectroscopy has assumed prime importance here because most other physical methods fail when applied to aqueous solutions of proteins and nucleic acids.[9]

# References

1. A. W. Overhauser, *Phys. Rev.* **92**, 411 (1953).
2. T. R. Carver and C. P. Slichter, *Phys. Rev.* **92**, 212 (1953).
3. C. D. Jeffries (Ed.), *Dynamic Nuclear Orientation*, Interscience, New York, 1963.
4. A. Abragam, J. Combrisson and I. Solomon, *Comptes Rendus Acad. Sci. Paris,* **247**, 2237 (1958).
5. K. F. Kuhlman and D. M. Grant, *J. Am. Chem. Soc.* **90**, 7355 (1968).
6. F. A. L. Anet and A. J. R. Bourn, *J. Am. Chem. Soc.* **87**, 5250 (1965).
7. R. A. Bell and J. K. Saunders, *Can. J. Chem.* **48**, 1114 (1970).
8. J. Jeener, B. H. Meier, P. Bachmann and R. R. Ernst, *J. Chem. Phys.* **71**, 4546 (1979).
9. K. Wüthrich, *NMR of Proteins and Nucleic Acids,* Wiley, New York, 1986.
10. I. Solomon, *Phys. Rev.* **99**, 559 (1955).
11. I. Solomon and N. Bloembergen, *J. Chem. Phys.* **25**, 261 (1956).
12. A. Abragam, *The Principles of Nuclear Magnetism,* Clarendon Press, Oxford, 1961.
13. J. D. Mersh and J. K. M. Sanders, *J. Magn. Reson.* **50**, 289 (1982).
14. V. Dötsch, G. Wider and K. Wüthrich, *J. Magn. Reson. A* **109**, 263 (1994).
15. J. H. Noggle and R. E. Schirmer, *The Nuclear Overhauser Effect,* Academic Press, New York, 1971.
16. D. Neuhaus and M. Williamson, *The Nuclear Overhauser Effect in Structural and Conformational Analysis,* VCH, New York, 1989.
17. G. Wagner and K. Wüthrich, *J. Magn. Reson.* **33**, 675 (1979).
18. S. Forsén and R. A. Hoffman, *J. Chem. Phys.* **39**, 2892 (1963).
19. R. R. Ernst, G. Bodenhausen and A. Wokaun, *Principles of Nuclear Magnetic Resonance in One and Two Dimensions,* Clarendon Press, Oxford, 1987.
20. H. J. Dyson and P. E. Wright, in *Two-Dimensional NMR Spectroscopy. Applications for Chemists and Biochemists* 2nd edn, Eds: W. R. Croasmun and R. M. K. Carlson, VCH, New York, 1994, Chapter 6.
21. S. Macura and R. R. Ernst, *Mol. Phys.* **41**, 95 (1980).
22. A. A. Bothner-By, R. L. Stephens, J-M. Lee, C. D. Warren and R. W. Jeanloz, *J. Am. Chem. Soc.* **106**, 811 (1984).
23. A. Bax and D. G. Davis, *J. Magn. Reson.* **63**, 207 (1985).
24. H. Kessler and S. Seip, in *Two-Dimensional NMR Spectroscopy. Applications for Chemists and Biochemists* 2nd edn, Eds: W. R. Croasmun and R. M. K. Carlson, VCH, New York, 1994, Chapter 5.

# 10

# In defence of noise

## 10.1 Introduction

Our reactions to random phenomena, for example games of chance such as roulette or blackjack, are often irrational and tainted by primitive superstition. Ask the man in the street whether a penny is more or less likely to fall heads if it has just shown three tails in a row. A science student will often have difficulty deciding just how to draw a straight-line graph through data showing a large scatter of the experimental points, even though he knows in principle about least-squares-fitting procedures. Noise is somehow regarded as wrong—something to be suppressed at all costs. We may feel guilty about publishing results with large error bands or spectra with prominent noise contributions. Perhaps we should have been more careful with our measurements? Perhaps there is a malevolent influence on the spectrometer today?

### 10.1.1 Thermal noise

Yet noise is very much a part of most NMR measurements, setting a limit on the sensitivity of the experiment, that is to say, the minimum amount of sample needed to give a detectable response. The situation is rather similar to that encountered in radar—however powerful the radar transmitter, there is always an extreme range at which the true radar returns are so weak as to merge into the background noise. Chemists often have to deal with samples where the intrinsic NMR response is at the limit of detection, either because of the unfavourable gyromagnetic properties of the chosen nucleus or through low concentration of material. The resulting signals may be so weak that they have to compete with the "Johnson" or thermal noise from the Brownian motion of the electrons in the receiver coil.

Thermal noise can be reduced by cooling the receiver coil, and over the years some heroic efforts[1] have been made to cool this coil with liquid helium, keeping it thermally isolated from the sample. The advantage arises from two independent effects. First, cooling slows down the thermal agitation of the electrons; the resulting noise voltage varies as the square root of the absolute temperature $T$. Then, if the coil is made superconducting, its dc resistance is zero and its radiofrequency resistance is greatly reduced. This gives rise to a very a high quality factor $Q$ and, since signal voltage increase in proportion to $Q$, whereas the noise voltage increases only as the square root of $Q$, there is a further improvement in sensitivity. Thus for a given sample volume, sensitivity increases in proportion to $(\eta Q/T)^{\frac{1}{2}}$, where $\eta$ is the filling factor. In practice, receiver coil $Q$ factors as high as 20 000 have been reported,[2] compared with $Q$ of about 250 for a standard room-temperature coil. The coil temperature might be typically 25 K rather than 300 K. These improvements are partially offset by a reduction in sensitivity due to the less favourable filling factor, imposed by the necessary thermal shielding of the coil from the sample. The potential net gain in sensitivity is approximately an order of magnitude. In practice an improvement of about five has been demonstrated[2] for a 5 mm standard proton test sample at 400 MHz. Unfortunately, practical difficulties still remain with aqueous samples (§ 11.1), where the dielectric loss degrades the $Q$ factor of the coil and tends to negate one of the advantages of cooling.

Amplifiers also contribute noise, principally shot noise from semiconductors. The most critical amplifier is the first (the preamplifier) and great care is taken to ensure that this introduces very little additional noise into the system. An electronic engineer would say that the noise figure of the preamplifier approaches the ideal value of unity. (With the superconducting receiver coil described above, care was taken to ensure that the preamplifier did not become the dominant source of noise, by cooling it in liquid nitrogen.) After preamplification, the noise is raised to a level where it dominates any shot-noise contributions from later amplification stages. Consequently, with proper spectrometer design, the noise which we observe in the final spectrum should be predominantly from the random motion of the electrons in the receiver coil.

It is in fact quite difficult to obtain a reproducible measure of the noise level in a spectrum. The peak-to-peak amplitude is not a suitable quantity since in principle it is indeterminate—if we search long enough we can always find a still larger noise excursion. The most convenient measure is the root-mean-square (rms) noise amplitude, analogous to quantifying random errors on an experimental quantity in terms of the standard deviation. Many spectrometer systems include a software routine to evaluate the rms noise from a signal-free region of the baseline. It is the accepted convention that signal-to-noise is defined as the ratio of the peak signal intensity to twice the rms value of the noise. If the noise is Gaussian, then 99% of the values fall within a range of $\pm 2.5$ times the rms noise.

In its pure form noise is white, that is to say, it contains all frequency components equally; any selected band of frequency space contains the same amount of noise. Now the NMR signals exist only in a very narrow frequency range, the spectral width, which is a function of the nuclear species under investigation (it spans a range of about 8 kHz for protons in a 800 MHz spectrometer). We are interested in what happens within this restricted window of frequencies, and we must be very careful to exclude noise from adjacent frequency bands. This is actually quite difficult to achieve because of aliasing, a phenomenon analogous to the stroboscopic effect that makes wagon wheels appear to turn much slower in a movie than in real life. The Nyquist theorem tells us that if we sample a time-domain signal at a rate $S$ kHz, we only obtain a faithful record of signals within a frequency window spanning the range $-\frac{1}{2}S$ kHz to $+\frac{1}{2}S$ kHz. After digitization, all signals and noise outside this range are aliased and appears to fall inside this frequency window. Normally there are no NMR signals outside the chosen window, but there is certainly a great deal of noise. The solution is to restrict the bandwidth of the detection system by introducing low-pass filters that cut off the noise beyond $\pm S/2$ kHz. In order to maintain an essentially flat response and uniform phase within the chosen spectral width, with a sharp cut-off just outside, digital filters are being increasingly used for this purpose. They usually operate in the time domain and require sampling at a rate that is several times faster than the Nyquist frequency (oversampling), followed by filtration and data compression. Provided that data storage is not a limitation, and provided that the Fourier transformation of the larger data table is fast enough, oversampling can be used without the digital filtration stage, simply by discarding all but the interesting central window. As we shall see below (§ 10.1.3), both oversampling schemes have the additional advantage that $t_1$ noise is also attenuated.[3]

## 10.1.2 **Digitization noise**

The act of digitizing the time-domain signal may introduce another type of spurious fluctuation usually called digitization noise, although the process is not strictly random. Digitization noise can be appreciable when the incoming NMR free induction decay spans a high dynamic range, comprising both very weak signals and very strong signals. Dilute aqueous solutions present a particularly serious case, and an entire field of solvent suppression methodology has been developed to alleviate the dynamic range problem (§ 11.2). The seat of the difficulty is the analog-to-digital converter (ADC) which is normally operated with the strongest signal scaled to be just less than full range (the most-significant bit). The ADC necessarily makes an error of the order of the least-significant bit when it digitizes the time-domain NMR signal. Suppose, for the sake of argument, that the conversion always triggers the next-highest bit, rounding up all the ordinates (Figure 10.1). At each conversion there is a small digitization error, shown as the small vertical bars. Since Fourier transformation is a linear operation we can consider the processing of the true signal and the digitization errors separately; the former would give the undistorted NMR spectrum, and the latter the digitization noise. These errors are not random,

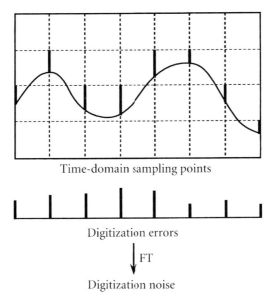

Time-domain sampling points

Digitization errors

$\downarrow$ FT

Digitization noise

**Figure 10.1** Digitization noise in an analog-to-digital converter (ADC). Each vertical unit represents one least-significant bit of the digital voltage. It is assumed that the ADC rounds up to the next-highest digit. The Fourier transform of these small digitization errors constitutes the digitization noise.

but after Fourier transformation, digitization noise closely resembles true noise, and it certainly interferes with the proper observation of the NMR spectrum. The possible remedies include the use of an analog-to-digital converter with a much higher dynamic range, oversampling of the time-domain signal to increase the effective dynamic range, and a massive armoury of solvent suppression techniques (§ 11.2).

During analog-to-digital conversion, it might be supposed that weak signals less than the least-significant bit would be lost altogether. This is where the true noise plays a positive role. Suppose there is a signal equal to (say) one-tenth of the least significant bit, repeated on each scan. Provided that the noise excursions are an order of magnitude larger, during time averaging the small signal will cause the noise to trigger the next bit in approximately 10% more scans than it would normally do. In this manner the noise carries the weak signal, which then makes its appropriate contribution to the output digital voltage after hundreds of scans have been accumulated.

### 10.1.3 **Things that go bump in the night—$t_1$ noise**

We also have to keep a close eye on other effects that could give rise to fluctuating responses that interfere with the detection of the true signal. These may not be random processes and the purist would not regard them as true noise. For example, some types of spurious fluctuations may be eliminated by difference spectroscopy or by a suitable phase cycle, processes that can have no influence on true random noise. In two-dimensional spectroscopy there is a new category, usually called $t_1$ noise—spurious baseline fluctuations running parallel to the $F_1$ frequency axis at $F_2$ ordinates where there are strong NMR signals.[4,5] Similar difficulties arise in conventional NMR spectra run in the difference mode. This new form of noise arises from unwanted modulation of the time-domain NMR signal from several possible sources. For example, there may be instabilities in the field:frequency ratio, variations in the field homogeneity, or variations in amplitude or phase of the radiofrequency pulses.[5] All these instabilities produce spurious fluctuations in the observed NMR signal. The resulting $t_1$ noise is essentially a set of sideband responses of the true signals and is therefore particularly serious when there is a strong solvent signal present; many two-dimensional spectra of aqueous solutions are marred by an intense band of $t_1$ noise at the frequency of the water signal (§ 11.1).

It may seem surprising, but even coherent 50 Hz (or 60 Hz) modulation from the main electrical supply can generate $t_1$ noise. If it is sampled at irregular time intervals in the $t_1$ dimension, it behaves like an incoherent fluctuation which is then Fourier transformed to give a response which resembles true noise. Unfortunately not all two-dimensional spectroscopy programmes make the effort to ensure that the sampling in the $t_1$ domain is exactly periodic, instead they indulge in housekeeping operations such as reading and writing onto hard disc at arbitrary times within the data-gathering phase of the program. This uncertainty in the timing chops up any sinusoidal modulation into an apparently random fluctuation. Most contributions to $t_1$ noise are in fact grossly undersampled, and therefore aliased many times in the $F_1$ frequency domain. Sampling at a higher rate in the $t_1$ dimension partly reverses this process[3] and spreads out the $t_1$ noise into a wider frequency band, part of which may then be discarded. This reduces the amount of $t_1$ noise but increases the number of $t_1$ increments. Normally any such increase in the number of sampling operations is to be avoided since it prolongs the experiment unnecessarily, but it can be advantageous for intrinsically weak signals, since it also acts as a signal averaging process.

Considerable progress has been made in reducing $t_1$ noise by replacing phase cycling with the appropriate pulsed field gradients.[6] Certain intense signals are then dispersed (rather than cancelled by difference spectroscopy) and since it is their modulation that causes much of the $t_1$ noise, the latter is suppressed along with the parent signals. Selection of coherence transfer pathways by pulsed field gradients (§ 6.2.2) rather than by phase cycling seems to afford many practical advantages, particularly for round-trip heteronuclear coherence transfer (§ 8.3.1).

### 10.1.4 Noisy artifacts

Any responses that appear noisy can interfere with the detection of very weak signals and thereby reduce the sensitivity of the NMR method. For example, broadband decoupling sequences based on repetitive cycles of spin inversion pulses generate cycling sidebands in the decoupled spectrum (§ 7.4.2). One way to decrease the amplitude of these undesirable sidebands is by systematically changing the basic decoupling cycle so that the cycling sidebands are chopped up into a set of much weaker responses. Unfortunately the resulting artifacts, though weaker and more widely distributed, now take on the character of noise and thus obscure very weak signals. If sensitivity is of key importance, it would be better to work with the original cycling sidebands, because their positions in the spectrum are predictable. Noisy artifacts can arise from many possible sources, for example, if the free induction signal is inadvertently clipped in the analog-to-digital converter, a set of spikes appears in the transformed spectrum.

## 10.2 Post-acquisition processing

When precautions have been taken to suppress the instrumental artifacts known as digitization noise and $t_1$ noise, the remaining component is the thermal noise. Efforts are often made to improve the situation still further. These are usually called post-acquisition methods. What is needed is a criterion that distinguishes signal from noise. Although we might imagine that there are several properties that could be utilized to filter signals from random noise, only two afford a genuine separation. These are considered below.

### 10.2.1 Smoothing

If the noise fluctuates faster than the NMR signal itself, then the signal-to-noise ratio can be improved by smoothing in the frequency domain, to take out the faster

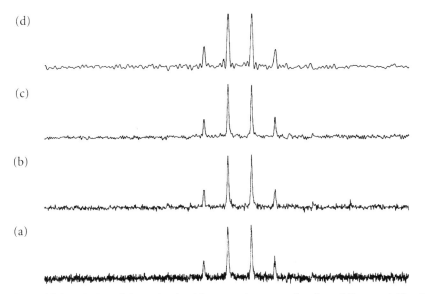

**Figure 10.2** Variation of the noise level in a spectrum as a function of the acquisition time: (a) 8 s, (b) 4 s, (c) 2 s, (d) 1 s. Shortening the acquisition time removes high-frequency components of the noise and increases the signal-to-noise ratio. However, no further improvement is observed for trace (d) since noise and signal fluctuate at about the same rate, and there is a perceptible broadening of the signal. Spectra courtesy of Dr Federico Del Rio-Portilla.

disturbances. If the process is carried too far, the NMR signal itself is significantly affected, and eventually too much smoothing degrades the signal-to-noise ratio. The extent to which sharp features appear in the spectrum (signal or noise) is determined by the acquisition time of the free induction decay. A long acquisition time retains more of the rapid fluctuations and generates a noisier spectrum. We can appreciate this phenomenon by comparing spectra derived from free induction decays acquired for different times (Figure 10.2). Clearly the noise is smoothed and reduced in amplitude as the acquisition time is reduced. Acquisition of the free induction decay over too long a period introduces noise without contributing any significant amount of signal, so the tail should be discarded. The limit is reached when truncation of the free induction signal appreciably distorts the lineshape and degrades the resolution. Figure 10.2(d) shows an example where the truncation is too severe.

Smoothing a spectrum in the frequency domain involves the process of convolution. Suppose that the spectrum of interest is represented by $g(f)$ and it is to be smoothed according to a function $h(f)$. This involves the convolution integral

$$g(f) * h(f) = \int_{-\infty}^{+\infty} g(f')h(f-f')\, df' \qquad \textbf{10.1}$$

where the asterisk denotes convolution. It is easier to visualize this process with digitized spectra, represented as histograms, and where the integral is replaced by a summation between finite limits.

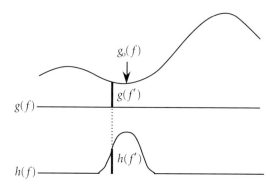

The smoothing function $h(f)$ is first reversed left-to-right (folded), but in many cases of interest it is symmetrical about its centre, so folding is irrelevant. We now consider a typical ordinate of the spectrum $g_0(f)$ and shift the entire smoothing function $h(f)$ so that it is centred at the frequency of this ordinate. Then we multiply each and every spectral ordinate $g(f')$ by the corresponding smoothing ordinate $h(f')$ at all points of the smoothing function histogram, and evaluate the sum. After normalization, this value replaces the original ordinate at $g_0(f)$. We then move on to the next point on the experimental spectrum and repeat the process, continuing until all points have been smoothed. (There are some slight problems with end effects where the smoothing function no longer overlaps the spectrum completely, but these can be circumvented, for example by wraparound.) After processing, points on the smoothed spectrum incorporate contributions from nearby points, and this has the effect of diminishing the more rapid fluctuations.

An exceptionally simple example would be the case where $h(f)$ is a 1:2:1 smoothing function. Each ordinate in the smoothed spectrum is then made up of 50% of the original ordinate in the unprocessed spectrum plus 25% contributions from the two immediate neighbours. This is a fast operation and can be repeated many times in cascade, giving the equivalent of convolution with the higher binomial functions (for example 1:4:6:4:1) and approaching a Gaussian smoothing function in the limit of many passes.[7]

Smoothing by convolution is used in many fields. The stock market analyst would like to identify real trends in the price of a certain stock (the signal) in the face of apparently meaningless short-term fluctuations (noise). He might address this problem by calculating the 30-day moving average of the stock price to smooth out the wilder fluctuations of the market. This is a rather crude convolution of the raw data with a rectangular function.

Now convolution in the frequency domain $g(f) * h(f)$ is equivalent to multiplication of the corresponding Fourier transforms $G(t)$ and $H(t)$ in the time domain. One way to reduce noise in the tail of the free induction decay is to multiply by an appropriate function that decreases with time, the so-called sensitivity enhancement weighting function. Exponential weighting in the time domain is equivalent to convolution with a Lorentzian function in the frequency domain. Sensitivity enhancement is very widely used in NMR spectroscopy; we see that it is equivalent to smoothing in the frequency domain. If the weighting function is too severe it begins to affect the time-domain NMR signal more than the noise and the signal-to-noise ratio is degraded. The optimum condition is the matched filter where the weighting function falls off at the same rate as the NMR signal itself. So there is a limit on the improvement to be achieved by smoothing and, unless the highest possible resolution is required, we normally seek to operate near the optimum matched filter condition. Note that with a proper sensitivity enhancement weighting of the free induction signal there is no longer a signal-to-noise penalty nor any resolution advantage associated with the use of a long acquisition time.

If we consider a region far down the tail of the free induction decay that has been strongly affected by the sensitivity enhancement function, it will be seen to contribute negligible signal and negligible noise. There is therefore little point in continuing to acquire the NMR signal beyond this point. There may however be an advantage in artificially extending the tail of the free induction decay by appending ordinates of zero intensity. This zero-filling operation has the effect of improving the digital definition of the spectrum (more data points per Hz) without significantly affecting its general form. This can improve the precision of computer peak-finding routines used to measure chemical shifts or coupling constants. Zero-filling is also convenient for extending the data table to be exactly $2^n$ points in preparation for the fast Fourier transform algorithm.

## 10.2.2 **Signal averaging**

The other property that distinguishes signals from noise is the fact that the signal is reproducible when the experiment is repeated, whereas random noise has a different form in each scan. By adding successive spectra with the proper frequency registration we can increase the signals at the expense of the noise. Noise increases as the square root of the number of scans $N$, whereas the signal is linearly proportional to $N$. This is the basis of the ubiquitous time averaging technique for improving the signal-to-noise ratio. Since Fourier transformation is a linear process, it does not matter whether the time averaging is performed in the time domain or the frequency domain. It is important to make a clear distinction between true random noise and other fluctuations that may be deterministic; the latter do not benefit in the same way from time averaging.

The dependence of the sensitivity enhancement on the square root of the number of scans means that time averaging yields diminishing returns. The difference between a typical overnight run and a week-end of data accumulation is only a factor of approximately two in signal-to-noise ratio. Beyond this point there is little to be gained by tying up an expensive high-resolution spectrometer in time averaging.

## 10.2.3 **Hadamard trick**

The principle of time averaging still applies even when a signal is changing with time. This is why two-dimensional spectra enjoy a sensitivity similar to their one-dimensional counterparts performed in the same total time (if no extra splittings are introduced). There is another way to take advantage of this principle. This is the Hadamard trick,[8–10] described in detail in § 5.5.4 in terms of the analogy with weighing several different objects on a balance. The upshot of the argument is that it is better to weigh all the objects together (in different combinations) than to weigh them one at a time. The accuracy then improves as the square root of the number of weighings.

Exactly the same principle can be used with one-dimensional NMR correlation experiments initiated by soft pulses.[11–13] Normally these correlations would be examined one-at-a-time in $N$ separate measurements. However, if $N$ experiments are carried out simultaneously and repeated $N$ times with a phase alternation scheme based on a Hadamard matrix, then the signal-to-noise ratio is improved by a factor $\sqrt{N}$. We might regard two-dimensional spectroscopy in the same light. It is a one-dimensional experiment repeated $N$ times, but instead of being coded with $+$ or $-$, phase (or amplitude) modulation is employed and Fourier transformation is used as the decoding procedure. The Hadamard trick has the advantage that we would choose to take the $N$ traces only at interesting regions of the spectrum, whereas the two-dimensional Fourier transform scheme samples indiscriminately across the entire spectrum. The Hadamard scheme thus enjoys a multiplex advantage at least as good as that of two-dimensional spectroscopy. Indeed it may well be superior because the selectively excited spectra are not split into several components in the second frequency domain.

## 10.2.4 **Reference deconvolution**

When we see a photograph of a beautiful model in a glossy magazine, our admiration may be tempered by the knowledge that the image has almost certainly been enhanced on a computer. What we see is not the raw experimental data but an ideal image from which any natural blemishes have been carefully removed; teeth have been whitened, spots have been suppressed. A comparable procedure may be imposed on experimental NMR data to strip away the effects of spectrometer instabilities, converting the experimental lineshape $E(f_2)$ into an ideal lineshape $T(f_2)$. We can identify the blemishes on $E(f_2)$ since they also appear on the reference line $R(f_2)$, because many of the common spectrometer instabilities act on all the lines of the spectrum in exactly the same manner. Examples include instability of the field:frequency controller, sidebands that arise from spinning the sample, and instabilities in the amplitude or phase of the radiofrequency pulses.[5] (Certain other perturbations, such as temperature-dependence of chemical shifts, can be site-specific; water is particularly susceptible to temperature fluctuations.) Random noise cannot, of course, be deconvoluted in this manner.

Reference deconvolution uses information about spectrometer instabilities residing in the shape of the signal from the reference compound (tetramethylsilane) to correct these perturbations on all the other lines.[14–18] In one-dimensional spectroscopy the result is a global replacement of an experimental lineshape $E(f_2)$ that carries noisy modulation components, with the ideal lineshape $T(f_2)$. The normal choice for $T(f_2)$ is a pure Lorentzian, but if resolution enhancement[19] is required, it may be a Gaussian that is deliberately made narrower than the instrumental linewidth.

The first step is to separate the reference line $R(f_2)$ from the rest of the experimental spectrum. It should have suitably extensive wings, so that no truncation artifacts are introduced. Here it is an advantage to derive the dispersion mode reference signal by Hilbert transformation of the absorption mode, to avoid the need to include long dispersion tails.[18] The reference line should also have a good signal-to-noise ratio so as not to introduce additional noise.

The reference line $R(f_2)$ is then back-transformed into the time domain to give $R(t_2)$. For this reason it is important that $R(f_2)$ be a singlet, so that there are no zero crossings in the time-domain signal $R(t_2)$. The time-domain correction function $C(t_2)$ is the (complex) ratio $T(t_2)/R(t_2)$, where $T(t_2)$ is the inverse transform of the ideal lineshape function $T(f_2)$. Multiplication of the experimental free induction decay by $C(t_2)$ corrects the amplitude and phase at each point, removing the effects of instrumental imperfections.

An analogous deconvolution process may be applied in two-dimensional spectroscopy in the $t_1$ domain, providing a powerful weapon against $t_1$ noise.[18] Since the longer-term spectrometer instabilities affect each signal in a given free induction decay in the same manner, the noise from these sources is highly correlated from one $f_1$ trace to another. Reference deconvolution using a correction function $C(t_1)$ applied to the entire time-domain data set $S(t_1, t_2)$ suppresses this correlated $t_1$ noise throughout the two-dimensional spectrum. A more complete solution to the problem also corrects the shorter-term fluctuations that occur during the free induction decays, as in the one-dimensional example described above. Phase cycling introduces a slight complication in that the reference deconvolution may have to be applied separately to two data sets before they are subtracted.

Two-dimensional round-trip coherence transfer experiments (§ 8.3.1) on low-abundance nuclei such as $^{13}$C or $^{15}$N provide a searching test of reference deconvolution, since it is the intense proton signals from $^{12}$C or $^{14}$N molecules that give rise to the spurious modulations, while it is the much weaker proton signals from $^{13}$C or $^{15}$N molecules that must be recorded. Gibbs et al [18] have successfully applied the deconvolution method to the long-range $^{13}$C–H correlation spectrum of a proanthocyanidin. In order to be able to acquire the reference signal (residual water in the solvent DMSO-d$_6$) the phase cycle was split into two parts, thus avoiding the cancellation that normally occurs for signals not involved in the round-trip coherence transfer. Figure 10.3(b) shows the appreciable improvement in the quality of the two-dimensional correlation spectrum achieved by reference deconvolution.

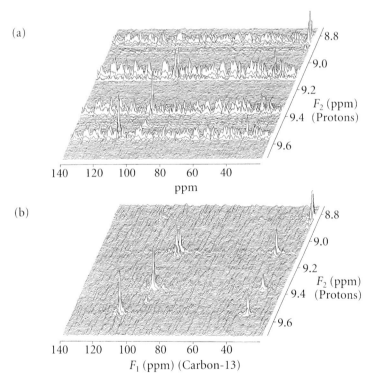

**Figure 10.3** Reference deconvolution applied to the heteronuclear multiple-bond correlation (HMBC) spectrum of a non-spinning sample of a proanthocyanidin. Interleaved acquisition was used to reduce $t_1$ noise from temperature fluctuations. (a) The conventional HMBC spectrum recorded in the absolute-value mode. (b) After reference deconvolution based on the residual water signal in dimethyl sulfoxide-d$_6$.

## 10.3  **Sensitivity and signal-to-noise ratio**

If we have ambitions to go further in suppressing random noise by post-acquisition processing, it is important to enquire carefully about the linearity of the process. An operation is linear if its effect on the sum of two signals $f(s_1 + s_2)$ is the same as the sum of the effects measured individually, $f(s_1) + f(s_2)$. To a good approximation, the spectrometer receiver behaves in a linear manner and, in particular, it amplifies noise in exactly the same way as it amplifies NMR signals. Once we relax this linearity requirement all kinds of bizarre consequences ensue.

Let us consider a very simple example concerning a properly phased pure absorption mode high resolution spectrum. All signals are positive, but the baseline noise goes both positive and negative. Why not improve sensitivity by a factor of two by eliminating all negative values? At first sight this seems a seductive idea and we might be tempted to carry it even further by resetting the zero point a little higher so the all baseline noise is eliminated and only signals appear on the display device.

The fallacy lies in a confusion between the concepts of sensitivity and signal-to-noise ratio. Sensitivity is defined in terms of the lowest concentration of sample required to give a detectable NMR signal. A processing technique can be said to enhance sensitivity if it can reveal the presence of a signal that was previously obscured by noise. Under conditions of linearity, the sensitivity is proportional to the signal-to-noise ratio; if we improve the latter, we improve the former. Unfortunately this is no longer true for non-linear data processing such as the treatment postulated above. A signal weaker than the noise may be lost altogether. The signal-to-noise ratio is only enhanced in cases where it is already adequate, a situation where we have no real need of improvement. Sensitivity is not increased; indeed if anything it is degraded.

Although not many spectroscopists will fall into such an obvious trap, there are several nonlinear post-acquisition processing schemes where the same catch is better disguised. Suppose we set out to fit the experimental spectrum with a set of (say) Lorentzian lines with adjustable frequencies, linewidths and amplitudes, using a least-squares optimization program. It is convenient to introduce the NMR lines one at a time, while monitoring the sum of squares of residuals between simulated and experimental spectra. A stage is normally reached when the introduction of further lines shows no significant reduction in the residuals, since the program is then attempting to fit noise excursions rather than true signals. If we stop there, then the apparent signal-to-noise ratio in the simulation is much higher than that in the unprocessed experimental spectrum. However, the procedure is nonlinear, weak signals below the general level of noise are ignored, and sensitivity is not improved.

Another seductive idea is to use prior knowledge of the symmetry properties of the spectrum to distinguish true signals from random noise. If several symmetrically related signals are combined into a single response, sensitivity is of course improved, and this kind of operation has been applied to enhance INADEQUATE spectra. On the other hand, if symmetry is employed merely to discriminate against noise, dangerous conclusions may be drawn. For example, we know that an isolated cross-peak in a two-dimensional correlation (COSY) spectrum must be antisymmetrical with respect to horizontal and vertical axes passing through its centre (the chemical shift co-ordinates). Once this centre has been located, it is possible to write a symmetrization program that tests the appropriate pairs of locations to see whether the intensities have opposite signs. If they have, these intensities are retained, if not, then they are attributed to random noise and both ordinates are replaced by zeroes. Genuine signals are unchanged but the baseplane noise is considerably reduced. (There is in fact a justifiable reason for processing the data in this manner, because any overlapping cross-peaks are suppressed by the symmetry filter.) Figure 10.4 illustrates the dramatic improvement in the appearance of an experimental COSY cross-peak achieved by such a symmetrization routine. However, there is no genuine improvement in sensitivity. There are two problems—noise that is riding on a true signal is not reduced at all, whereas signals that are hidden in the baseplane noise are suppressed. The improvement is merely cosmetic.

## 10.3.1 **Maximum Entropy**

Post-acquisition data processing is used in many situations in experimental science—for example radioastronomy or photography. In many cases the experimental data are in some way defective. In radioastronomy, part of the antenna array may not be functioning; in photography the object may be moving while the shutter is open. Maximum entropy reconstruction[20,21] offers a method for suppressing some of these undesirable artifacts. Consider the application to NMR spectroscopy, given a noisy free induction decay as the raw data. Instead of performing the usual Fourier transformation, we could employ an inverse procedure. We might postulate the existence of a small family of different NMR spectra, each having the property that its inverse Fourier transform fits the experimental free induction decay within experimental error (the noise). There would appear to be no reason to prefer one such spectrum over the others, and the safest choice would be to select the spectrum with the lowest information content, the one least likely to contain features for which there is no firm experimental evidence. To the extent that entropy represents the negative of the information content, this is the maximum entropy solution (§ 6.5.1).

We see at once the power of the method. If the experimental data set contains an instrumental artifact and if we have some knowledge about its source, then the maximum entropy method can be used to suppress that artifact. Take the example of a time-domain NMR signal that has been prematurely truncated, as often happens in two-dimensional spectroscopy. Straightforward Fourier transformation would introduce sinc function artifacts into the spectrum; apodization severe enough to smooth the step function would seriously compromise the resolution. In the maximum entropy treatment, the trial time domain trial signal can be truncated so that it can be directly compared with the similarly truncated experimental free-

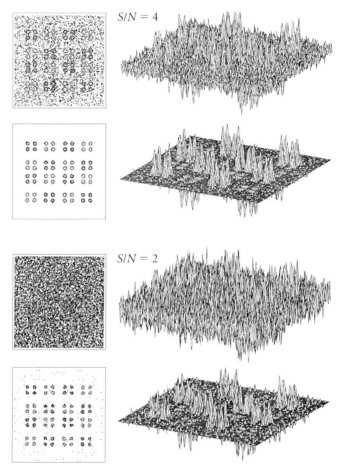

**Figure 10.4** Suppression of baseplane noise on a COSY cross-peak, starting with a signal-to-noise ratios of 2 and 4. The processing works on the assumption that the intensities of genuine signals must be antisymmetric with respect to vertical and horizontal axes passing through the chemical shift coordimates. Intensities at corresponding pairs of locations are compared; if the signs are opposite, these intensities are retained, if not, they are replaced by zeroes. The apparent improvement (between upper and lower spectra) is merely cosmetic. There is no true sensitivity enhancement.

induction decay (§ 6.5.1). Consequently the spectrum predicted by the maximum entropy treatment shows no sinc function artifacts.

We can see how an analogous procedure might be used to improve resolution, by imposing a sensitivity enhancement weighting function onto the trial free-induction decay before comparison with the experimental signal.[22] Note that the crux of the method is the inverse nature of the processing, not any intrinsic virtue of the maximum entropy method itself. We must be careful not to ascribe to this technique magical properties to deal with any undesirable features of the spectrum.

Noise cannot be regarded as an instrumental artifact in the same category as those described above. Maximum entropy is a nonlinear operation, and although it can indeed improve the signal-to-noise ratio, it does so only in circumstances where the signal is already emerging from the noise, a situation where the cruder nonlinear operations would achieve the same end. We could have reached this conclusion from the very definition of the maximum entropy method—that it specifically excludes all features for which there is no firm evidence in the experimental data. The only cases where we really need to improve the signal-to-noise ratio are those where the signal amplitude is below the noise, and the maximum entropy method can be of no use here. Examples where improvements in sensitivity appear to have occurred are often cases where the experimental NMR signal has not been properly smoothed beforehand; the maximum entropy method is then acting as an unnecessarily complicated filter.

The lesson to be learned is that nonlinear processing can present serious dangers, and that not all improvements in signal-to-noise ratio are actually productive. We must accept noise as an inevitable and even useful part of a typical spectrum. It is useful in the sense that it serves as a measure of the reliability of NMR intensities, provided that all amplification and processing methods are linear. When baseline noise is 10% of the signal height, then the intensity carries an intrinsic error of 10%, whether we like it or not. By contrast, artificial reduction of the baseline noise by nonlinear data processing could mislead us into believing that the signal intensities are more reliable than they actually are. We should not strive to minimize the noise when the signal-to-noise ratio is already adequate for our purposes, this would only be a pointless cosmetic exercise. In situations where the signal-to-noise ratio is already good, smoothing should also be used sparingly, since it begins to broaden the lines near the matched filter condition.

## 10.4  Stochastic excitation

Noise may be used as a method for exciting nuclear spins. The natural reaction of most spectroscopists is to be horrified at the idea of deliberately introducing systematic noise into a process that is already beset with severe problems of thermal noise, but we shall see below that there is a simple trick that allows us to remove the systematic noise at the end of the experiment. Stochastic excitation is one of the three principal methods for interrogating the spin system, although it has received far less attention than slow-passage continuous-wave methods and pulse excitation.

Imagine you are an automobile engineer concerned with eliminating undesirable oscillations (squeaks, rattles etc.) from a particular structural design. You might test the prototype by deliberately shaking the model sinusoidally, slowly increasing the frequency to see where any resonances might occur. This is the equivalent of the slow-passage continuous-wave method in NMR, the method of choice in the early NMR spectrometers. Alternatively, you might subject the structure to a single sharp impulse and study the transient response that followed. This would correspond to the well-known pulse-Fourier transform technique that most modern spectrometers use. The essential equivalence of these frequency-domain and time-domain experiments has been widely discussed. There is, however, a third approach. If the car is driven fast over a very rough road the squeaks and rattles soon become apparent; we might say that this is equivalent to exposing the structure to random noise—stochastic excitation.[23–28]

Theoretically the ideal would be to excite with pure "white" noise, where the intensity is uniform as a function of frequency, at least over the entire NMR frequency band of interest. When applied to the system (the nuclear spins) this white noise acquires a "colour" that is a property of the system itself and which eventually will tell us about the high resolution NMR spectrum. But what about the danger of introducing systematic noise into the spectrometer? The key is that we always know the exact form of the noise excitation. For convenience this is usually pseudorandom noise—a sequence of radiofrequency pulses modulated according to a specific code which approximates a random process but which is in fact deterministic. We might liken the process to the operation of a scrambler telephone. Information to be transmitted is chopped up in what appears to be an unpredictable noisy manner so as to be incomprehensible to any eavesdroppers (random number generation is in fact the basis of much of the science of cryptology). Upon reception, the message is unscrambled by a device that employs an exact knowledge of the original scrambling code, deconvoluting the scrambling function from the output signal. If this procedure can be carried out efficiently, then the systematic noise introduced by the stochastic excitation can be reduced to a level well below that of the inevitable thermal noise.

Stochastic excitation is not therefore an exercise in masochism. Indeed it has some interesting advantages over the other two techniques, combining some of the advantages of continuous-wave excitation (low transmitter power) with advantages of pulsed NMR (rapid data acquisition). The simplest implementation would be a regular sequence of weak radiofrequency pulses chosen to have 0° or 180° phase according to a pseudorandom code. The system may be driven in a regime where the response is nonlinear. After the initial transient response has had time to settle down into a steady state, signal acquisition takes place (once) in each interval between pulses. The main trick in generating a suitable pseudo-random sequence by electronic methods is to avoid repeating the pattern, or at least to delay repetition for as long a time as possible. Usually a shift register is used for this purpose, one with a sufficiently large number of stages that it accomplishes many many steps before it repeats itself.

The detected signal is thus a very long string of apparently random ordinates, but it actually contains NMR information in addition to the systematic noise. Separation involves the process of cross-correlating this signal with the excitation function. It is not merely a question of removal of some pseudorandom modulation imposed upon the desired NMR time-domain signal, and in this sense the scrambler telephone analogy could be misleading. At any given point in time, the NMR signal represents the sum of the responses to all previous radiofrequency pulses. Naturally the spins 'forget' about the effect of the very early pulses but are progressively more sensitive to pulses closer in time, reflecting the spin–spin relaxation time $T_2$ and (in the absence of echo effects) the instrumental parameter $T_2^*$.

Suppose, for simplicity, that the excitation is weak and that only the first-order response is appreciable. It may be represented by $y(t)$, which is the convolution of the stochastic excitation function $x(t)$ with the system impulse response $k_1(\tau)$:

$$y(t) = \int_0^\infty k_1(\tau)\, x(t-\tau)\, \mathrm{d}\tau \qquad\qquad \textbf{10.2}$$

The spins retain a phase memory of all the pulses that occurred in a period of the order of $T_2^*$. Consequently $k_1(\tau)$ is equivalent to the well-known free induction decay. Cross-correlation is the process of deconvolution of $x(t)$ from $y(t)$ to give the linear impulse response $k_1(\tau)$. In practice this is computed as a summation (rather than an integral) and it continues to a finite limit $t_m$, rather than to infinity. These restrictions can adversely affect the efficiency with which the systematic noise is removed, but the situation is greatly improved by oversampling the experimental data using a much finer time grid than that imposed by the Nyquist sampling criterion. Figure 10.5 shows how systematic noise in the stochatically-excited proton spectrum of ethanol is progressively reduced as a function of the degree of oversampling of the data.[28] When the sampling rate is increased by a factor 16, the signal-to-noise improves approximately fourfold. The thermal noise is essentially negligible in comparison with the systematic noise in this situation.

As NMR spectrometers are designed to operate at higher and higher magnetic fields, the chemical shift ranges expand accordingly and the excitation scheme needs to cover a broader frequency band. With the widely used pulse method, limits will eventually be reached because of radiofrequency voltage breakdown in the probe and the practical power limitations of the transmitter amplifier. Although these

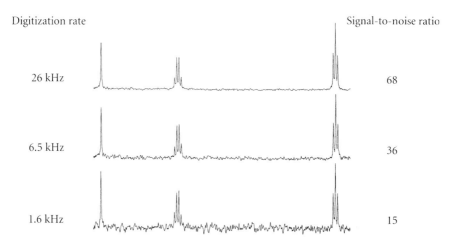

Digitization rate                                                                                    Signal-to-noise ratio

26 kHz                                                                                                           68

6.5 kHz                                                                                                          36

1.6 kHz                                                                                                          15

**Figure 10.5** Oversampling used to reduce the systematic noise introduced by stochastic excitation. Spectra of ethanol obtained with different digitization rates. The signal-to-noise ratio increases approximately as the square-root of the digitization rate.

problems may be alleviated to some degree by the use of composite pulses, it would be desirable to be able to combine the low-power excitation implicit in the continuous-wave method with the multiplex advantage enjoyed by the more usual pulse-Fourier transform experiment. Stochastic excitation enjoys both these advantages.

Stochastic excitation would normally operate in a regime where the NMR signal is partly saturated and where multiple-quantum and other nonlinear processes are appreciable. This introduces some exciting new possibilities, which have not yet been realized. For example it should be feasible to detect coherence transfer processes equivalent to the well-known two-dimensional correlation spectroscopy (COSY). This could conceivably offer a fresh approach to the methodology of NMR spectroscopy. One could imagine a complete separation of the act of data gathering from the spin engineering practised by modern NMR spectroscopists to extract various kinds of structural information. Instead of returning to the spectrometer to perform additional experiments to clarify some structural point, the spectroscopist could obtain all this additional information from a single stochastic excitation experiment. The different types of information would be extracted by processing the experimental data in appropriate fashions, off-line from the spectrometer. This would divorce the act of data gathering from that of data manipulation. An organic chemist could then have second thoughts about a particular molecular structure problem months after the actual NMR experiment had been performed, return to the stored data and extract (say) correlation information without any further recourse to the sample or to the spectrometer.

## 10.5  Noise decoupling

The broadband nature of stochastic excitation was recognized at an early stage and exploited for the purpose of decoupling, the removal of C–H splittings in $^{13}$C spectroscopy.[29] The basic problem is that all proton resonances must be affected essentially equally, irrespective of their chemical shifts. Decoupling not only simplifies carbon spectra but also improves sensitivity, partly through the collapse of the multiplet structure, and partly through the nuclear Overhauser effect. Noise decoupling is only effective if a sufficiently high decoupler power is employed. At intermediate powers some of the pseudorandom noise applied at the proton frequency is transmitted through the scalar coupling to the $^{13}$C spins and contributes to the general noise level. Carbon-13 spin echoes are particularly sensitive to interference attributable to noise transferred from the proton decoupler through the Hartmann–Hahn effect.[30]

For many years noise decoupling was used almost universally, until new deterministic pulse sequences were discovered (§ 7.2.3) that make a more efficient use of the decoupler power and minimize sample heating. In this case there is a certain satisfaction in replacing a pseudorandom process with a calculated pulse sequence, well-grounded on theoretical principles.

## 10.6 **Relaxation**

One of the most useful stochastic processes for high resolution NMR spectroscopy is the random reorientational motion of molecules in a liquid that gives rise to relaxation. The noisy motion of surrounding magnetic species provides this much-needed coupling to the environment. If we were to imagine a world where these interactions were reduced to the point that relaxation times were (say) two orders of magnitude longer, the spectroscopist's life would be impossibly difficult. Indeed the first attempts to discover magnetic resonance in bulk matter were thwarted by a most unfortunate choice of a sample that had an abnormally long spin–lattice relaxation time.[31]

How does one describe this reorientational motion? If we think of it as noise, it must be band-limited, since the molecules in a liquid are constrained by inertial and viscosity effects that entail a cut-off frequency for molecular rotation. It is usual to describe this 'Fourier spectrum of the molecular motion' in terms of a spectral density function, a graph representing the number of molecules rotating at a given frequency, versus frequency (§ 9.1.2). For want of a detailed knowledge of the true form of this function it is often assumed to be Lorentzian, corresponding to a single reorientational correlation time $\tau_c$. For small molecules in relatively nonviscous liquids, $1/\tau_c$ is very much faster than the nuclear Larmor precession frequency. It is the component of this noise at the Larmor frequency that gives rise to spin–lattice relaxation.

These fluctuating (radiofrequency) fields from surrounding magnetic species cause resonance transitions in the spin under investigation. But in contrast to a radiofrequency field applied from an external transmitter, they induce net downward transitions, bringing spin populations to Boltzmann equilibrium. The difference is that the radiation from the radiofrequency transmitter is hot, whereas the (noisy) radiation from the magnetic environment is at the sample temperature and acts to bring the spin system into thermal equilibrium with its surroundings. Here is one noisy process for which we should be profoundly thankful.

Relaxation has much in common with slow chemical exchange. In the latter, a molecule flips into a new conformation or an atom jumps to another molecule in a randomly timed event. Analogous experimental techniques can be used to study cross-relaxation and slow chemical exchange by labelling the mobile spins with a population disturbance, usually a population inversion. Both phenomena may be displayed as cross-peaks in two-dimensional spectra (NOESY and EXSY).

## 10.7 **Spin noise**

We are taught that spontaneous emission is entirely negligible for nuclear magnetic dipole transitions since the effect is proportional to $\nu^3\mu^2$, where the emitted frequency $\nu$ is very low and the magnetic moment $\mu$ is very small. For a Larmor frequency of 600 MHz, the rate of spontaneous emission is roughly $10^{24}$ times smaller than that for an electric dipole transition in the visible region of the spectrum. However, the nuclear magnetic dipoles in question are not in free space but confined inside a tuned circuit of high $Q$ factor and this can enhance the effect. Even so, detection seems quite impractical even for (say) $10^{22}$ spins.

Undeterred by these very unfavourable odds, Hahn and co-workers[32] set out to observe this effect in the nuclear quadrupole resonance of $^{35}Cl$ in $NaClO_3$ and $KClO_3$ at liquid helium temperatures. His trump card was a very low noise device—a superconducting quantum interference detector (SQUID). The spins are first saturated so there is no nuclear polarization. The spectral density of the noise is measured as a function of frequency, showing a broad response with a maximum at the resonance frequency of the tuned circuit, with a superimposed weak bump at the nuclear Larmor frequency. The bump is caused by nuclear spin noise, tiny random fluctuations in the transverse component of the nuclear magnetization. This spontaneous emission of radiation is incoherent, a most unusual situation in magnetic resonance. In this application spin noise can only be regarded as a useful property for it betrays the presence of nuclear quadrupole resonance without the need for an excitation source.

## 10.8 **Discussion**

We may conclude that noise is an inevitable feature of magnetic resonance experiments and that we should be very wary of cosmetic procedures aimed to make it go away. The intensity of the NMR signal provides us with certain quantitative information; the associated noise indicates the reliability of that information. True signals can only be distinguished from noise by their reproducibility (time averaging) or their frequency response (smoothing). Be very wary of magical sensitivity enhancement schemes. If nonlinear processing is employed, signals are enhanced at the expense of noise only in situations where the signal-to-noise ratio is already adequate; they do not improve the sensitivity of the experiment at all. Finally it should be recognized that noise can often be a useful commodity, for example as a method of wideband excitation or decoupling of nuclear spins.

## **References**

1. P. Styles, N. F. Soffe, C. A. Scott, D. A. Cragg, F. Row, D. J. White and P. C. J. White, *J. Magn. Reson.* **60**, 397 (1984).
2. W.A. Anderson, W. W. Brey, A. L. Brooke, B. Cole, K. A. Delin, J. F. Fuks, H. D. W. Hill, M. E. Johanson, V. Y. Kotsubo, R. Nast, R. S. Withers and W. H. Wong, *Bull. Magn. Reson.* **17**, 98 (1995).
3. J. M. Nuzillard and R. Freeman, *J. Magn. Reson. A* **110**, 252 (1994).
4. K. Nagayama, P. Bachman, K. Wüthrich and R. R. Ernst, *J. Magn. Reson.* **31**, 133 (1978).
5. A. F. Mehlkopf, D. Korbee, T. A. Tiggleman and R. Freeman, *J. Magn. Reson.* **58**, 315 (1984).
6. R. Hurd, *J. Magn. Reson.* **87**, 472 (1990).
7. P. Xu, X. L. Wu and R. Freeman, *J. Magn. Reson.* **95**, 132 (1991).
8. J. Hadamard, *Bull. Sci. Math.* **17**, 240 (1893).

9. R. J. Ordidge, A. Connelly and J. Lohman, *J. Magn. Reson.* **66**, 285 (1986).

10. L. Bollinger and J. S. Leigh, *J. Magn. Reson.* **80**, 162 (1988).

11. H. R. Bircher, C. Müller and P. Bigler, *J. Magn. Reson.* **89**, 146 (1990).

12. V. Blechta and R. Freeman, *Chem. Phys. Lett.* **215**, 341 (1993).

13. V. Blechta, F. del Rio-Portilla and R. Freeman, *Magn. Reson. Chem.* **32**, 134 (1994).

14. J. M. Wouters and G. A. Petersson, *J. Magn. Reson.* **28**, 81 (1977).

15. J. M. Wouters, G. A. Petersson, W. C. Agosta, F. H. Field, W. A. Gibbons, H. Wyssbrod and D. J. Cowburn, *J. Magn. Reson.* **28**, 92 (1977).

16. G. A. Morris, *J. Magn. Reson.* **80**, 547 (1988).

17. G. A. Morris and D. Cowburn, *Magn. Reson. Chem.* **27**, 1085 (1989).

18. A. Gibbs, G. A. Morris, A. G. Swanson and D. Cowburn. *J. Magn. Reson. A* **101**, 351 (1993).

19. R. R. Ernst, R. Freeman, B. Gestblom and T. R. Lusebrink, *Mol. Phys.* **13**, 283 (1967).

20. C. E. Shannon, *Bell Syst. Tech. J.* **27**, 623 (1948).

21. P. J. Hore, *Maximum Entropy in Action,* Eds: B. Buck and V. A. Macaulay, Oxford University Press, 1993.

22. S. J. Davies, C. J. Bauer, P. J. Hore and R. Freeman, *J. Magn. Reson.* **76**, 476 (1988).

23. R. R. Ernst, *J. Magn. Reson.* **3**, 10 (1970).

24. R. Kaiser, *J. Magn. Reson.* **3**, 28 (1970).

25. B. Blümich and D. Ziessow, *J. Chem. Phys.* **78**, 1059 (1983).

26. B. Blümich and R. Kaiser, *J. Magn. Reson.* **54**, 486 (1983).

27. B. Blümich, *Prog. NMR Spectrosc.,* **19**, 331 (1987).

28. J. Paff, R. Freeman and B. Blümich, *J. Magn. Reson. A* **102**, 332 (1993).

29. R. R. Ernst, *J. Chem. Phys.* **45**, 3845 (1966).

30. R. Freeman and H. D. W. Hill, *J. Chem. Phys.* **54**, 3367 (1971).

31. C. J. Gorter and L. J. F. Broer, *Physica* **9**, 591 (1942).

32. T. Sleator, E. L. Hahn, C. Hilbert and J. Clarke, *Phys. Rev. B* **36**, 1969 (1987).

# 11

## Water

## 11.1 Water, water, everywhere

In spite of the amazing abundance of chemical information offered by NMR, it might still be argued the most important sample is water, since the vast majority of medical applications focus on the proton signal from water in the human body. Water gives one of the strongest signals in the liquid phase, being 110 molar in protons. Historically, water seemed the natural choice to the Stanford group for their first NMR experiments (rather surprisingly, the Harvard group decided to go out and buy a sample of paraffin wax from the local hardware store). In the days when scientists built their own NMR spectrometers, water was the obvious choice for locating an NMR response for the first time. Even today, if something has gone seriously wrong with a high resolution spectrometer and the resonance condition has been lost, any fresh search for the correct field usually focuses on the proton signal from water, or on the deuterium signal from heavy water.

Today water has become the villain of the piece for the chemist. It is ubiquitous in biological experiments, it has a very intense signal that taxes the dynamic range of the analog-to-digital converter and the linearity of the receiver, it is broadened by radiation damping and temperature gradients, and it readily undergoes chemical exchange. Since it is the most intense signal in many two-dimensional spectra, it causes the most trouble with baseline distortion and $t_1$ noise (§ 10.1.3) and it can completely obscure weak signals lying under its skirts. Substitution of heavy water may not be a practical option in many cases because of slow chemical exchange with NH groups that could be of interest. However, most other solvents with high proton densities are usually used in their deuterated form, leaving water as the principal offender.

Sensitivity is a prime concern for NMR spectroscopy of biochemical molecules. One of the most promising recent schemes for sensitivity enhancement is cryogenic cooling for the receiver coil by liquid helium (§ 10.1.1). This greatly increases the quality factor $Q$ of the coil and reduces the noise arising from the random motion of electrons in the wire (Johnson noise). Both effects improve sensitivity quite markedly. Unfortunately this technique is ineffective for aqueous solutions, particularly if ions are present, because the dielectric losses adversely affect the $Q$ factor.

## 11.2  Water suppression

The most serious of the practical problems of aqueous solutions arises from the limited dynamic range of the analog-to-digital converter. In Fourier transform NMR it is not permissible to clip the intense water signal, and consequently the receiver gain must be adjusted so that the maximum voltage excursion in the free induction decay just triggers the most-significant bit of the analog-to-digital converter. At the other end of the range, the weak solute signals have amplitudes which may only just trigger the least significant bit of the analog-to-digital converter, and as a consequence, they suffer severe digitization errors (§ 10.1.2). We may imagine decomposing the experimental free induction decay into a part that contains only true signals and thermal noise, and another part that contains only the digitization errors. The Fourier transform of the latter is the spectrum of the digitization noise, which often exceeds the thermal noise and sets a limit on the sensitivity of the method. (Note that slow-passage continuous-wave NMR does not have this problem; the intense water signal may be safely truncated.)

The magnitude of this problem is reflected in the vast and still growing literature[1] on solvent suppression techniques. This should warn us that no single solvent suppression procedure has yet been generally accepted. Note that we are concerned here with schemes to suppress the water signal before analog-to-digital conversion, in order to avoid digitization noise and spurious harmonics; cosmetic removal of water signals from the final spectrum is quite another question. The spectroscopist has two options—termination or isolation.

### 11.2.1  Killing the water peak

One simple expedient for suppressing the water signal is to set the transmitter frequency on the water resonance so that its signal is carried at frequencies very close to zero, and then introduce a low-frequency rejection filter into the detector, just before analog-to-digital conversion. Since the water resonance is not too far from the centre of many proton spectra, the necessary displacement of the transmitter frequency is small and so is any concomitant increase in the spectral width. Naturally, signals close to the water resonance are lost or severely attenuated, but the remainder of the spectrum is recorded with undistorted intensities.

Presaturation with a long weak pulse or repeated train of hard pulses is a widely used method for suppressing the water signal.[2] It is simple and reasonably effective but has the disadvantages that it can bleach nearby proton resonances and can also affect other signals through cross-relaxation or chemical exchange.[3] The saturating field is usually supplied by the proton decoupler, and is switched off during signal acquisition to avoid Bloch–Siegert shifts.[4] The duration and intensity of the saturation field are chosen as a compromise between the requirements of speed and selectivity. Part of the suppression of the water signal comes about because of the divergence of isochromats in the spatially inhomogeneous saturating field. It can therefore be advantageous to introduce a phase shift at some point in the presaturation pulse, so that this divergence is spread over all directions over a sphere, rather than in just one plane. Since the water signal is so intense, appreciable contributions arise from unexpected regions of the sample which may be in very inhomogeneous parts of the $B_0$ field. For example, a small part of the total signal may be excited by the stray radiofrequency field from the leads to the receiver coil.[5] The main advantages of presaturation are its simplicity and versatility; it is the favoured method both in conventional and two-dimensional spectroscopy. However, it is difficult to use in experiments that employ polarization transfer from protons to a heteronuclear species (§ 4.2.6).

We may also take advantage of differences in relaxation rates. The protons of macromolecules usually have significantly shorter spin–lattice relaxation times than water. Consequently, a population inversion followed by a suitable delay can catch the water signal at the null condition when the read pulse is applied. Since the solute protons relax much faster, they are well on the way towards thermal equilibrium at that time of the read pulse, and give a strong signal. Theoretically the null occurs at a time $T_1 \ln 2$, where $T_1$ is the spin–lattice relaxation time of the water, but in practice the delay is adjusted by trial and error since the inversion pulses are imperfect, and radiation damping[6] also affects the recovery rate (§ 11.3). This water-eliminated Fourier transform (WEFT) technique[7] offers the possibility of detecting signals very close to the water resonance, but it provides only limited suppression, and it inevitably introduces intensity distortions in the spectrum. To minimize these distortions a selective inversion pulse seems to be preferable to broadband inversion.[8]

Differences in spin–spin relaxation may also be exploited.[9] Addition of a suitable relaxation reagent can affect water preferentially, thus greatly attenuating the water contribution to a spin echo.[10] The Carr–Purcell–Meiboom–Gill technique (§ 4.1.6) is used with a relatively high pulse repetition rate to avoid $J$ modulation of the echo. Another way to reduce the water spin–spin relaxation time is by addition of a high concentration (typically 0.5 M) of ammonium chloride, which has exchangeable protons.[11] These methods are not applicable to solutes with short spin–spin relaxation times.

As described in § 6.2, pulsed field gradient methods can be very useful for solvent suppression.[12] Many modern spectrometers have a facility for applying intense $B_0$ gradients along all three axes. The water isochromats can be dispersed very effectively, along with the associated $t_1$ noise that can otherwise wreak havoc in a two-dimensional spectrum. Advantage may be taken of the fact that water diffuses more rapidly than the (larger) solute molecules under investigation; pulsed field gradients can then be used to discriminate against the water signal, because the phase divergence caused by molecular diffusion cannot be refocused.[12] Double-quantum filtration (§ 6.3.1) employing matched field gradients acts in a similar manner, discrimination against the singlet water resonance.

Another pulsed-gradient technique, called WATERGATE,[13] employs a hard $\pi$ pulse at the same time as a selective $\pi$ pulse for the water resonance, sandwiched between two equal field gradient pulses of the same polarity.

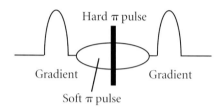

All the coherences that are dispersed by the first gradient pulse are refocused by the second, except for the water signal which experiences no net refocusing pulse. This gradient-recalled spin-echo technique is now one of the most popular schemes for water suppression because if offers a high suppression ratio and can be used as a module that may be inserted into existing pulse sequences. However, because it detects a spin echo, the recorded signals may be distorted by $J$ modulation and attenuated by spin–spin relaxation.

## 11.2.2 Selective excitation

The second category of water suppression techniques avoids excitation of the water signal or, alternatively, returns the water signal to the $+z$ axis after some specific manipulation. Redfield's 214 composite pulse[14] was designed to have a relatively broad (zero- and first-order) null at the water resonance frequency. It may be expressed as

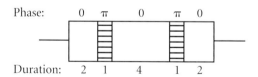

Although it induces frequency-dependent phase shifts in the spectrum, these can be corrected with the appropriate computer program. Various combinations of hard and soft excitation pulses can also be devised to excite the main spectrum but take the water magnetization back to the $+z$ axis. The effect of a self-calibrating soft pulse can even be engineered by augmenting the radiation damping signal from the water resonance (§ 11.3.2).

Broadband excitation with a rejection notch at the water frequency appears to offer another attractive approach, since it is designed to retain the correct intensities in the remainder of the spectrum. Tailored excitation[15] has been suggested for this purpose; it starts by defining the requisite frequency-domain excitation profile and then derives a pulse sequence to achieve it; unfortunately it relies on a Fourier transform relationship between the time-domain modulation and the frequency-domain excitation profile, and this is only approximately valid.

## 11.2.3 Polychromatic pulses

Polychromatic pulses[16] (§ 5.5.1) offer the possibility of uniform excitation across the entire proton spectrum, except for a rejection notch at the water frequency. Several soft pulse elements are arrayed in a uniform comb of frequencies across the spectral width, and their relative intensities are suitably adjusted. This has the advantage that the frequency domain excitation profile may be tailored by hand, the width and shape of the rejection notch being controlled to suit the application, either for a high degree of suppression, or to favour the detection of signals close to the water resonance. The time-domain envelope of the soft pulse is defined as a histogram made up of many narrow segments. The frequency of a particular element is offset from the transmitter frequency by incrementing the radiofrequency phase from one segment to the next. A large number of soft pulses of slightly different frequencies can be combined by vector addition at each segment. These experiments employ a digitally controlled waveform generator or interleaved DANTE sequences (§ 5.3.5).

An example of a polychromatic pulse[16] specifically constructed for water suppression is PC(19):

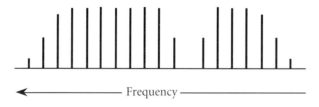

Frequency

It has a flip angle of 45° to ensure an approximately linear response. It has basically 19 frequency elements, except that one element has zero intensity and creates an off-centre notch. The gradual attenuation of the elements at all four edges improves the uniformity of the response in the two excitation bands by avoiding sharp discontinuities. An experimental test of the frequency-domain excitation pattern is mapped out in Figure 11.1. The transmitter frequency was moved in steps of 25 Hz through the range of proton frequencies, recording the signal from residual HDO in a heavy water sample. There are two reasonably flat regions (900 Hz and 500 Hz wide) flanking a rejection notch of approximately 100 Hz at half height.

Rather better performance (with a more linear phase gradient) is achieved by incorporating more soft pulse elements. This provides additional flexibility for

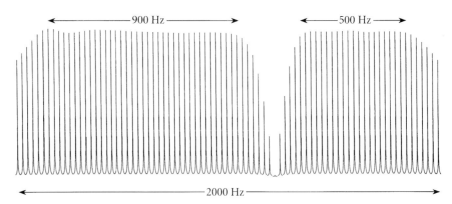

**Figure 11.1** Experimental examination of the frequency domain profile of a polychromatic pulse PC(19) designed to have a uniform excitation except for a notch at the water frequency. A weak proton signal (absolute-value mode) from the residual HDO in heavy water was moved in 25 Hz steps across a 2000 Hz region. The width of the notch at half height is approximately 100 Hz.

shaping the rejection notch. One such pulse, PC(54) was used to study the 400 MHz proton spectrum of vancomycin in DMSO-$d_6$ (Figure 11.2). Conventional pulse excitation gives an intense water peak at 3.5 ppm and a weak water harmonic at 8.5 ppm, attributed to the effect of receiver nonlinearity. In addition, there is an appreciable level of digitization noise arising from the analog-to-digital converter. After

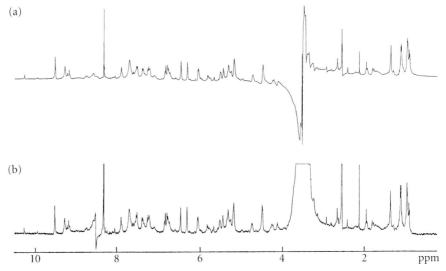

**Figure 11.2** Water suppression by introduction of a rejection notch into an otherwise uniform excitation scheme.[16] The polychromatic pulse PC(54) was designed to have a notch at 3.5 ppm. The water signal in the conventional spectrum of vancomycin (b) has been truncated at 0.22% of its peak height. Note the high degree of suppression of the water signal in (a) together with its weak harmonic at 8.3 ppm. The digitization noise is significantly reduced.

excitation with the PC(54) pulse, the water signal and the much weaker harmonic are effectively suppressed, and the digitization noise is eliminated (Figure 11.2(a)). More importantly, signals outside the rejection notch are of uniform intensity and in pure absorption, something that is very difficult to achieve with many other solvent suppression schemes.

### 11.2.4 Binomial pulses

Binomial pulses belong to the category of schemes that return the water magnetization to the $z$ axis. The germ of the idea can be traced back to the jump-and-return experiment of Plateau and Guéron.[17] With the transmitter set at the water frequency, they apply a hard pulse sequence,

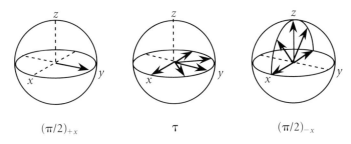

$$(\pi/2)_{+x} \qquad\qquad \tau \qquad\qquad (\pi/2)_{-x}$$

The water signal undergoes no precession during $\tau$ and is carried back to the $z$ axis by the second pulse, but other signals in the spectrum precess to some degree and therefore retain an $x$ component after the second pulse. Near resonance the signal amplitudes increase with offset from the transmitter frequency, and change sign according to whether they are up-field or down-field of the water signal. Unlike some other schemes for water suppression, the jump-and-return method does not introduce frequency-dependent phase shifts into the spectrum. It is simple and effective, and can be incorporated into two-dimensional experiments. When it forms part of a phase cycle in a two-dimensional experiment, care must be taken to eliminate radiation damping effects (§ 11.3.2) which vary from one stage of the phase cycle to the next. For reasons which will soon become clear, we now describe the jump-and-return scheme as the 1:$\bar{1}$ sequence.

The principal drawback of the 1:$\bar{1}$ sequence arises from the width of the water signal, which is broadened by $B_0$ inhomogeneity and possibly by thermal gradients acting on the chemical shift. The excitation profile of the 1:$\bar{1}$ sequence is rather too sharp to remove the tails of the water signal efficiently. An improvement in the width of the null condition is obtained with the 1:$\bar{2}$:1 hard pulse sequence,[18] and it was soon realized[19,20] that there is an entire family of binomial pulse sequences that can be put to work on the case. Probably the most useful is the 1:$\bar{3}$:3:$\bar{1}$ sequence, which has an even broader null and is also rather insensitive to spectrometer imperfections such as inhomogeneity in the radiofrequency field $B_1$, or poor pulse calibration.[21]

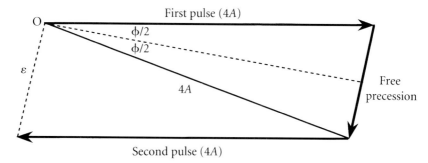

**Figure 11.3** Typical magnetization trajectory for the 1:$\bar{1}$ binomial sequence shown in projection on the *xy* plane for the case that all excursions are small with respect to the radius of the unit sphere. For the purpose of illustration, the excursions due to free precession have been exaggerated and their curvature is not shown. The origin (O) is the $+z$ axis. At exact resonance (no free precession) the terminus is also the origin, but at small offsets it is displaced by a small distance $\varepsilon$. Simple trigonometry gives the expression $\varepsilon = 8A\sin(\phi/2)$.

It is interesting to examine the shapes of the frequency-domain excitation in the vicinity of the null condition for the three different sequences 1:$\bar{1}$, 1:$\bar{2}$:1 and 1:$\bar{3}$:3:$\bar{1}$. This problem can be analyzed geometrically to a first-order approximation by assuming a very small offset such that the free precession angle $\phi$ between pulses is very small. If we also assume that the pulse flip angles are small, all magnetization vectors remain close to the $+z$ axis and it is permissible to draw their trajectories as a plane projection (parallel to the *xy* plane) rather than on the surface of the unit sphere. Maxima in the excitation profile occur at offsets $\pm n/(2\tau)$, where the water signal precesses through an odd multiple of 180° in the $\tau$ intervals and the radio-frequency pulses have a cumulative effect. The individual flip angles are therefore normalized to give the same overall flip angle at these offsets, corresponding to an arc of length 8A. Thus the radiofrequency pulses in the 1:$\bar{1}$ sequence have arcs of length 4A.

Figure 11.3 shows trajectories for the 1:$\bar{1}$ sequence. Since these effects are small, the vertical scale (free precession) has been grossly exaggerated in comparison with the horizontal scale (pulses) in all of these diagrams. The origin (O) is the $+z$ axis of the rotating frame, and after two pulses separated by a period of free precession, the trajectory returns within a distance $\varepsilon$ of the origin. This represents the residual signal amplitude at offsets close to the null condition:

$$\varepsilon = 8A\sin(\phi/2) \qquad\qquad \textbf{11.1}$$

Figure 11.4 demonstrates that the 1:$\bar{2}$:1 sequence, where the pulses have arcs of length 2A, 4A and 2A, leaves a much smaller displacement $\varepsilon'$, giving a residual signal amplitude that follows the expression

$$\varepsilon' = 4A(1 - \cos\phi) = 8A\sin^2(\phi/2) \qquad\qquad \textbf{11.2}$$

Note that a sine-squared curve corresponds to a broader null than a sine curve.

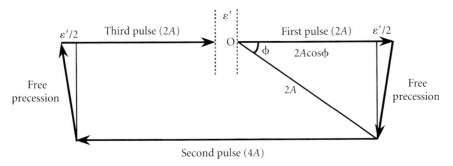

**Figure 11.4** Typical magnetization trajectory for the 1:$\bar{2}$:1 binomial sequence shown in projection on the *xy* plane and with the approximations used in Figure 11.3. The increased symmetry of the overall trajectory brings the terminus much closer to the origin (O). Simple trigonometry gives the expression $\varepsilon' = 8A\sin^2(\phi/2)$.

Figure 11.5 maps the trajectories for the 1:$\bar{3}$:3:$\bar{1}$ sequence, where the pulses have arcs of length $A$, $3A$, $3A$, and $A$. This path has a high degree of symmetry but does not quite return to the origin; the approximate displacement $\varepsilon''$ is given by

$$\varepsilon'' = 2A(3 - \cos\phi)\tan(\phi/2) - 2A\sin\phi$$
$$= 4A\{[1 + \sin^2(\phi/2)]\tan(\phi/2) - \sin(\phi/2)\cos(\phi/2)\}$$
$$= 8A\sin^3(\phi/2)/\cos(\phi/2) \qquad\qquad \textbf{11.3}$$

With the approximation $\cos(\phi/2) = 1$, which is justified in this limit, the deviation is

$$\varepsilon'' \approx 8A\sin^3(\phi/2) \qquad\qquad \textbf{11.4}$$

We see that in this approximation the responses of three types of binomial pulse sequence follow sine, sine-squared and sine-cubed curves in the vicinity of the null, as indeed would be predicted from Fourier transform considerations. These response curves are sketched out in Figure 11.6. The null condition may be shifted to approximately $\pm 1/(2\tau)$ by adopting the in-phase sequences 1:1, 1:2:1 and 1:3:3:1, which have their maximum excitation at the transmitter frequency.

The choice of which binomial sequence to use depends on many instrumental factors—sensitivity to $B_0$ and $B_1$ inhomogeneities and radiofrequency switching transients, ease of calibration, internal compensation for relaxation and radiation damping. The phase-alternating sequences tend to be superior for some of these features. We are also concerned to have a high suppression ratio, a broad null, a wide band of near-uniform excitation off-resonance, and a phase gradient that is not too severe and which can be compensated with the usual linear phase correction routine. Because the residual water signal has broad tails, a high degree of phase distortion is undesirable since after the application of a phase correction it gives rise to a rolling baseline. The NERO soft pulse sequences,[22,23] which are reminiscent of the binomial pulse schemes, give a much broader band of uniform excitation with only small ($\pm 10\%$) phase variations across this band.

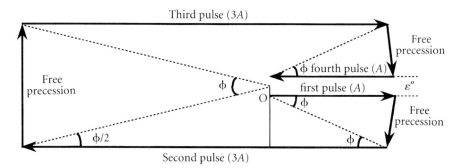

**Figure 11.5** Typical magnetization trajectory for the $1:\bar{3}:3:\bar{1}$ binomial sequence shown in projection on the $xy$ plane and with the approximations used in Figure 11.3. There is a high degree of compensation and the terminus is very close to the origin, being given by the expression $\varepsilon'' = 8A\sin^3(\phi/2)$.

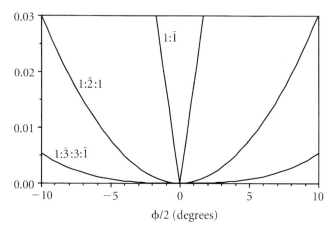

**Figure 11.6** Calculated excitation profiles near the null condition for the binomial pulse sequences assuming a linear NMR response and infinitely short pulses. The $1:\bar{1}$ (jump-and-return) sequence follows $|\sin(\phi/2)|$, the $1:\bar{2}:1$ sequence follows $\sin^2(\phi/2)$ and the $1:\bar{3}:3:\bar{1}$ sequence follows $|\sin^3(\phi/2)|$. The angle $\phi = 2\pi\Delta F\tau$ where $\Delta F$ is the offset from the transmitter frequency and $\tau$ is the interval between pulses. The sense of the signal is not indicated.

## 11.2.5 Residual water signals

Even after water suppression methods have been applied, there is often a quite intense residual water peak in the spectrum that can make it difficult to detect weak responses in the immediate vicinity. This is one of the reasons for preferring the double-quantum-filtered COSY experiment[24] over the primitive COSY technique, since it attenuates all singlet responses and makes it possible to detect signals close to the diagonal. The long tails from the residual water signal also cause baseline distortions that are particularly undesirable in two-dimensional spectra. An entire family of baseline correction routines has evolved to improve this situation; see, for example, the article by Waltho and Cavanagh.[25]

Perhaps the most serious legacy of the intense water signal is its contribution to $t_1$ noise, since this creates noise bands that run right across the two-dimensional spectrum, obscuring all in their path. Some improvement can be achieved for spectra such as COSY or NOESY, where there is an inherent symmetry about the principal diagonal. The data matrix may be processed by comparing pairs of symmetrically related ordinates, putting the lower intensity in both locations.[26] Although this is a nonlinear operation, it has little effect on true signals but strongly attenuates the $t_1$ noise, which lacks the appropriate symmetry. Reference deconvolution[27,28] which is described in more detail in § 10.2.4, offers another mode of attack on the water $t_1$ noise. By monitoring the effects of instability on the reference line, that is to say, by comparing its shape with that of an ideal line, these imperfections can be removed from all the lines of the two-dimensional spectrum. In particular, the spurious modulations that constitute $t_1$ noise are very effectively suppressed.

However, the most promising development of all is the gradual replacement of difference methods and phase cycles by single-scan methods employing matched field gradients.[29] If the water signal is eliminated in a single scan, it generates no $t_1$ noise. The effectiveness of the method is also enhanced by the relatively fast diffusion of water molecules in an applied field gradient, dispersing the isochromats in a manner that precludes refocusing. An example is presented in § 6.2.2 that demonstrates a remarkable degree of suppression of water $t_1$ noise.

## 11.3  Radiation damping

The intensity of the water response makes it the prime source of radiation damping.[6] This phenomenon (§ 2.4.1) is often overlooked by high resolution spectroscopists but it can have insidious effects because it interferes with free precession. Any nuclear magnetization that precesses in the $xy$ plane induces a small radiofrequency current in the receiver coil; this is the signal that is amplified in the receiver, detected, transformed and recorded. If the induced current is sufficiently intense, it generates a radiofrequency field $B_{rad}$ strong enough to influence the nuclear spins. The sense of this induced field $B_{rad}$ is opposite to that of the applied field $B_1$ and it therefore acts to drive the nuclear magnetization vector towards the $+z$ axis, causing a more rapid decay of the free induction signal than would be expected on the basis of the values of $T_2$ or $T_2^*$. Recently schemes have been devised to suppress the radiation damping field $B_{rad}$ by electronic feedback circuitry[30] or by rapid Q switching.[31]

### 11.3.1  Line broadening

The magnitude of the effect is usually quantified in terms of a radiation damping time constant:

$$\tau_{rad} = (2\pi\eta Q\gamma M)^{-1} \qquad\qquad \mathbf{11.5}$$

where $\eta$ is the coil filling factor (approximately unity), $Q$ is the coil quality factor (approximately 250) and $M$ is the transverse nuclear magnetization (with a maximum value of $M_0$). Only samples with very high spin densities (such as water) exhibit appreciable radiation damping (a short time constant $\tau_{rad}$), but note that the effect increases in high-field spectrometers since the equilibrium magnetization $M_0$ increases with the strength of the applied field $B_0$. The $Q$ factor reflects the enhancement of the induced NMR signal by the tuned resonant circuit. This enhancement may easily be reduced by detuning the receiver coil, thus moving to a lower point on the resonance curve of the tuned circuit. This is the simplest test for radiation damping (but it has a disastrous effect on the signal-to-noise ratio).

The most obvious symptom of radiation damping is a spurious broadening of an intense line in the spectrum, attributable to the more rapid decay of that component in the free induction signal. The effect falls off rapidly as a function of offset from the intense line since any nearby resonance would be acted on by an effective field:

$$B_{eff} = [\Delta B^2 + (B_{rad})^2]^{\frac{1}{2}} \qquad \textbf{11.6}$$

tilted away from the $xy$ plane through an angle $\theta$ given by

$$\tan\theta = \Delta B/B_{rad} \qquad \textbf{11.7}$$

The main effect on such a neighbour line is a slight phase shift towards dispersion, rather than a line broadening. This dephasing can be observed (by difference spectroscopy) on the long-range $^{13}C$ satellites of a proton signal if the parent ($^{12}C$) proton signal is very intense.[32]

## 11.3.2 Undesirable torques

An interesting situation arises when a $\pi$ pulse is applied to an intense resonance such as water.[33] In principle, if the spin inversion is exact ($M$ aligned along $-z$), there is no induced radiation damping field $B_{rad}$. In practice, this condition can be approached quite closely by employing an intense pulsed field gradient to purge any residual transverse components of magnetization. But as soon as any transverse signal is generated, however slight, it induces a radiation damping field that starts to rotate the intense magnetization vector away from the $-z$ axis in an escalating process where $B_{rad}$ increases rapidly with time. The consequent free-induction signal grows to a maximum and then decays, leading to a marked distortion of the lineshape after Fourier transformation. This interference by the radiation damping field clearly has serious consequences for spin–lattice relaxation measurements.[34]

The complex manipulations of spin systems that are now commonplace in high resolution NMR are particularly susceptible to perturbation by radiation damping if water is present. What should have been a period of *free* precession degenerates into an unexpected rotation about the tilted effective field defined by Equations 11.6 and 11.7. One example is the jump-and-return sequence[17] that is often incorporated into two-dimensional experiments to suppress the water signal:

$$(\pi/2)_{+x} - \tau - (\pi/2)_{-x} \; \text{acquire} \qquad \textbf{11.8}$$

During the $\tau$ period there is a strong transverse water signal with its attendant radiation damping field, and this can interfere with the suppression, since it tends to rotate the water magnetization vector towards the $+z$ axis, instead of leaving it in the $xy$ plane. When phase cycling is used (for example, for axial peak suppression), the situation becomes even more complicated because the severity of radiation damping varies between the different stages of the cycle.

One of the dangers inherent in manipulating the water signal in aqueous solutions of proteins is that magnetization can be transferred to other protons by chemical exchange. Often these are NH resonances that carry important structural information. Consequently, the popular technique of selective population inversion of the water signal must be used with some care. Electronic feedback to the receiver coil can be used to control radiation damping in this situation. Negative feedback diminishes the NMR signal picked up on the receiver coil, thus reducing the radiation damping effect, while positive feedback increases the induced NMR signal and enhances radiation damping. Difference spectroscopy incorporating both effects can be put to good use as a test for chemical exchange between water and other protons.[35] A soft $\pi$ population inversion pulse is applied to the water resonance, and two spectra are acquired. In the first, the radiation damping is eliminated by strong negative feedback, thus retaining the inverted water resonance and diminishing the signals of the exchangeable protons. In the second, radiation damping is enhanced by strong positive feedback, rapidly returning the water magnetization to the $+z$ axis where it has no effect on the signals of the exchangeable protons. In effect, the radiation damping signal, with positive feedback, acts as a self-calibrating soft pulse to return the water signal to the $z$ axis.

## 11.4 Nuclear susceptibility

When a sample is placed in an applied magnetic field, the internal field is slightly reduced (diamagnetism) or increased (paramagnetism) according to the bulk susceptibility of the material. This well-known electronic effect slightly changes the field experienced by the nuclei, but it is nowadays widely ignored since the Larmor frequencies are measured with respect to the internal reference standard (tetramethylsilane) and therefore small changes in the total applied field are largely irrelevant. The operating field is still determined by the choice of the radiofrequency used to excite the deuterium field/frequency lock, and the tiny diamagnetic field shift is compensated.

### 11.4.1 Frequency shifts

There is also a small contribution to the bulk susceptibility due to the nuclear polarization, as demonstrated by Lazarew and Schubnikow[36] for solid hydrogen at 2 K (the effect is greatly enhanced at low temperatures). The possibility that this effect might shift NMR frequencies was first considered by Dickinson.[37] At that time (1951), the magnetic fields of NMR spectrometers were much less intense, and

Dickinson concluded that the nuclear contribution to bulk susceptibility could safely be neglected. At present-day field strengths this is no longer the case; the nuclear susceptibility of materials with a high proton concentration (such as water) is large enough to shift the Larmor frequency by a measurable amount.[38] Unlike the electronic susceptibility, the nuclear susceptibility is not constant but changes in proportion to the $z$-component of nuclear magnetization. Strictly speaking, suscept-ibility is an equilibrium property; magnetization might be a better term to use in this context, but we follow the custom among NMR spectroscopists. Sometimes the term "demagnetization field" is employed but this could be misleading. Thus satur-ation or population inversion causes transient shifts of all the NMR frequencies, but they return to normal as spin–lattice relaxation restores Boltzmann equilibrium. For a sample that approximates to an infinite cylinder, the shift relative to the usual resonance frequency $f_0$ is given by:

$$\Delta f / f_0 = -\tfrac{1}{3}(1 - \cos\alpha)\,\chi_N \qquad\qquad \textbf{11.9}$$

where $\chi_N$ is the nuclear susceptibility, and $\alpha$ is the flip angle of the radiofrequency pulse. For saturation $\cos\alpha = 0$ while for a population inversion, $\cos\alpha = -1$.

Usually the shift is masked by the fact that the deuterium reference signal experi-ences an identical displacement of the magnetic field (due to the *proton* susceptibility) and the field/frequency regulator compensates accordingly, but several spectro-scopists have reported transient disturbances of the locking system when water is suddenly saturated. The nuclear magnetic susceptibility is a paramagnetic effect, increasing the internal magnetic field. Saturation or population inversion reduces this contribution, causing a shift to lower internal fields and lower Larmor frequencies.

The magnitude of the bulk nuclear susceptibility $\chi_N$ has been calculated by Edzes[38] to be

$$\chi_N = 1000\, C_A\, N_A\, \mu_0\, g_N^2 \beta_N^2 I(I+1)/3kT \qquad\qquad \textbf{11.10}$$

where $C_N$ is the concentration of spins in moles $dm^{-3}$, $N_A$ is Avogadro's number, $\mu_0$ is the conversion factor for SI units, $g_N$ is the nuclear Landé factor, $\beta_N$ is the nuclear magneton, $I$ is the nuclear spin quantum number, $k$ is the Boltzmann factor and $T$ is the absolute temperature. This reduces to:

$$\chi_N = +3.63 \times 10^{-11}\, C_N \qquad\qquad \textbf{11.11}$$

Water has $C_N = 110$ which suggests a maximum shift of $-0.0027$ ppm after a population inversion. At 750 MHz this is a shift of $-2$ Hz, which is readily detectable if the field:frequency regulator is disabled.

If pulse-excitation is used (rather than continuous saturation) there is also a dis-tortion[38] of the resonance lineshape attributable to the change in Larmor frequency as recovery of $z$ magnetization causes the nuclear susceptibility to vary during the free induction decay. Of course, unless special precautions are taken, this phenomenon is overshadowed by the effects of radiation damping.

Fortunately these bulk susceptibility shifts are usually masked by the action of the field regulator locked to the deuterium signal, since this experiences an identical shift in the field due to the proton susceptibility. Furthermore, the common solvents dimethyl sulphoxide, acetone and benzene (which would otherwise have high molar concentrations of protons) are used in the perdeuterated form and do not induce appreciable shifts. Water is the exception because in many situations chemical exchange with NH protons precludes the use of heavy water. Clearly, some care should be exercised in experiments where the water signal is saturated or inverted, particularly in very-high field NMR spectrometers.

## 11.4.2 Multiple Hahn echoes

The nuclear susceptibility phenomenon is also responsible for an unusual effect that occurs when Hahn spin echoes are excited in a sample of high proton density placed in a linear field gradient.[39–41] At first sight one would have expected only a single spin echo, but in practice a train of multiple echoes is observed, although only one refocusing pulse is employed. The pulse sequence may be written as

$$\pi/2 - \tau - \pi/2 - \tau - \text{echo } (E_1) \qquad \textbf{11.12}$$

Suppose that there is a linear field gradient $G$ along the $z$ axis. During the first $\tau$ period (Figure 11.7(b)) a typical isochromat precesses through an angle:

$$\phi = \gamma G z \tau \qquad \textbf{11.13}$$

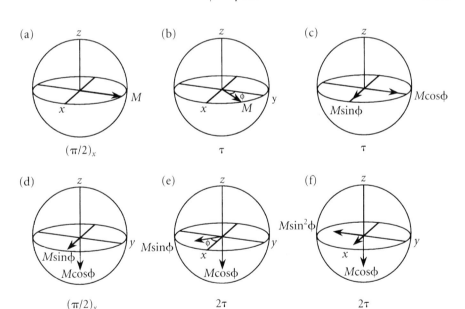

**Figure 11.7** Formation of the Hahn echo by consideration of the motion of a typical isochromat $M$, (a) initially aligned along $+y$ which (b) precesses through an angle $\phi$ during the $\tau$ interval. (c) This vector is then resolved along the $x$ and $y$ axes. (d) The second pulse $(\pi/2)_x$ rotates the $y$ component into the $-z$ axis. (f) Free precession of the transverse component for a further period $\tau$ generates an echo of amplitude $M\sin^2\phi$, but leaves the longitudinal magnetization $M\cos\phi$ unchanged. This causes the nuclear susceptibility shift. For a symmetrical distribution of isochromats the $x$ components vanish.

where $z$ is the distance from the point where the gradient is zero. Immediately after the second pulse the $y$ component of magnetization is converted into longitudinal magnetization given by:

$$M_z = -M\cos(\phi) \tag{11.14}$$

while a component $M\sin\phi$ remains along the $x$ axis (Figure 11.7(d)). Further precession of this transverse component for a period $\tau$ gives an echo of amplitude $M\sin^2\phi$ along the $-y$ axis (Figure 11.7(f)). For a symmetrical distribution of isochromats, the $x$ components of magnetization vanish.

If there is a strong water signal, the component $M\cos\phi$ along the $-z$ axis induces an appreciable modulation of the nuclear susceptibility along the sample, in addition to the linear dependence caused by the applied field gradient. Since we are only concerned with the changes (temporal and spatial) in the field $B^*$ due to nuclear susceptibility, it is convenient to define $B^*$ as zero for equal spin populations ($M_z = 0$). Thus:

$$\gamma B^*/2\pi = \Delta f \tag{11.15}$$

where the relative frequency shift is now given by

$$\frac{\Delta f}{f_0} = -\tfrac{1}{3}(\cos\phi)\,\chi_N \tag{11.16}$$

where $\phi$ is the precession angle during $\tau$. The internal field $B^*$ thus oscillates along the length of the sample (Figure 11.8). We neglect the effects of relaxation and diffusion for the moment.

First we consider the conventional Hahn echo $E_1$ in the absence of any susceptibility field $B^*$. The amplitude of the echo (Figure 11.7(f)) is given by

$$I = \tfrac{1}{2}M_0\sin^2\phi \tag{11.17}$$

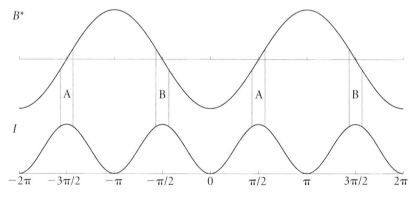

**Figure 11.8** Schematic diagram showing the spatial variation of the field due to nuclear susceptibility $B^*$, and the intensity of the Hahn echo ($I$). Most of the observed echo originates from the zones A and B. Within the B zones, the gradient $dB^*/dz$ opposes the applied gradient $G$ and creates isolated pockets of high field homogeneity. When the NMR signal is very intense (strong $B^*$) this causes multiple Hahn echoes.

Consequently most of the echo arises from values of $\phi$ near to the condition

$$\phi = (2n+1)\pi/2 \qquad \qquad \textbf{11.18}$$

where $n$ is an integer. This defines a set of active zones that contribute most of the signal in the Hahn echo (Figure 11.8). They are situated at evenly spaced intervals along the $z$ direction:

$$z = (2n+1)\, z_0 \qquad \qquad \textbf{11.19}$$

where $z_0$ is the distance of the first zone from the point where the applied gradient is zero

$$\gamma G z_0 \tau = \pi/2 \qquad \qquad \textbf{11.20}$$

It is convenient to label as A the zones with $n$ even and as B the zones with n odd.

The scheme for multiple echo generation is set out in Figure 11.9. A linear gradient $G$ is applied along the $z$ axis, creating a field $Gz$ throughout the experiment. The Hahn echo is $E_1$ at time $2\tau$. Our task is to explain the generation of the unexpected echoes $E_2$, $E_3$ ... at times $3\tau$, $4\tau$ ... etc. The additional field due to nuclear suscept-ibility $(B^*)$ begins at the end of the first $\tau$ interval and continues indefinitely (if spin–lattice relaxation is neglected). The consequent frequency modulation (at $1/\tau$) is given by Equation 11.16. It gives rise to a Fourier spectrum made up of a centre-band at the Larmor frequency $f_0$ and a symmetrical set of modulation sidebands at frequencies $f_0 \pm m/\tau$ Hz, where $m$ is an integer. (The situation is analogous to that for spinning sidebands.) If the modulation index is low, only the first-order side-bands $(m = \pm 1)$ have appreciable intensity. The centreband response gives rise to the expected strong Hahn echo $E_1$ at time $2\tau$.

Magnetization vectors corresponding to the modulation sidebands rotate clock-wise and counterclockwise with respect to the centreband vector at $f_0$, all coming back into coincidence whenever the precession angles reach $\pm 2k\pi$, thus setting the

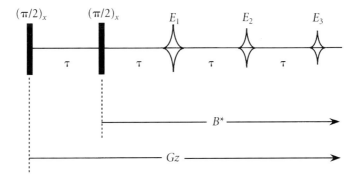

**Figure 11.9** Formation of multiple Hahn echoes in an applied field gradient G when there is an intense NMR signal that creates a periodic field B* due to the change in nuclear susceptibility along the sample. The echo $E_1$ is the ordinary Hahn echo, but $E_2$, $E_3$ etc are extraordinary echoes.

stage for further echoes at times $k\tau$ ($k = 3, 4, 5$, etc). For example, in the simple case where only the first-order sidebands ($\alpha$ and $\beta$) are appreciable, we have the evolution:

At first sight this appears sufficient to account for the multiple echoes, but we must remember that if these isochromats are allowed to evolve in the applied gradient $G$ they would normally experience a progressive phase divergence and could not form any subsequent echoes. Only a single Hahn echo would be observed. However, there is a rather unexpected compensation effect. Since the field $B^*$ varies cosinusoidally along the sample, there are sample regions where the gradient $dB^*/dz$ is almost linear and opposed to the applied gradient $G$. These are the B zones of Figure 11.8 (the A zones have the two gradients in the same sense). They are situated at regular intervals along the $z$ direction according to Equation 11.19 with $n$ odd.

If the nuclear susceptibility is high enough, the two gradients cancel, creating isolated pockets of almost uniform magnetic field. Isochromats from these active zones are not subject to the usual progressive divergence and thus give rise to the extra echoes $E_2$, $E_3$ etc. In the general case where the gradient $dB^*/dz$ is not large enough to match $G$ exactly, only partial cancellation occurs, giving weaker echoes. Imperfect matching leads to a progressive attenuation of the echoes with time, in addition to losses due to diffusion and relaxation (both spin–spin and spin–lattice). Consequently the number of echoes observed depends on the strength of the nuclear susceptibility effect (which is high at high spectrometer fields) and also upon the strength of the applied gradient. Isochromats from the B zones of the sample that form these multiple echoes precess at the mean frequencies $(2n+1)\pi/(2\tau)$ with $n$ odd. This introduces a progressive phase shift of $-90°$ for each subsequent echo, as observed by Bowtell et al.[40]

### 11.4.3 Anomalous correlation effects

One of the most disconcerting consequences of the bulk nuclear susceptibility of water (and of other samples with high spin density) is the appearance of anomalous cross-peaks in two-dimensional correlation spectra. This subject has given rise to prolonged controversy in the literature[42–45] because it can be treated by two entirely different theories, a classical description based on spurious modulation of the precession frequencies in a two-dimensional COSY experiment, and a quantum-mechanical treatment that invokes long-range multiple-quantum coherences. A recent review article by Levitt[45] presents an admirably clear comparison of the two approaches to this curious phenomenon. The effect is important because it challenges the usual assumption that a cross peak in a COSY spectrum must indicate $J$ coupling between the two species involved, yet in these instances one of the species is water!

The classical description is easier to visualize than the multiple-quantum hypothesis. As we saw in § 2.6.1, the COSY experiment involves modulation of the $z$-component of magnetization of the $I$ spins after an evolution period $t_1$. In the case of $J$ coupling to a second species $S$, there are two $I$-spin resonances and it is the differential population effect that eventually leads to polarization transfer from $I$ to $S$, but in the case of the singlet resonance of water there is, of course, only a net effect. Suppose that the vector representing the water signal precesses during the evolution period through an angle $\phi = 2\pi\delta t_1$; its $z$ magnetization after the second $\pi/2$ pulse is modulated according to $M_z = -M_0 \cos\phi$, leading to a variable frequency shift, because the macroscopic nuclear susceptibility is similarly modulated. This additional wobbling of the precession frequency of water gives rise to a set of modulation sidebands in the $F_1$ dimension. Not only do they appear on the water signal, but also on the signal of any solute that may be present, because the nuclear susceptibility of water affects the entire sample. (All the sidebands disappear if the transmitter is set exactly at the water frequency, because $\phi = 0$.) These spurious cross peaks could be interpreted as the result of multiple-quantum coherence, and they have exactly the behaviour expected of true multiple-quantum coherence with respect to phase cycling (§ 6.2.2) and applied field gradients (§ 6.3.1).

Satisfying the tests for multiple-quantum behaviour lends some credence to the arguments of Warren and co-workers[42,43] that the anomalous off-diagonal responses could arise from genuine multiple-quantum effects generated by dipole–dipole interactions between spins at very large distances. For spins at macroscopic separations, the random translational diffusion of the molecules in a liquid does not quite average the dipolar interactions to zero, and since there is a very large number of such neighbours at a distance $r$ when $r$ is very much larger than the molecular diameter, the overall dipolar effect can be appreciable. The anomalous cross-peaks arise from very large numbers of extremely small long-range multiple-quantum coherences, rendered observable by the action of long-range dipolar interactions. The theory must therefore encompass interactions between all pairs of spins in the sample, except for very close neighbours. The usual picture of an isolated molecular spin system has to be abandoned and the appropriate density operator acquires astronomical dimensions. Because of the large numbers of spins involved, it is difficult to derive quantitative results by this treatment, but it is self-consistent and has the advantage of restoring linearity to the spin dynamics. It may well be that the two alternative treatments of this phenomenon are both valid.[45] However, the nuclear susceptibility description appears simpler, being based on the frequency shift effects reported in § 11.4.1.

## 11.5 Discussion

Water gives rise to the dynamic range problem—one of the few serious drawbacks of pulse-Fourier transform NMR that has not been satisfactorily resolved to date. It seems that water suppression techniques will be required for the foreseeable future, but that the progressive introduction of pulsed field gradient schemes will improve the situation, particularly with respect to $t_1$ noise. The other instrumental difficulties with water—radiation damping and the nuclear susceptibility effect—are merely a nuisance in high resolution studies, providing little information of interest. Indeed, the introduction of anomalous cross peaks in COSY spectra is an alarming feature for routine high resolution studies, and it is reassuring to note that the effect can be eliminated by the simple expedient of setting the transmitter at the water frequency. The real danger is that these complications, and the distortions and spurious signals that they can cause, might be overlooked during multi-pulse experiments on aqueous samples.

# References

1. P. J. Hore, *Method. Enzymol.* **176**, 64 (1989).
2. A. G. Redfield and R. K. Gupta, in *Advances in Magnetic Resonance*, Ed. J. S. Waugh, Academic Press, New York, 1971, Vol. **5**.
3. I. D. Campbell, C. M. Dobson and R.G. Ratcliffe, *J. Magn. Reson.* **27**, 455 (1977).
4. F. Bloch and A. Siegert, *Phys. Rev.* **57**, 522 (1940).
5. R. W. Dykstra, *J. Magn. Reson.* **72**, 162 (1987).
6. N. V. Bloembergen and R. V. Pound, *Phys. Rev.* **95**, 8 (1954)
7. S. L. Patt and B. D. Sykes, *J. Chem. Phys.* **56**, 3182 (1972).
8. R. K. Gupta, *J. Magn. Reson.* **24**, 461 (1976).
9. D. L. Rabenstein and A. A. Isab *J. Magn. Reson.* **36**, 281 (1979).
10. R. G. Bryant and T. M. Eads, *J. Magn. Reson.* **64**, 312 (1985).
11. D. L. Rabenstein, S. Fan and T. T. Nakashima, *J. Magn. Reson.* **64**, 541 (1985).
12. P. C. M. van Zijl and C. T. W. Moonen, *J. Magn. Reson.* **87**, 18 (1990).
13. M. Piotto, V. Saudek and V. Sklenár, *J. Biomed. NMR* **2**, 661 (1992).
14. A. G. Redfield, S. D. Kunz and E. K. Ralph, *J. Magn. Reson.* **19**, 114 (1975).
15. B. L. Tomlinson and H. D. W. Hill, *J. Chem. Phys.* **59**, 1775 (1973).
16. Ē. Kupče and R. Freeman, *J. Magn. Reson. A* **103**, 358 (1993).
17. P. Plateau and M. Guéron, *J. Am. Chem. Soc.* **104**, 7310 (1982).
18. V. Sklenár and Z. Starčuk, *J. Magn. Reson.* **50**, 495 (1982).
19. D. L. Turner, *J. Magn. Reson.* **54**, 146 (1983).
20. P. J. Hore, *J. Magn. Reson.* **54**, 539 (1983).
21. P. J. Hore, *J. Magn. Reson.* **55**, 283 (1983).
22. M. H. Levitt and M. F. Roberts, *J. Magn. Reson.* **71**, 576 (1987).
23. M. H. Levitt, *J. Chem. Phys.* **88**, 3481 (1988).
24. U. Piantini, O. W. Sørensen and R. R. Ernst, *J. Am. Chem. Soc.* **104**, 6800 (1982).
25. J. P. Waltho and J. Cavanagh, *J. Magn. Reson. A* **103**, 338 (1993).
26. R. Baumann, G. Wider, R. R. Ernst and K. Wüthrich, *J. Magn. Reson.* **44**, 402 (1981).
27. G. A. Morris, *J. Magn. Reson.* **80**, 547 (1988).
28. A. Gibbs, G. A. Morris, A. G. Swanson and D. Cowburn, *J. Magn. Reson. A* **101**, 351 (1993).
29. R. Hurd, *J. Magn. Reson.* **87**, 472 (1990).
30. P. Broekaert and J. Jeener, *J. Magn. Reson. A* **113**, 60 (1995).
31. C. Anklin, M. Rindlisbacher, G. Otting and F. H. Laukien, *J. Magn. Reson. B* **106**, 199 (1995).

32. R. Freeman and W. A. Anderson, *J. Chem. Phys.* **42**, 1199 (1965).
33. A. Szöke and S. Meiboom, *Phys. Rev.* **113**, 585 (1959).
34. X. A. Mao, J. X. Guo and C. H. Ye, *Chem. Phys. Lett.* **222**, 417 (1994).
35. J-Y. Lallemand, A. Louis-Joseph and D. Abergel, *La RMN, un Outil pour la Biologie*, Institut Pasteur, Paris, 1996.
36. B. Lasarew and L. Schubnikow, *Phys. Z. Sowjet.* **11**, 445 (1937).
37. W. C. Dickinson, *Phys. Rev.* **81**, 717 (1951).
38. H. T. Edzes, *J. Magn. Reson.* **86**, 293 (1990).
39. G. Deville. M. Bernier and J. M. Delrieux, *Phys. Rev. B* **19**, 5666 (1979).
40. R. Bowtell, R. M. Bowley and P. Glover, *J. Magn. Reson.* **88**, 643 (1990).
41. R. P. O. Jones, G. A. Morris and J. C. Waterton, *J. Magn. Reson.* **98**, 115 (1992).
42. Q. He, W. Richter, S. Vathyam and W. S. Warren, *J. Chem. Phys.* **98**, 6779 (1993).
43. W. S. Warren, W. Richter, A. H. Andreotti and B. T. Farmer, *Science* **262**, 2005 (1993).
44. J. Jeener, A. Vlassenbroek and P. Broekaert, *J. Chem. Phys.* **103**, 1309 (1995).
45. M. H. Levitt, *Concepts in Magnetic Resonance,* **8**, 77 (1996).

# 12

# Measurement of coupling constants

## 12.1 Introduction

Vicinal spin–spin coupling constants have proved to be of key importance for the determination of molecular conformation in solution. The foundation of all such studies is the Karplus equation[1,2] which relates the three-bond coupling constant to the dihedral angle $\phi$.

$$^3J = A + B\cos\phi + C\cos2\phi \tag{12.1}$$

Despite the fact that we are usually forced to employ empirical values for the factors $A$, $B$ and $C$, this relationship has proved to be extraordinarily useful in practice. This chapter examines the practical methods available for measuring spin–spin splittings. For first-order spectra, these splittings are equal to the corresponding coupling constants; for strongly coupled spectra, a full analysis of the spin system is normally required to extract the coupling constants. Coupling constants carry a sign (§ 8.2.3) but these are rarely required when the coupling is used for diagnostic purposes, and sign determination is not considered in this chapter.

## 12.2 **Direct measurements**

When the splitting of interest is well-resolved, it is a relatively simple matter to measure it directly, using a peak-finding software routine if necessary. There is then no need to consider the more sophisticated manipulations described below. Precautions should be taken to ensure fine digitization in the frequency domain, either by using a long acquisition time, or by zero-filling, or (in extreme cases) by performing a multiplet-selective excitation experiment in order to limit the spectral width. In two-dimensional correlation spectra there is the additional complication that the splitting may be in-phase or anti-phase.

The difficulties arise when the multiplet under investigation is poorly resolved. For an in-phase doublet the apparent splitting is then smaller than the coupling constant, because the overlapping component lines begin to merge. For an anti-phase doublet the apparent splitting becomes larger than the coupling constant, owing to mutual cancellation of signals in the overlap region. The remainder of this chapter is mainly devoted to schemes to circumvent these problems.

Heteronuclear two-dimensional correlation spectroscopy can sometimes offer a way to avoid overlap difficulties. Groups of lines separated by a homonuclear splitting that is normally unresolved, are displaced in the second frequency dimension by a large heteronuclear splitting, allowing the homonuclear separation to be measured directly.[3]

## 12.3 **Resolution enhancement**

Improvement in the resolving power is the next obvious recourse. The natural decay of the free induction signal caused by spin–spin relaxation and magnet inhomogeneity is partially reversed by the application of a weighting function that increases with time; the Fourier transform then gives narrower lines and small splittings are resolved. If this process is carried too far, it introduces an appreciable step function at the end of acquisition, distorting the transformed line shape, so the free induction decay must be brought smoothly down to zero at the end (apodization). Analogous deconvolution processes can be devised in the frequency domain.[4] Maximum entropy reconstruction[5] can also be used as a resolution enhancement technique[6,7] by virtue of its inverse nature (§ 6.5.1). There is an inevitable loss in signal-to-noise with these resolution enhancement methods, and this eventually limits their applicability.

An alternative approach[8] is to restrict the effective sample volume by selective excitation in an applied field gradient, followed by acquisition of the free induction decay in the absence of that gradient (§ 5.4.5). The gradient is normally applied along the spinning axis ($z$ axis). This limits the effective sample length in the $z$ direction so that the natural $z$ gradients have little effect. Although one might have anticipated a similar improvement in resolution by an actual reduction of the physical size of the sample, this is not observed in practice because discontinuities in magnetic susceptibility at the walls of the sample container distort the applied magnetic field. These effects are relatively more severe for very small samples. Recently this problem has been attacked by spinning small samples about an axis inclined with respect to the $z$ axis at the magic angle (54.7°) in a specially-constructed high resolution probe.

## 12.4  Fitting procedures

Another common method is to apply a least-squares fitting procedure to a multiplet extracted from the one-dimensional spectrum or from a suitable section through a two-dimensional cross-peak, using the linewidth and coupling constant as adjustable parameters. We may adopt an appropriate lineshape, for example a Lorentzian, or we may borrow an experimental lineshape from a known singlet in the same spectrum.[4] Fitting may be performed either in the time-domain or the frequency domain, with comparable efficiencies. Studies of a large number of vicinal proton–proton couplings in human lysozyme from two-dimensional correlation spectra confirm the reliability of such curve-fitting methods (and also of the Karplus relationship) by showing that the derived coupling constants are in good agreement with values calculated from the coordinates of the X-ray crystal structure.[9]

## 12.5  J spectroscopy

Resolution can also be enhanced by spin echo methods, in which the broadening due to magnet inhomogeneity is reversed by a $\pi$ refocusing pulse (§ 4.2.9). In the resulting two-dimensional $J$ spectrum, the resolution in the $F_1$ frequency dimension is limited not by $T_2^*$ but (in principle) by the spin–spin relaxation time $T_2$, which is usually considerably longer than $T_2^*$. Unfortunately instrumental instabilities still contribute to the observed linewidth and the improvement in resolution is then rather less than expected. Complications arise in the case of strong coupling. In homonuclear $J$ spectra, the undesirable phase-twist lineshape[10] makes it difficult to exploit the resolution advantage, but in heteronuclear $J$ spectroscopy pure absorption-mode lines can be recorded. The main drawback here is that it may be necessary to set up an entire two-dimensional experiment simply to measure a single coupling constant of interest.

## 12.6  Comparison methods

In these, and in many other procedures that follow, the spin multiplet under investigation must first be separated from the remainder of the high resolution spectrum by discarding the unwanted frequency-domain data. For this reason the chosen multiplet must be well separated from adjacent signals. Often it is then zero-filled (for finer digitization) and back-transformed into the time domain. When overlapping multiplets are involved, there are techniques available for separating them into the basic components (§ 5.4.4 and § 8.2.4).

## 12.6.1 Absorptive and dispersive multiplets

The splitting in a poorly resolved anti-phase doublet is not equal to the corresponding coupling constant since there are serious interference effects in the region where the two lines overlap. Kim and Prestegard[11] have shown that if an anti-phase doublet is composed of pure Lorentzian lines, the separation between the extrema of the absorption mode doublet and the separation between the extrema of the corresponding dispersion-mode doublet can be related to the coupling constant, independent of the linewidth. This relationship involves a cubic equation in $(J_{IS})^2$ and it can be solved analytically. The scheme has been tested by simulation for $J = 8.0$ Hz and a range of linewidths from 0.3 to 16 Hz, yielding values of $J = 8.2 \pm 0.5$ Hz. To be successful, this method requires the lineshape to be close to a pure Lorentzian.

## 12.6.2 Coupled and decoupled multiplets

Decoupling has long been used as a diagnostic for small unresolved splittings by observing the slight reduction in linewidth when a neighbour spin is decoupled. An extension of this idea provides an estimate of the actual spin–spin coupling constant, provided that the decoupling is efficient (negligible residual splitting). Naturally, this method is not applicable to anti-phase multiplets, since they would suffer self-cancellation when decoupled. Well-digitized coupled and decoupled versions of the time-domain signal are obtained by back transformation of the corresponding multiplets extracted from the full spectra. The missing splitting in the decoupled time-domain signal is then reintroduced by multiplying by $\cos(\pi J^* t)$, where $J^*$ is a trial coupling constant. The result is compared with the coupled signal in a least-square fitting routine. When $J^* = J_{IS}$ there is a global minimum in the sum of the squares of the residuals. This fitting procedure may also be carried out in the frequency domain. When further splittings are present in the multiplet under investigation, partial fits are also observed, giving rise to subsidiary minima at frequencies equal to sums and differences of the splittings. However, these minima are always significantly weaker than the principal minimum so there is usually no ambiguity. Accuracies of the order of $\pm 0.03$ Hz can be achieved in favorable cases.[12] The principal application is the measurement of poorly resolved couplings, where the coupling information resides in the distorted line shapes in the coupled multiplet.

There is an alternative scheme for extracting the coupling constant from coupled and decoupled multiplets that does not involve a least-squares fitting procedure.[13] Assuming no other perturbation by the decoupler, the time-domain signals from coupled $S_c$ and decoupled $S_d$ multiplets may be written as

$$S_c = A\cos(2\pi\delta_S t) \cos(\pi J_{IS} t) \exp(-t/T_2^*) \prod_R \cos(\pi J_{SR} t) \qquad \textbf{12.2}$$

$$S_d = A\cos(2\pi\delta_S t) \exp(-t/T_2^*) \prod_R \cos(\pi J_{SR} t) \qquad \textbf{12.3}$$

where the terms in $J_{SR}$ represent all the passive couplings to the observed spin $S$. The ratio $S_c/S_d$ can be evaluated by dividing corresponding ordinates of the time-domain signals, leaving only a single term $\cos(\pi J_{IS} t)$, which gives the coupling constant of interest. (If necessary, an exponential weighting function can be applied to this cosine wave and the result transformed into the frequency domain to give a doublet of splitting $J_{IS}$.) However, there are obvious complications in the division process when there are zero-crossings in $S_d$. They can be circumvented by pre-processing $S_d$ to replaces all values less than a given threshold $\theta$ with the value of $\theta$ itself, retaining the sign. A surprisingly wide range of choices of the threshold can be accommodated.[13]

### 12.6.3 In-phase and anti-phase splittings

There are several experimental techniques that generate two forms of a given doublet —one in-phase, and characterized by a time-domain modulation $\cos(J_{IS}t)$, and another anti-phase, modulated by $\sin(\pi J_{IS}t)$. Many coherence transfer schemes fall into this category. For example, a cross-peak in a homonuclear correlation (COSY[14]) spectrum has an anti-phase active splitting, whereas the corresponding diagonal peak exhibits the same splitting in-phase. Related cross-peaks in COSY and TOCSY[15] spectra of the same molecule display an anti-phase and an in-phase splitting respectively.

The DISCO technique[16,17] appears to have been the first to exploit this phenomenon. Two traces are extracted from the two-dimensional spectrum, one with the active splitting ($J_{IS}$) in-phase and the other anti-phase, and the amplitudes of the signals in these traces are adjusted to be equal. If the active splitting is small and if there is also a larger passive splitting ($J_{SR}$) then the two traces can be sketched as $a$ and $b$:

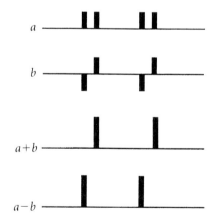

Addition or subtraction of the two traces gives doublets with the (larger) passive splitting $|J_{SR}|$, but there is a relative displacement equal to the (smaller) active splitting $|J_{IS}|$, which may then be measured by the direct method. Overlap effects, which would normally degrade the accuracy of the measurement, are considerably reduced in these submultiplets. The only limitations are that the two versions of the multiplet should be properly phased and appropriately scaled.

The DISCO method is particularly well-adapted to the selective relayed coherence transfer experiment[16] in a three-spin ($ISR$) system, where only the $I$ resonance is excited. The relayed multiplet ($I \rightarrow S \rightarrow R$) contains two anti-phase splittings ($J_{IS}$ and $J_{SR}$), and its time domain signal can be described by:

$$M_R = M_0 \cos(2\pi\delta_R t) \cos(\pi J_{IR} t) \sin(\pi J_{IS} t) \sin(\pi J_{SR} t) \exp(-t/T_2^*) \qquad \textbf{12.4}$$

The back-transfer peak ($I \rightarrow S \rightarrow I$), involving two inversions with respect to $J_{IS}$, has all the splittings in phase, and may be represented by

$$M_B = M_0 \cos(2\pi\delta_I t) \cos(\pi J_{IR} t) \cos(\pi J_{IS} t) \cos(\pi J_{SR} t) \exp(-t/T_2^*) \qquad \textbf{12.5}$$

With the proper registration ($\delta_I = \delta_S = 0$), subtraction of the corresponding frequency-domain multiplets produces a multiplet with the splittings $|J_{IR}|$ and $|J_{IS} + J_{SR}|$, both in phase, represented by the time-domain modulation

$$S(t) = M_0 \cos(\pi J_{IR} t) \cos [\pi(J_{IS} + J_{SR})t] \exp(-t/T_2^*) \qquad \textbf{12.6}$$

Addition of the same two traces leaves a multiplet with the splittings $|J_{IR}|$ and $|J_{IS} - J_{SR}|$, both in phase, corresponding to the time-domain modulation

$$S'(t) = M_0 \cos(\pi J_{IR} t) \cos [\pi(J_{IS} - J_{SR})t] \exp(-t/T_2^*) \qquad \textbf{12.7}$$

This allows all three coupling constants to be extracted, even though this information may be obscured in the conventional spectrum by overlap effects.

Titman and Keeler[18] have attacked the same problem by a least-squares fitting procedure, starting with an anti-phase doublet from a COSY cross-peak and an in-phase doublet from the corresponding TOCSY cross-peak. The time-domain signals may be written as

$$S_{\text{COSY}} = A \cos(2\pi\delta_S t) \sin(\pi J_{IS} t) \exp(-t/T_2^*) \qquad \textbf{12.8}$$

$$S_{\text{TOCSY}} = B \cos(2\pi\delta_S t) \cos(\pi J_{IS} t) \exp(-t/T_2^*) \qquad \textbf{12.9}$$

where $J_{IS}$ is the active splitting, and passive splittings have been omitted for simplicity. A trial splitting $J^*$ is now introduced by multiplying these time domain signals by $\cos(\pi J^* t)$ and $\sin(\pi J^* t)$ respectively:

$$S'_{\text{COSY}} = A\cos(2\pi\delta_S t) \sin(\pi J_{IS} t) \cos(\pi J^* t) \exp(-t/T_2^*) \qquad \textbf{12.10}$$

$$S'_{\text{TOCSY}} = B\cos(2\pi\delta_S t) \cos(\pi J_{IS} t) \sin(\pi J^* t) \exp(-t/T_2^*) \qquad \textbf{12.11}$$

If $J^* = J_{IS}$ and $A = B$, these two expressions become identical, irrespective of the lineshape and the presence of passive splittings on both cross-peaks. A least-squares fitting procedure adjusts the relative amplitudes $A$ and $B$ and searches for the condition $J^* = J_{IS}$. Accuracies of $\pm 0.3$ Hz have been achieved for a typical coupling $J_{IS} = 7.0$ Hz.

A variant of this experiment[19] has been used to solve a tricky problem that arises in the measurement of long-range $^{13}$C–H spin–spin couplings. A round-trip coherence transfer experiment (§ 8.3.1) is used to optimize sensitivity. The resulting cross-peaks have complex phase properties because they are modulated by proton chemical shifts and proton–proton couplings during the initial fixed delay Δ. The required anti-phase splitting $J_{IS}$ is so distorted by these effects and by the instrumental line broadening that it is virtually impossible to extract $J_{IS}$ by inspection. The first step is to introduce the same phase distortion into a $^{13}C$ satellite of the proton spectrum by inserting an identical delay Δ before acquisition. This is used as a reference multiplet. It is then convoluted with an anti-phase trial doublet, equivalent to multiplication of the time-domain signal by $\sin(\pi J^{*}t)$. The result is compared with the experimental multiplet from the heteronuclear correlation experiment in a least-squares fitting program that searches for the condition $J^{*} = J_{IS}$. The relative amplitudes of the reference and experimental multiplets is the second adjustable parameter in this two-dimensional search. The procedure tolerates reasonable amounts of noise and line broadening without making significant errors.

## 12.7 J extension

Pairs of in-phase and anti-phase multiplets can be exploited in a quite different manner to obtain an accurate value for a poorly-resolved splitting $J_{IS}$ without the necessity for a search routine at all. The idea is to extend the experimental splitting $J_{IS}$ by adding a well-defined dummy splitting $J_0$, thus making the extended splitting $|J_{IS} + J_0|$ amenable to direct measurement, even when there are several other splittings involved. If we multiply the time-domain in-phase doublet $S_{ip}$ by $\cos(\pi J_0 t)$ and the time domain anti-phase doublet $S_{ap}$ by $\sin(\pi J_0 t)$ then the difference represents a new in-phase doublet of splitting $|J_{IS} + J_0|$:

$$
\begin{aligned}
S_{ip}\cos(\pi J_0 t) - S_{ap}\sin(\pi J_0 t) &= A\cos(2\pi\delta_S t)\,\cos(\pi J_{IS}t)\,\cos(\pi J_0 t)\,\exp(-t/T_2^{*}) \\
&\quad - A\cos(2\pi\delta_S t)\,\sin(\pi J_{IS}t)\,\sin(\pi J_0 t)\,\exp(-t/T_2^{*}) \\
&= A\cos(2\pi\delta_S t)\,\cos[\pi(J_{IS} + J_0)t]\,\exp(-t/T_2^{*}) \qquad \textbf{12.12}
\end{aligned}
$$

The adjustment of the relative intensities is most easily performed by comparing the frequency-domain doublets convoluted by in-phase and anti-phase stick doublets. J extension can also be achieved through the alternative trigonometrical identity

$$
\cos(\pi J_{IS}t)\,\sin(\pi J_0 t) + \sin(\pi J_{IS}t)\,\cos(\pi J_0 t) = \sin\left[\pi(J_{IS} + J_0)t\right] \qquad \textbf{12.13}
$$

which gives an anti-phase splitting $|J_{IS} + J_0|$.

An example of the application of J extension is provided by the complex unresolved multiplet of proton H-14 in the 400 MHz spectrum of strychnine. Homonuclear equivalents[20] of the INEPT[21] and refocused INEPT[22] techniques were used to obtain in-phase and anti-phase versions of this multiplet. The in-phase version is a broad featureless resonance due to many small unresolved couplings (Figure 12.1(a)). The anti-phase version exhibits a splitting of 7.1 Hz, but this is evidently grossly exaggerated by the well-known self-cancellation effect in the overlap region (Figure 12.1(b)). J-extension with a value $J_0 = 40.0$ Hz gives a splitting of 44.8 Hz (Figures

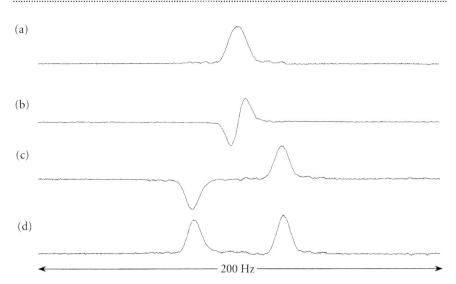

**Figure 12.1** Selectively excited spectrum of proton H-14 in strychnine at 400 MHz: (a) with an unresolved in-phase splitting; (b) with an anti-phase splitting of 7.1 Hz, exaggerated because of self-cancellation in the overlap region. $J$ extension has been used to add a dummy splitting of 40.0 Hz, giving observed splittings (c) and (d) of 44.8 Hz, indicating that the actual coupling constant is 4.8 Hz.

12.1(c) and (d)), indicating that the true value of $J_{IS}$ is 4.8 Hz. This is probably the most straightforward technique when the in-phase and anti-phase versions of the multiplet are available.

## 12.8  J deconvolution

### 12.8.1  Time domain

The first report of direct deconvolution of a spin–spin splitting appears to have been the paper of Bothner–By and Dadok.[23] They extracted a suitable spin multiplet from the full spectrum, applied the reverse Fourier transform into the time domain, and then divided this by $\cos(\pi J^* t)$, where $J^*$ is a trial coupling. They note that a 1:2:1 triplet would require division by $\cos^2(\pi J^* t)$, and that a 1:3:3:1 quartet would require division by $\cos^3(\pi J^* t)$. Successive deconvolution processes would be cascaded by dividing by the appropriate cosine functions one after the other. However, they observed a serious drawback of this procedure occasioned by division with near-zero values of the cosine function, an operation that introduces unacceptably large spikes into the deconvoluted time-domain signal.

One way to avoid this pitfall is through judicious choice of sampling intervals on the cosine function, so that there is never any occasion to divide by a value close to zero.[24,25] Anti-phase and in-phase splittings require different treatments.

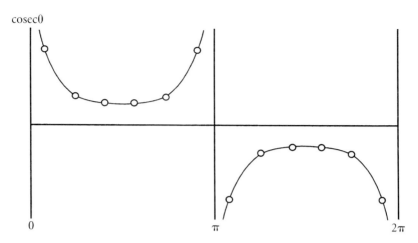

cosec θ

**Figure 12.2** A cosecant function sampled at regular intervals θ = (2k−1)π/12 so as to avoid the singularities. This is equivalent to sampling the corresponding sinewave so as to avoid the zero-crossings.

## 12.8.2 Anti-phase splittings

Many coherence transfer experiments generate anti-phase splittings, represented by a modulation term $\sin(\pi J_{IS}t)$ in the time domain or by an up-down pattern in the frequency domain. Rather than dividing by $\sin(\pi J^*t)$, it is convenient to think in terms of multiplication by $\mathrm{cosec}(\pi J^*t)$, sampled at regular intervals, but avoiding the singularities (Figure 12.2). For deconvolution of the anti-phase splitting, $\mathrm{cosec}(\theta)$ would be sampled at intervals $t$ such that

$$t = \frac{(2k-1)}{4aJ^*} \tag{12.14}$$

which, with $\theta = \pi J^*t$, is equivalent to

$$\theta = \frac{(2k-1)\pi}{4a} \tag{12.15}$$

where $4a$ is the number of samples per cycle and $k$ is an integer. Thus in Figure 12.2 where $a=3$, the cosecant is sampled at π/12, 3π/12, 5π/12, 7π/12, 9π/12, 11π/12, 13π/12, 15π/12, 17π/12, 19π/12, 21π/12 and 23π/12.

The deconvoluted time domain signal may be written as

$$S' = S_0 \cos(2\pi\delta_S t) \sin(\pi J_{IS}t) \, \mathrm{cosec}(\pi J^*t) \exp(-t/T_2^*) \tag{12.16}$$

where $J^*$ is a trial coupling constant. Fourier transformation of this expression generates a frequency-domain pattern we shall call the test spectrum. It is not obvious at first sight what form this test spectrum will take, but it can be visualized if we make a digression to find a simpler replacement for the term $\mathrm{cosec}(\pi J^*t)$.

Consider the trigonometrical series

$$\text{Sum} = 2\sin\theta \sum_{p=1}^{a} \sin[(2p-1)\theta)] = 2\sin\theta\{\sin\theta + \sin3\theta + \sin5\theta + \sin7\theta + \ldots\} \quad \textbf{12.17}$$

Each product of sines can be written in terms of cosines of sums and differences, giving

$$\text{Sum} = \cos0 - \cos2\theta + \cos2\theta - \cos4\theta + \cos4\theta - \ldots - \cos(2a\theta) = 1 - \cos(2a\theta) \quad \textbf{12.18}$$

since the intermediate terms cancel. If we now invoke the sampling restriction of Equation 12.15, then $\cos(2a\theta) = 0$, and Sum $= 1$. Consequently we may write

$$\text{cosec}\theta = 2\sum_{p=1}^{a} \sin[(2p-1)\theta)] \quad \textbf{12.19}$$

Thus the choice of sampling has had an important simplifying rôle. With the substitution $\theta = \pi J^* t$, the deconvoluted time-domain signal may be written as

$$S' = 2S_0 \cos(2\pi\delta_S t) \exp(-t/T_2{}^*) \sin(\pi J_{IS} t) \sum_{p=1}^{a} \sin[(2p-1)\pi J^* t)] \quad \textbf{12.20}$$

In this form it is much easier to visualize the Fourier transform. For example, if $a=1$, the deconvoluted time-domain signal is simply

$$S' = 2S_0 \cos(2\pi\delta_S t) \exp(-t/T_2{}^*) \sin(\pi J_{IS} t) \sin(\pi J^* t) \quad \textbf{12.21}$$

In the frequency domain this is an anti-phase doublet of splitting $|J_{IS}|$ convoluted with another anti-phase splitting $|J^*|$, represented (for simplicity) as the stick pattern shown in Figure 12.3(b). When more samples are used ($a=2$) then the deconvoluted time-domain signal becomes

$$S' = 2S_0 \cos(2\pi\delta_S t) \exp(-t/T_2{}^*) \sin(\pi J_{IS} t) [\sin(\pi J^* t) + \sin(3\pi J^* t)] \quad \textbf{12.22}$$

This transforms into the eight-line pattern shown in Figure 12.3(c). In the general case there are $4a$ component lines in the test spectrum. This might appear quite surprising—the number of lines in the test spectrum is always equal to the number of sampling points on one cycle of the cosecant function.

The two central lines coalesce to a singlet when $J^* = J_{IS}$, but all the other lines disappear through mutual cancellation. (This is also true of the two lines at the extreme edges of the spectrum because one-half of each line is aliased and inverted.) This self-cancellation the key to the determination of $J_{IS}$. If we monitor $\Sigma$, the sum of the moduli of all the ordinates in the test spectrum, it passes through a well-defined minimum when $J^* = J_{IS}$. Although the central in-phase doublet coalesces to a singlet at this condition, this does not affect $\Sigma$. This behaviour is illustrated by the simulations of Figure 12.4 which show the effect of varying $J^*$ in small increments in the vicinity of the actual splitting $J_{IS} = 1.20$ Hz. Clearly $\Sigma$ is a sensitive test for the condition $J^* = J_{IS}$.

However, this is not the only minimum in $\Sigma$ recorded as a function of $J^*$. When $J^*$ is reduced to $J_{IS}/2$ there is a new interference effect, attributable to mutual cancel-

(a)

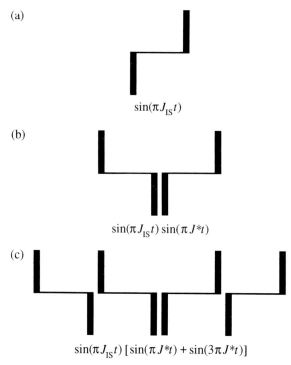

$$\sin(\pi J_{IS} t)$$

(b)

$$\sin(\pi J_{IS} t) \sin(\pi J^* t)$$

(c)

$$\sin(\pi J_{IS} t) [\sin(\pi J^* t) + \sin(3\pi J^* t)]$$

**Figure 12.3** (a) Schematic frequency-domain spectra of an anti-phase doublet of splitting $|J_{IS}|$. (b) The same, convoluted with an anti-phase doublet $|J^*|$, slightly larger than $|J_{IS}|$. (c) The same, convoluted with a further anti-phase splitting $3|J^*|$. Such spectra arise by Fourier transformation of a time-domain function $\sin(\pi J_{IS} t) \operatorname{cosec}(\pi J^* t)$ when the latter is sampled (b) four times per cycle, (c) eight times per cycle.

lation of next-nearest anti-phase neighbours, and when $J^*$ reaches $J_{IS}/3$ there is another anti-phase coincidence between next-but-one nearest neighbours. The minima at these subharmonic conditions are less pronounced than the principal minimum at $J^* = J_{IS}$. The total number of subharmonic minima is given by $(a-1)$, where $4a$ is the number of samples taken on the cosecant function. Figure 12.5 shows a simulation of $\Sigma$ versus $J^*$ for an anti-phase splitting $J_{IS} = 1.20$ Hz, comprising the principal minimum and three subharmonic minima. It also illustrates how added Gaussian noise (10%) and a Lorentzian line broadening of 0.16 Hz affect the determination.

Of course in practical cases we are not normally dealing with a simple anti-phase doublet but with a multiplet containing several other passive splittings. The introduction of a passive splitting $J_{SR}$ onto all the lines of the test spectrum complicates the graph of $\Sigma$ against $J^*$. It converts the principal minimum into a 1:2:1 triplet at frequencies $J_{IS} - J_{SR}$, $J_{IS}$, and $J_{IS} + J_{SR}$. The subharmonic minima are similarly affected. We shall see that this triplet effect is a common feature of several other methods described below.

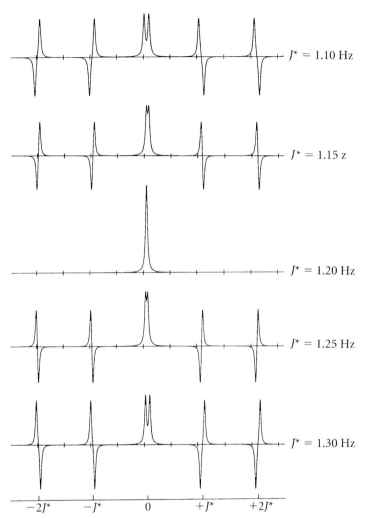

**Figure 12.4** Simulated test spectra for $J$ deconvolution as $J^*$ is varied in small increments through the value $J_{IS} = 1.20$ Hz. At the condition $J^* = J_{IS}$, the central in-phase doublet coalesces to a singlet and the anti-phase doublets vanish. This is a sensitive test for measuring $J_{IS}$.

## 12.8.3 **In-phase splittings**

A doublet with an in-phase splitting is represented in the time domain by

$$S = S_0 \cos(2\pi\delta_S t) \cos(\pi J_{IS} t) \exp(-t/T_2^*)$$  **12.23**

Instead of dividing by $\cos(\pi J^* t)$ we multiply by $\sec(\pi J^* t)$, sampled at regular intervals but avoiding the singularities

$$t = \frac{(2k-1)}{4aJ^*}$$  **12.24**

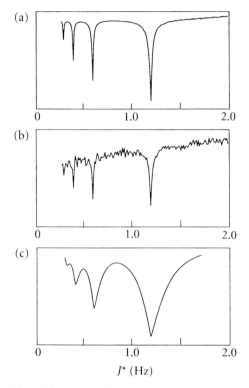

**Figure 12.5** (a) Simulation of the graph of $\Sigma$ against $J^*$ for an anti-phase splitting $J_{IS}$ = 1.20 Hz. Subharmonic responses are observed at $J_{IS}/2$, $J_{IS}/3$, and $J_{IS}/4$. (b) When 10% of Gaussian noise is added to the time-domain signal, or (c) when a Lorentzian line broadening of 0.16 Hz is imposed, the positions of the minima are scarcely affected.

which, with $\theta = \pi J^* t$, is equivalent to

$$\theta = \frac{(2k-1)\pi}{4a} \qquad \textbf{12.25}$$

where $4a$ is the number of samples per cycle, and $k$ is an integer.

In this case we need to find a replacement for $\sec\theta$. Consider the trigonometrical series

$$\text{Sum} = 2\cos\theta \sum_{p=1}^{a} \beta\cos[(2p-1)\theta)] = 2\cos\theta\{\cos\theta - \cos3\theta + \cos5\theta - \cos7\theta \dots\} \quad \textbf{12.26}$$

where $\beta$ is a sign alternation term given by $(-1)^{p+1}$. Each product of cosines can be written in terms of cosines of sums and differences, giving

$$\text{Sum} = \cos0 + \cos2\theta - \cos2\theta - \cos4\theta + \cos4\theta + \dots - \cos(2a\theta) = 1 - \cos(2a\theta)$$

$$\textbf{12.27}$$

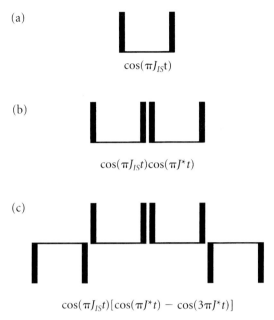

(a)

$\cos(\pi J_{IS}t)$

(b)

$\cos(\pi J_{IS}t)\cos(\pi J^*t)$

(c)

$\cos(\pi J_{IS}t)[\cos(\pi J^*t) - \cos(3\pi J^*t)]$

**Figure 12.6** Schematic frequency-domain spectra representing (a) an in-phase doublet of split-ting $|J_{IS}|$. (b) The same, convoluted with an in-phase splitting $|J^*|$, slightly larger than $|J_{IS}|$. (c) The same, convoluted with a further (inverted) in-phase splitting $3|J^*|$. Such spectra arise by Fourier transformation of a time-domain function $\cos(\pi J_{IS}t)\,\sec(\pi J^*t)$ when the latter is sampled (b) four times per cycle, (c) eight times per cycle.

since the intermediate terms cancel. If we now invoke the sampling restriction of Equation 12.25, then $\cos(2a\theta) = 0$, and Sum $= 1$. Consequently we may write

$$\sec\theta = 2\sum_{p=1}^{a} \beta\cos[(2p-1)\theta)] \qquad \textbf{12.28}$$

Thus the deconvoluted time-domain signal may be rewritten as

$$S' = 2S_0 \cos(2\pi\delta_S t) \exp(-t/T_2^*) \cos(\pi J_{IS}t) \sum_{p=1}^{a} \beta\cos[(2p-1)\pi J^*t)] \qquad \textbf{12.29}$$

As before this allows us to visualize the Fourier transform, the test spectrum. For example, if $a=1$, then we have a test spectrum of four lines all in the same sense (Figure 12.6(b)), obtained by convolution of an in-phase splitting $|J_{IS}|$ with another in-phase splitting $|J^*|$:

$$S' = 2S_0 \cos(2\pi\delta_S t) \exp(-t/T_2^*) \cos(\pi J_{IS}t) \cos(\pi J^*t) \qquad \textbf{12.30}$$

For $a=2$, the convolution may be written as

$$S' = 2S_0 \cos(2\pi\delta_S t) \exp(-t/T_2^*) \cos(\pi J_{IS}t) [\cos(\pi J^*t) - \cos(3\pi J^*t)] \qquad \textbf{12.31}$$

which gives the eight-line spectrum shown in Figure 12.6(c) where the inverted lines arise because of the phase-alternation factor $\beta = (-1)^{p+1}$.

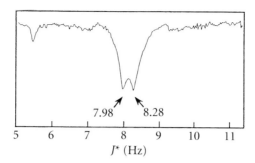

**Figure 12.7** Graph of $\Sigma$ versus $J^*$ for $J$ deconvolution of a section through a cross-peak in the 400 MHz proton spectrum of $m$-bromonitrobenzene, showing the resolution of two in-phase splittings only 0.30 Hz apart. The weak minimum near 5.4 Hz is a subharmonic response.

The sum of the moduli of all ordinates in the test spectrum $\Sigma$ exhibits a well-defined minimum when $J^* = J_{IS}$ through mutual cancellation of anti-phase signals. However, in contrast to the anti-phase case treated above, the summation only produces subsidiary minima at the odd subharmonics. This simplifies the determination somewhat.

Deconvolution is useful not only for measuring small coupling constants but also for resolving two large splittings that have almost the same value. This is the situation for one of the correlation cross-peaks of $m$-bromonitrobenzene where there are two in-phase (passive) splittings representing *ortho* couplings. The graph of $\Sigma$ against $J^*$ shows two clearly resolved minima, only 0.30 Hz apart, indicating splittings of 7.98 and 8.28 Hz (Figure 12.7). There is also a weak minimum at 5.42 Hz from a subharmonic response. We can appreciate how this technique would lend itself to automated measurements of coupling constants.

### 12.8.4 **Frequency domain**

An equivalent deconvolution may be performed in the frequency domain. However, although convolution in the frequency domain with a doublet of delta functions is a straightforward operation, deconvolution requires that we find a reciprocal sequence of delta functions. In the case of an in-phase doublet (1,1) this is the sequence $\{1, -1, 1, -1, 1, -1, 1, -1, 1, -1, \dots\}$. Deconvolution by this method is less straightforward than the time-domain scheme described above, and suffers from complications with baseline offsets and in the treatment of noise.[26,27]

Not all applications of deconvolution are required to operate on poorly resolved multiplets. Deconvolution may be used in an automated program to measure resolved splittings and then remove them from the spectrum. For example, it is sometimes advantageous to reduce a COSY cross-peak to a singlet at the chemical shift coordinates. A variant of deconvolution in the frequency domain[28] achieves this result in a very simple manner, using nonlinear processing. Suppose we have a well-resolved, in-phase splitting $J_{IS}$. We employ an in-phase doublet of delta func-

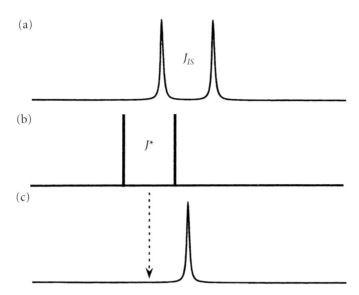

**Figure 12.8** Deconvolution in the frequency domain with a nonlinear operation. The delta function doublet is moved in small steps across the experimental trace (a). Pairs of ordinates in the experimental trace are compared and the lower value placed at the mid-point (indicated by the arrow). There is an appreciable signal in the processed trace (c) only when $J^* = J_{IS}$ and the doublets (a) and (b) coincide. The result is to remove the splitting $J_{IS}$ and to halve the total intensity.

tions with a trial splitting $J^*$ and move it step-by-step through the experimental multiplet. We then compare the two ordinates of the experimental spectrum that coincide with the two delta functions and place the lower (absolute) value at the mid-point (Figure 12.8). This scheme is based on an idea proposed by Baumann et al.[29] for reducing $t_1$ noise in two-dimensional spectra. The result is near-zero ordinates everywhere except for frequencies close to the chemical shift, *and* when $J^*$ is close to $J_{IS}$. The sum of all the ordinates in the processed spectrum ($\Sigma$) passes through a maximum when $J^* = J_{IS}$. In this manner we simultaneously measure $J_{IS}$ and remove it from the spectrum. The technique fails for unresolved splittings because the algorithm merely attenuates the response when $J_{IS}$ is smaller than the experimental line width. On the other hand, the method has the advantage over true $J$ deconvolution that the graph of $\Sigma$ against $J^*$ does not have maxima at the subharmonic frequencies. However, there are subsidiary responses at the sums and differences of the splittings in the multiplet under investigation. The actual splittings generate the most intense maxima; those corresponding to sums or differences are normally reduced by at least a factor of two. Occasionally two subsidiary maxima coincide and exhibit a stronger maximum. The situation is made easier if there are approximate estimates of the splittings available; then accurate values can be extracted for all the splittings from a single plot of $\Sigma$ versus $J^*$.

An example is provided by the 400 MHz proton spectrum of nicotine. The spin

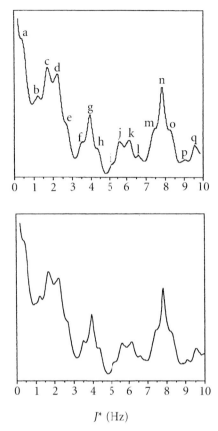

**Figure 12.9** A plot of $\Sigma$ against $J^*$ for one of the proton multiplets of nicotine subjected to $J$ deconvolution by the nonlinear process described in the text. The four principal maxima give the coupling constants (a) 0.45 Hz, (c) 1.71 Hz, (d) 2.24 Hz, and (n) 7.85 Hz. The other maxima represent sums or differences of coupling constants. Occasionally two subsidiary peaks coincide; for example (g) is the superposition of peaks at 3.90 and 3.95 Hz. The lower graph is a simulation based on the measured coupling constants.

multiplet from proton $M$ is quite complex but the plot of $\Sigma$ against $J^*$ reveals four principal maxima, indicating the splittings $J_{MZ} = 0.45$ Hz, $J_{MP} = 1.71$ Hz, $J_{MX} = 2.24$ Hz and $J_{AM} = 7.85$ Hz (Figure 12.9). There are also 18 subsidiary maxima at sums and differences of splittings. Although the graph is quite complicated, it is in excellent agreement with the simulated curve based on the known values of the coupling constants. The accuracy is estimated to be better than $\pm 0.03$ Hz in this example. This technique offers distinct possibilities for computer analysis of complex multiplets.

The progressive removal of splittings from the corresponding COSY cross-peak of nicotine is illustrated in Figure 12.10. First, the active splitting $J_{AM} = 7.85$ Hz is removed from each frequency dimension, using an anti-phase doublet of delta

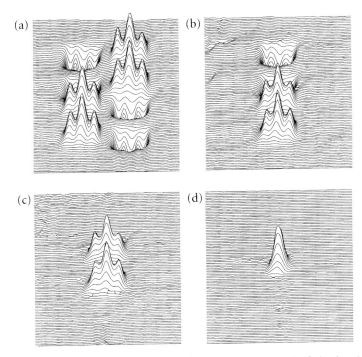

**Figure 12.10** Progressive removal of splittings from a COSY cross-peak of nicotine. (a) Original cross-peak (24 × 24 Hz). (b) After removal of an anti-phase splitting of 7.85 Hz from the $F_1$ (horizontal) dimension. (c) After removal of this same splitting from the $F_2$ (vertical) dimension. (d) After removal of all the in-phase (passive) splittings.

functions. Then the passive splittings are removed one at a time using an in-phase doublet of delta functions. The result is a singlet at the chemical shift coordinates. Note, however, that because of the nonlinear nature of the processing, intensity is sacrificed at each stage of the deconvolution; the final singlet has the same height as the individual lines of the original two-dimensional multiplet instead of representing the sum of these components. This would be a disadvantage only in situations of marginal signal-to-noise.

## 12.8.5 **Maximum entropy**

Because it is an inverse operation, the maximum entropy reconstruction (§ 6.5.1) can be used to remove a spin–spin splitting $J_{IS}$ from an experimental NMR trace if $J_{IS}$ is known beforehand.[30] The splitting, which may be in-phase or anti-phase, is imposed on the trial spectra by convolution with the appropriate stick doublet. After Fourier transformation, the result is compared with the time-domain signal derived from the isolated experimental multiplet, and the entropy is maximized subject to the usual $\chi^2$ constraint. This method of $J$ deconvolution works well if we have prior knowledge of the splitting $J_{IS}$. On the other hand, if the intention is to determine an unknown splitting, another search dimension must be introduced[30] to allow variation of a trial coupling $J^*$. Unfortunately such a procedure fits all the splittings of the experimental multiplet indiscriminately.[31] This is because incomplete deconvolution gives a higher entropy than complete deconvolution; a trial spectrum with many weak lines has a higher entropy than a spectrum of fewer lines with the same total intensity. Jones et al.[31] have demonstrated that in such cases it is better to abandon maximum entropy reconstruction altogether and proceed instead by simply minimizing $\chi^2$.

## 12.9 J doubling

### 12.9.1 Time domain

$J$ extension solves the overlap problem in a poorly resolved multiplet by increasing the active splitting $J_{IS}$ by a known amount, much larger than the line width and all the passive splittings. A similar result can be achieved by doubling the splitting, either once, or many times, in a cascade of doubling operations.[32] This exploits the standard trigonometrical expression for $\sin 2\theta$. Consider the case of an experimental in-phase doublet ($J_{IS}$), represented in the time domain by

$$S = S_0 \cos(2\pi\delta_S t)\, \cos(\pi J_{IS} t)\, \exp(-t/T_2^*) \qquad \textbf{12.32}$$

If this is multiplied by a term involving a trial coupling constant $J^*$, we have

$$S' = S_0 \cos(2\pi\delta_S t)\, \exp(-t/T_2^*)\, \sin(\pi J^* t)\, \cos(\pi J_{IS} t) \qquad \textbf{12.33}$$

After Fourier transformation we obtain a test spectrum that is the four-line pattern illustrated in Figure 12.11(b). Clearly if $J^* = J_{IS}$, the two central lines mutually cancel while the separation of the (anti-phase) outer lines becomes $2|J_{IS}|$.

A second stage of doubling may be imposed by multiplication of the time domain signal by $\cos(2\pi J^* t)$, giving

$$S' = S_0 \cos(2\pi\delta_S t)\, \exp(-t/T_2^*)\, \sin(\pi J^* t)\, \cos(2\pi J^* t)\, \cos(\pi J_{IS} t) \qquad \textbf{12.34}$$

In the frequency domain, the spectrum now has eight lines (Figure 12.11(c)). A third stage of doubling involves multiplication by $\cos(4\pi J^* t)$ giving a 16-line spectrum (Figure 12.11(d)). When $J^* = J_{IS}$ all except the extreme outer lines vanish.

In all these cases the sum of the moduli of all ordinates in the test spectrum $\Sigma$ passes through a well-defined minimum when $J^* = J_{IS}$, where there is self cancellation of all the inner lines. Multiple doubling ensures that the remaining two lines are so well-separated that there is no danger of overlap, even though there may be several other passive splittings present. Although each doubling stage halves the signal-to-noise ratio in the test spectrum, what actually matters is the signal-to-noise ratio in the graph of $\Sigma$ against $J^*$, and this is actually rather better for multiple doubling because a higher proportion of the lines cancel at the condition $J^* = J_{IS}$.

When an in-phase splitting is doubled twice, there is further minimum in the plot of $\Sigma$ versus $J^*$ at the condition $J^* = J_{IS}/3$. This subharmonic minimum is less

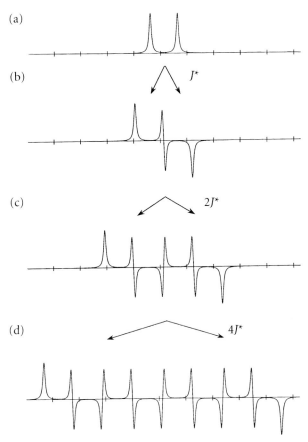

**Figure 12.11** Simulation of J doubling, represented in the frequency domain. (a) The original in-phase splitting $J_{IS}$. (b) One stage of J doubling with an anti-phase splitting $J^*$. (c) Two stages of J doubling with an additional in-phase splitting $2J^*$. (d) Three stages of J doubling with an additional in-phase splitting of $4J^*$. When $J^* = J_{IS}$, all the inner lines vanish by mutual cancellation.

pronounced than the principal minimum since it is caused by mutual cancellation of the anti-phase lines 2 and 7 of Figure 12.11(c). There are progressively weaker minima at the odd subharmonics $J_{IS}/5$, $J_{IS}/7$ etc, if more doubling operations are employed.

When the multiplet under investigation contains further splittings, the plot of $\Sigma$ against $J^*$ exhibits subsidiary minima at the sum and difference frequencies. As with J-deconvolution, these form 1:2:1 triplets. In the absence of accidental coincidences, subsidiary minima are less pronounced than the principal minimum. Traces through cross-peaks in two-dimensional correlation spectra normally contain one anti-phase splitting (the active coupling) and several in-phase splittings (the passive couplings). Multiplication by $\sin(\pi J^* t)$ picks out the in-phase splittings but ignores the anti-phase splitting. A good example is provided by a section through one of the

(a)

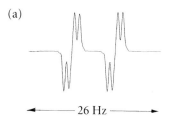

26 Hz

(b)

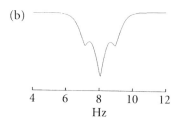

4     6     8    10    12
Hz

**Figure 12.12** (a) A section through one of the cross-peaks of the COSY spectrum of *m*-bromo-nitrobenzene, showing one active and two passive splittings. (b) The lower trace is a plot of $\Sigma$ against the trial coupling $J^*$, and was obtained by the $J$ doubling procedure. This gives the two in-phase (passive) splittings of 7.99 ± 0.04 Hz and 0.99 ± 0.04 Hz.

COSY cross-peaks of *m*-bromonitrobenzene shown in Figure 12.12(a). The graph of $\Sigma$ against $J^*$ (Figure 12.12(b)) shows a deep minimum at $J^* = 7.99 \pm 0.04$ Hz flanked by two shallow minima displaced by $\pm 0.99 \pm 0.04$ Hz. These represent good estimates of the two passive couplings.

The procedure for measuring anti-phase splittings runs along analogous lines. The experimental time-domain signal is obtained by back-transformation of a short section through a cross-peak in a correlation spectrum. It may be written as

$$S = S_0 \cos(2\pi\delta_S t)\, \sin(\pi J_{IS} t)\, \exp(-t/T_2^*) \tag{12.35}$$

Doubling is achieved by multiplication by cosine terms, repeated if necessary:

$$S' = S_0 \cos(2\pi\delta_S t)\, \exp(-t/T_2^*)\, \sin(\pi J_{IS} t) \prod_{n=0}^{k} \cos(2^n \pi J^* t) \tag{12.36}$$

The corresponding test spectra are similar to those shown in Figure 12.11, except that the central anti-phase doublets no longer alternate in sign. As a consequence, subharmonic minima are observed at $J_{IS}/2$, $J_{IS}/3$, $J_{IS}/4$, ... etc, not just at the odd subharmonics. If the multiplet contains further splittings, subsidiary minima are observed at sum and difference frequencies.

$J$ doubling is thus seen to have many features in common with $J$ deconvolution, but it is a more straightforward operation, and is to be preferred for this reason. Both methods are eventually subject to limitations imposed by the spectrometer resolution and the signal-to-noise ratio. If the line width is too broad in comparison with $J_{IS}$, the minimum in the curve of $\Sigma$ against $J^*$ becomes broad and ill-defined; if the noise level is too high, there are problems locating the true minimum. Nevertheless, with a reasonable signal-to-noise ratio, splittings can be measured that are several times smaller than the full linewidth.

## 12.9.2 Frequency domain

There are practical advantages in applying the doubling procedure in the frequency domain; indeed this facilitates more general schemes for $J$ tripling or $J$ quintupling, should these be deemed appropriate.[33] A trace with an in-phase splitting $J_{IS}$ is simply convoluted with a regular array of alternating-phase delta functions (a stick spectrum) where the splittings are equal to the trial splitting $J^*$. Figure 12.13 illustrates the case of three-stage $J$ doubling. Exactly as in the corresponding time-domain procedure, we obtain a 16-line pattern that degenerates into an anti-phase doublet of splitting $8J_{IS}$ when $J^* = J_{IS}$. Experimental anti-phase splittings are processed by convolution with a regular array of delta functions all in the same sense.

In order to avoid difficulties with edge effects during the convolution, the experimental frequency-domain trace should have adequate sections of signal-free baseline on each side of the multiplet under investigation. If these are not available, a baseline correction is applied and the baseline is artificially extended by appending zeroes. Fine digitization is ensured by the appropriate choice of the acquisition time, and by zero-filling if necessary. The digitization steps should be small,

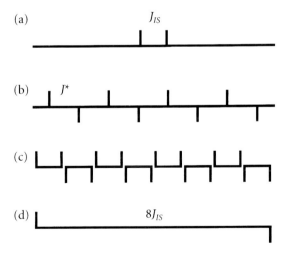

**Figure 12.13** Schematic representation of $J$ doubling in the frequency domain. The experimental doublet (a) is convoluted with a regular array of alternating-phase delta functions of separation $J^*$ (b). The result (c) is a 16-line spectrum which degenerates (d) into a two-line spectrum of splitting $8J_{IS}$ when $J^* = J_{IS}$, because all save the extreme outer lines mutually interfere at this condition.

comparable with the expected precision of the measurement, and the same digitization steps should be used for the convoluting stick spectrum. Whereas $J$ doubling in the time domain requires a fresh Fourier transformation for each increment of $J^*$, processing in the frequency domain avoids multiple transformations, and is therefore much faster to implement.

## 12.10 Zero crossing

Suppose a poorly resolved doublet $J_{IS}$ has been excised from the full spectrum or from a trace through a two-dimensional cross-peak. It may be represented by the time-domain signal

$$S'(t) = S_0 \cos(2\pi\delta_S t) \cos(\pi J_{IS} t)\, L(t) \qquad\qquad \textbf{12.37}$$

where $L(t)$ is some general form of decay function representing an arbitrary lineshape in the frequency domain. If the doublet is itself symmetrical, and if it is carefully centred at the mid-point of the chosen frequency range, we may set $\delta_S = 0$ and write the inverse Fourier transform as

$$S(t) = S_0 \cos(\pi J_{IS} t)\, L(t) \qquad\qquad \textbf{12.38}$$

Since the chemical shift modulation has been removed, the only zero crossings of this time-domain signal arises from the spin–spin coupling term. Stonehouse and Keeler[34] have used this to evaluate $J_{IS}$. Identification of the zero-crossing point has some practical advantages, because its position is quite unaffected by the shape of the decaying time-domain envelope (the Fourier transform of the lineshape).

For poorly resolved splittings only the first zero crossing is readily accessible, and in the presence of noise and line-broadening it is no simple matter to pinpoint it directly, but a search algorithm can be devised that evaluates the integral

$$|I(J^*)| = \int S(t)\cos(\pi J^* t))\, dt \qquad\qquad \textbf{12.39}$$

where, as usual, $J^*$ is a trial coupling. This integral has the important restriction that it is only evaluated over time ranges for which the integrand $S(t)\cos(\pi J^* t)$ is negative. The process resembles a Fourier transformation, but in a modification that acts principally on the first negative excursion of the $J$-modulation. If $J^* = J_{IS}$, the integral is zero, since the integrand is then $\cos^2(\pi J_{IS} t)$, which is necessarily positive. This condition corresponds to a minimum in $|I(J^*)|$, which serves as an indirect indication of the first zero crossing of $S(t)$.

In practice the integral is evaluated as a digital summation, and it is therefore important to ensure fine digitization of the experimental signal. As the line broadening is increased in comparison with $J_{IS}$, the minimum in the graph of $|I(J^*)|$ against $J^*$ becomes shallower and less well defined, but splittings less than half the line width have been measured.[34] Note that this scheme relies implicitly on the initial centering process, and therefore involves an iterative algorithm prior to the search for the zero crossing.

## 12.11 **J modulation**

The ultimate limitation on the accuracy with which a coupling constant can be measured is the natural width of the NMR lines, determined by the reciprocal of the spin–spin relaxation time, $T_2$. This suggests that spin-echo modulation should provide the most accurate values for homonuclear coupling constants if we are prepared to take the trouble to set up an experiment specifically for this measurement.[35] The technique employs soft radiofrequency $\pi$ pulses (§ 5.4.7) for spin inversion and echo refocusing. The first set of measurements follows the time-dependence of the echo modulation with the sequence:

S spins:         hard $(\pi/2)_x$ — $\tau$ — soft $(\pi)_x$ — $\tau$ — acquire

I spins                            soft $(\pi)_x$           acquire      **12.40**

The selectivity of the soft $\pi$ pulses is set for inversion of an entire spin multiplet, with negligible perturbation of adjacent multiplets. Gaussian-shaped pulses[36] (§ 5.2.1) are convenient for this application. After Fourier transformation of the second half of each spin echo, the time dependences of the $I$-spin and $S$-spin multiplets are recorded as a function of $\tau$. There is no need for particularly high resolution of these signals; sometimes it is preferable to switch off the sample spinner and deliberately broaden the multiplet to obtain more reliable values of the peak intensities. The time dependence can be represented as

$$S(t) = S_0 \cos(\pi J_{IS} t) \exp(-t/T_2) \qquad \textbf{12.41}$$

where $t = 2\tau + t_p$, and $t_p$ is the duration of the soft $\pi$ pulse. Note that none of the other couplings in the molecule are involved in the modulation, provided that the inversion of the $I$ spins is sufficiently selective. A three-parameter ($S_0$, $J_{IS}$, $T_2$) fitting routine may then be used to extract the coupling constant from this modulated decay curve. A rather better procedure avoids the use of a fitting program altogether. The experiment is repeated with the omission of the spin inversion pulse on the $I$ spins, giving an unmodulated exponential decay curve:

$$S'(t) = S_0 \exp(-t/T_2) \qquad \textbf{12.42}$$

Division of $S(t)$ by $S'(t)$ at each time increment gives a cosine curve with a frequency $\frac{1}{2}J_{IS}$ Hz. Note that the later experimental points on this curve become increasing unreliable when $S(t)$ and $S'(t)$ become small. Omitting the $\pi$ pulse on the $S$ spins but retaining the $\pi$ pulse on the $I$ spins gives the spin–spin relaxation curve of the $I$ spins, and thereby an independent determination of the same coupling constant. The spin–spin relaxation times obtained by this single-echo technique may be artificially shortened by molecular diffusion through the naturally occurring field gradients, but this does not invalidate the measurement of the coupling constant.

Figure 12.14 shows modulated and unmodulated decay curves obtained for the spin echo responses of proton $H_2$ of $m$-bromonitrobenzene, designed to measure the small coupling constant to the *para* proton $H_5$. Division of the modulated curve

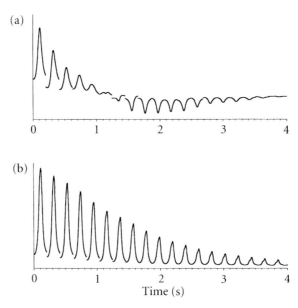

**Figure 12.14** Determination of the *para* coupling in *m*-bromonitrobenzene by recording the spin echo modulation using multiplet-selective $\pi$ pulses. (a) The modulated decay curve of the *S* spins, obtained with simultaneous soft $\pi$ pulses on both the *I* and *S* spins. (b) The unmodulated decay obtained when the *I*-spin pulses were omitted. The result gives $J_{IS} = 0.35 \pm 0.02$ Hz.

by the unmodulated decay gives data that can be fitted to a cosine wave, suggesting $J_{IS} = 0.35 \pm 0.02$ Hz.[37] This is in excellent agreement with other measurements.

Most of this chapter has deliberately concentrated on methods for measuring small, poorly resolved splittings recorded in one- or two-dimensional spectroscopy. In experiments on macromolecules, where natural line widths are broad and often coarsely digitized, even quite large coupling constants have to be estimated by more indirect methods, such as the time dependence of the intensity of a cross-peak in a correlation spectrum due to echo modulation effects. Sometimes this information is only qualitative or is expressed relative to another coupling in the same spectrum. In other situations the experiment is specifically designed to follow the *J* modulation and the determination can be more quantitative. There is a growing literature[38–42] on this methodology and a recent review article.[43]

## 12.12 Discussion

The simplest methods are invariably the best. If it is at all possible, the measurement of a spin–spin coupling constant should be performed by the direct method, using proven software routines to evaluate the relevant frequency splitting. Resolution enhancement can help, and fine digitization is advisable. Only when the straightforward methods run into trouble, should we have recourse to the more sophisticated manipulations outlined above. Many of these come straight out of a high-school textbook of elementary trigonometry—identities for the sines and cosines of sums and differences of angles. Underpinning it all is the convolution theorem, which states that the product of two functions $g(t)\, h(t)$ in the time domain corresponds to the convolution of their Fourier transforms $G(f) \star H(f)$ in the frequency domain.

The direct method tends to underestimate poorly resolved in-phase splittings or even miss them entirely, whereas it overestimates poorly resolved anti-phase splittings, because of the self-cancellation that occurs in the overlap region. If both in-phase and anti-phase versions of a given splitting are available, then $J$ extension is a safe and robust procedure. More commonly we are presented with a single version of the splitting of interest. Then $J$-doubling appears to hold the advantage. This procedure is rather simpler to implement in the frequency domain than in the time domain. Accuracies of the order of $\pm 0.03$ Hz can be achieved in favourable cases.

# References

1. M. Karplus, *J. Chem. Phys.* **30**, 11 (1959).
2. M. Karplus, *J. Am. Chem. Soc.* **85**, 2870 (1963).
3. S. D. Emerson and G. T. Montelione, *J. Am. Chem. Soc.* **114**, 354 (1992).
4. R. R. Ernst, R. Freeman, B. Gestblom and T. R. Lusebrink, *Mol. Phys.* **13**, 283 (1967).
5. P. J. Hore, in *Maximum Entropy in Action*, Eds: B. Buck and V. A. Macaulay, Oxford University Press, 1991.
6. F. Ni, G. C. Levy and H. Scheraga, *J. Magn. Reson.* **66**, 385 (1986).
7. S. J. Davies, C. J. Bauer, P. J. Hore and R. Freeman, *J. Magn. Reson.* **76**, 476 (1988).
8. A. Bax and R. Freeman, *J. Magn. Reson.* **37**, 177 (1980).
9. C. Redfield and C. M. Dobson, *Biochemistry* **29**, 7201 (1990).
10. G. Bodenhausen, R. Freeman, R. Niedermeyer and D. L. Turner, *J. Magn. Reson.* **26**, 133 (1977).
11. Y. Kim and J. H. Prestegard, *J. Magn. Reson.* **84**, 9 (1989).
12. F. del Rio-Portilla and R. Freeman, *J. Magn. Reson. A* **104**, 358 (1993).
13. F. del Rio-Portilla and R. Freeman, *J. Chem. Soc., Faraday Trans.* **89**, 4275 (1993).
14. W. P. Aue, E. Bartholdi and R. R. Ernst, *J. Chem. Phys.* **64**, 2229 (1976).
15. L. Braunschweiler and R. R. Ernst, *J. Magn. Reson.* **53**, 521 (1983).
16. H. Oschkinat and R. Freeman, *J. Magn. Reson.* **60**, 164 (1984).
17. H. Kessler, A. Müller and H. Oschkinat, *Magn. Reson. Chem.* **23**, 844 (1985).
18. J. J. Titman and J. Keeler, *J. Magn. Reson.* **89**, 640 (1990).
19. J. J. Titman, D. Neuhaus and J. Keeler, *J. Magn. Reson.* **85**, 111 (1989).
20. X. Miao and R. Freeman, *J. Magn. Reson. A* **117**, 128 (1995).
21. G. A. Morris and R. Freeman, *J. Am. Chem. Soc.* **101**, 760 (1979).
22. D. P. Burum and R. R. Ernst, *J. Magn. Reson.* **39**, 163 (1980).
23. A. A. Bothner-By and J. Dadok, *J. Magn. Reson.* **72**, 540 (1987).
24. J. M. Le Parco, L. M. McIntyre and R. Freeman, *J. Magn. Reson.* **97**, 553 (1992).
25. R. Freeman and L. M. McIntyre, *Israel J. Chem.* **32**, 231 (1992).
26. P. Huber and G. Bodenhausen, *J. Magn. Reson. A* **102**, 81 (1993).
27. P. Huber and G. Bodenhausen, *J. Magn. Reson. A* **104**, 96 (1993).

28. F. Del Rio-Portilla and R. Freeman, *J. Magn. Reson. A* **108**, 124 (1994).
29. R. Baumann, G. Wider, R. R. Ernst and K. Wüthrich, *J. Magn. Reson.* **44**, 402 (1981).
30. M. A. Delsuc and G. C. Levy, *J. Magn. Reson.* **76**, 306 (1988).
31. J. A. Jones, D. S. Grainger, P. J. Hore and G. J. Daniell, *J. Magn. Reson. A* **101**, 162 (1993).
32. L. M. McIntyre and R. Freeman, *J. Magn. Reson.* **96**, 425 (1992).
33. F. Del Rio-Portilla, V. Blechta and R. Freeman, *J. Magn. Reson. A* **111**, 132 (1994).
34. J. Stonehouse and J. Keeler, *J. Magn. Reson. A.* **112**, 43 (1995).
35. R. Freeman and H. D. W. Hill, *J. Chem. Phys.* **54**, 301 (1971).
36. C. J. Bauer, R. Freeman, T. Frenkiel, J. Keeler and A. J. Shaka, *J. Magn. Reson.* **58**, 442 (1984).
37. H-W. Jia unpublished results.
38. D. Neri, G. Otting and K. Wüthrich, *J. Am. Chem. Soc.* **112**, 3663 (1990).
39. M. Billeter, D. Neri, G. Otting, Y. Q. Qian and K. Wüthrich, *J. Biomol. NMR* **2**, 257 (1992).
40. A. Bax, D. Max and D. Zax, *J. Am. Chem. Soc.* **114**, 6923 (1992).
41. G. W. Vuister, F. Delaglio and A. Bax, *J. Am. Chem. Soc.* **114**, 9674 (1992).
42. F. Fogolari, G. Esposito, S. Cauci and P. Viglino, *J. Magn. Reson. A.* **102**, 49 (1993).
43. C. Biamonti, C. B. Rios, B. A. Lyons and G. T. Montelione, *Adv. Biophys. Chem.* **4**, 51 (1994).

# *Appendix: Acronyms*

The variety and complexity of the pulse sequences used for high resolution NMR have engendered a large vocabulary of acronyms intended as a concise method of referencing. Acronyms should be short, apt, descriptive and preferably amusing. These are the acronyms used in this book for sequences that are cited several times; there are many more in circulation among NMR spectroscopists.

| | |
|---|---|
| **BIRD** | Bilinear rotation decoupling.[1] |
| **BURP** | Band-selective, uniform response, pure-phase pulses.[2] |
| **CAMELSPIN** | Overhauser experiment in the rotating frame.[3] |
| **COSY** | Correlation spectroscopy.[4] |
| **ψ-COSY** | Pseudo-correlation spectroscopy.[5] |
| **CPMG** | Carr–Purcell–Meiboom–Gill spin echo sequence.[6] |
| **CYCLOPS** | Cyclically-ordered phase sequence for quadrature detection.[7] |
| **DANTE** | Delays alternating with nutation for tailored excitation.[8] |
| **DEPT** | Distortionless enhancement by polarization transfer.[9] |
| **DIPSI** | Decoupling in the presence of scalar interactions.[10] |
| **DISCO** | Differences and sums in correlation spectroscopy.[11] |
| **DQ-COSY** | Double-quantum filtered correlation spectroscopy.[12] |
| **E-COSY** | Exclusive correlation spectroscopy.[13] |
| **EXORCYCLE** | Phase cycle to suppress phantoms and ghosts.[14] |
| **EXSY** | Two-dimensional exchange spectroscopy.[15] |
| **FLOPSY** | Flip-flop spectroscopy.[16] |
| **GARP** | Globally-optimized, alternating-phase rectangular pulses for broadband decoupling.[17] |
| **HMBC** | Heteronuclear multiple-bond correlation.[18] |
| **HMQC** | Heteronuclear multiple-quantum correlation.[19] |
| **HOHAHA** | Homonuclear Hartmann–Hahn spectroscopy (TOCSY).[20] |
| **HSQC** | Heteronuclear single-quantum correlation spectroscopy.[21] |
| **INADEQUATE** | Incredible natural-abundance double-quantum transfer experiment.[22] |
| **INEPT** | Insensitive nuclei enhanced by polarization transfer.[23] |
| **MLEV** | Malcolm Levitt's decoupling cycle.[24] |
| **NERO** | Nonlinear excitation with rejection on resonance,[25] |
| **NOESY** | Two-dimensional nuclear Overhauser spectroscopy.[15] |
| **RELAY** | Relayed coherence transfer.[26] |

| | |
|---|---|
| **ROESY** | Rotating-frame Overhauser enhancement spectroscopy.[27] |
| **SEMUT** | Subspectral editing using a multiple-quantum trap.[28] |
| **TANGO** | Testing for adjacent nuclei with a gyration operator.[29] |
| **TOCSY** | Total correlation spectroscopy (HOHAHA).[30] |
| **TPPI** | Time-proportional phase incrementation.[31] |
| **TSETSE** | Double resonance two-spin effect.[32] |
| **WALTZ** | Decoupling sequence: $\{1\bar{2}3\ 1\bar{2}3\ \bar{1}2\bar{3}\ \bar{1}2\bar{3}\}$.[33] |
| **WATERGATE** | A water suppression sequence.[34] |
| **WEFT** | Water-eliminated Fourier transform.[35] |
| **WURST** | Decoupling sequence using adiabatic pulses.[36] |

# References

1. J. R. Garbow, D. P. Weitekamp and A. Pines, *Chem. Phys. Lett.* **93**, 514 (1982).
2. H. Geen and R. Freeman, *J. Magn. Reson.* **93**, 93 (1991).
3. A. A. Bothner-By, R. L. Stephens, J-M. Lee, C. D. Warren and R. W. Jeanloz, *J. Am. Chem. Soc.* **106**, 811 (1984).
4. W. P. Aue, E. Bartholdi and R. R. Ernst, *J. Chem. Phys.* **64**, 2229 (1976).
5. S. Davies, J. Friedrich, and R. Freeman, *J. Magn. Reson.* **75**, 540 (1987).
6. H. Y. Carr and E. M. Purcell, *Phys. Rev.* **94**, 630 (1954) and S. Meiboom and D. Gill, *Rev. Sci. Instr.* **29**, 688 (1958).
7. D. I. Hoult and R. E. Richards, *Proc. Roy. Soc. (London) A.* **344**, 311 (1975).
8. G. A. Morris and R. Freeman, *J. Magn. Reson.* **29**, 433 (1978).
9. D. M. Doddrell, D. T. Pegg and M. R. Bendall, *J. Magn. Reson.* **48**, 323 (1982).
10. A. J. Shaka, C. J. Lee and A. Pines, *J. Magn. Reson.* **77**, 274 (1988).
11. H. Kessler, A. Müller and H. Oschkinat, *Magn. Reson. Chem.* **23**, 844 (1985).
12. U. Piantini, O. W. Sørensen and R. R. Ernst, *J. Am. Chem. Soc.* **104**, 6800 (1982).
13. C. Griesinger, O. W. Sørensen and R. R. Ernst, *J. Am. Chem. Soc.* **107**, 6394 (19985).
14. G. Bodenhausen, R. Freeman, R. Niedermeyer and D. L. Turner, *J. Magn. Reson.* **26**, 133 (1977).
15. J. Jeener, B. H. Meier, P. Bachmann and R. R. Ernst, *J. Chem. Phys.* **71**, 4546 (1979).
16. A. J. Shaka, C. J. Lee and A. Pines, *J. Magn. Reson.* **77**, 274 (1988).
17. A. J. Shaka, P. B. Barker and R. Freeman, *J. Magn. Reson.* **64**, 547 (1985).
18. A. Bax and M. F. Summers, *J. Am. Chem. Soc.* **108**, 2093 (1986).
19. A. Bax, R. H. Griffey and B. L. Hawkins, *J. Magn. Reson.* **55**, 301 (1983).
20. A. Bax and D. G. Davis, *J. Magn. Reson.* **65**, 355 (1985).
21. A. Bax, M. Ikura, L. E. Kay, D. E. Torchia and R. Tschudin, *J. Magn. Reson.* **86**, 304 (1990).
22. A. Bax, R. Freeman and T. A. Frenkiel, *J. Am. Chem. Soc.* **103**, 2102 (1981).
23. G. A. Morris and R. Freeman, *J. Am. Chem. Soc.* **101**, 760 (1979).
24. M. H. Levitt and R. Freeman, *J. Magn. Reson.* **43**, 502 (1981).
25. M. H. Levitt, *J. Chem. Phys.* **88**, 3841 (1988).
26. G. W. Eich, G. Bodenhausen and R. R. Ernst, *J. Am. Chem. Soc.* **104**, 3731 (1982).
27. A. Bax and D. G. Davis, *J. Magn. Reson.* **63**, 207 (1985).
28. H. Bildsøe, S. Dønstrup, H. K. Jakobsen and O. W. Sørensen, *J. Magn. Reson.* **53**, 154 (1983).
29. S. Wimperis and R. Freeman, *J. Magn. Reson.* **58**, 348 (1984).

30. L. Braunschweiler and R. R. Ernst, *J. Magn. Reson.* **53**, 521 (1983).
31. G. Drobny, A. Pines, S. Sinton, D. Weitekamp and D. Wemmer, *Symp. Faraday Soc.* **13**, 49 (1979).
32. Ē. Kupče, J. M. Nuzillard, V. S. Dimitrov and R. Freeman, *J. Magn. Reson. A.* **107**, 246 (1994).
33. A. J. Shaka, J. Keeler, T. Frenkiel and R. Freeman, *J. Magn. Reson.* **52**, 335 (1983).
34. M. Piotto, V. Saudek and V. Sklenár, *J. Biomed. NMR.* **2**, 661 (1992).
35. S. L. Patt and B. D. Sykes, *J. Chem. Phys.* **56**, 3182 (1972).
36. Ē. Kupče and R. Freeman, *J. Magn. Reson. A.* **115**, 273 (1995).

# *Index*

Page numbers in bold refer to main sections of the book